Springer Monographs in Mathematics

This series publishes advanced monographs giving well-written presentations of the "state-of-the-art" in fields of mathematical research that have acquired the maturity needed for such a treatment. They are sufficiently self-contained to be accessible to more than just the intimate specialists of the subject, and sufficiently comprehensive to remain valuable references for many years. Besides the current state of knowledge in its field, an SMM volume should ideally describe its relevance to and interaction with neighbouring fields of mathematics, and give pointers to future directions of research.

Jean-Pierre Bourguignon

Variational Calculus

 Springer

Jean-Pierre Bourguignon
Institut des Hautes Études Scientifiques
Bures-sur-Yvette, France

ISSN 1439-7382   ISSN 2196-9922   (electronic)
Springer Monographs in Mathematics
ISBN 978-3-031-18309-6   ISBN 978-3-031-18307-2   (eBook)
https://doi.org/10.1007/978-3-031-18307-2

This Springer imprint is published by the registered company Springer Nature Switzerland AG
The registered company address is: Gewerbestrasse 11, 6330 Cham, Switzerland

LA VISION[1]

D'où est sorti ce livre

Il n'est pas de brouillards, comme il n'est point d'algèbres,
Qui résistent, au fond des nombres ou des cieux,
A la fixité calme et profonde des yeux ;
Je regardais ce mur d'abord confus et vague,
Où la forme semblait flotter comme une vague,
Où tout semblait vapeur, vertige, illusion ;
Et, sous mon œil pensif, l'étrange vision
Devenait moins brumeuse et plus claire, à mesure
Que ma prunelle était moins troublée et plus sûre.

Victor HUGO

in *La Légende des Siècles*

---

[1] This is an excerpt from the introduction of the famous (long!) collection of poems by Victor Hugo called "*La légende des siècles*", created in the period 1859–1883, aiming at presenting his poetary vision of the history and evolution of humanity (note that the title and subtitle of the excerpt are due to him):

THE VISION

From where this book comes

There is no fog, as there is no algebra,
Which persist, in the depths of numbers or of the skies,
To the quiet and deep gaze of the eyes;
I looked at this wall, at first confused and vague,
Where form seemed to float like a wave,
Where all seemed to be mist, vertigo, illusion;
And, under my pensive eye, the strangest vision
Became less foggy and clearer,
As my eye was less troubled and surer.

Editor's Preface

As pointed out in the preface to this book, the Calculus of Variations underlies many of the applications of mathematics to the natural sciences. In particular, the *Euler–Lagrange equations* are the universal form of the specific equations that govern essentially all the entities of fundamental physics, from the elementary constituents of matter to the forces that hold them together to the structure of space and time. They are also the starting point for the most natural approach to the quantum theory of fields. It is a monumental achievement of humanity to have discovered them in the concise and elegant form that they now have. The readers of this book will encounter their theory and use at the hands of a master and thereby come to appreciate their profundity for themselves.

Jean-Pierre Bourguignon has become widely known and respected in the global mathematical community as a great leader on account of the distinguished service roles he has most generously assumed over the years. These include his directorship of the IHES, one of the truly great research institutions in the world, and his presidency of the European Research Council. His prominence in the socio-politics of mathematics may lead some to momentarily forget his outstanding contributions as a researcher and his remarkable skills as a teacher. He is a world-leading expert on the geometric variational equations of mathematical physics, while several generations of excellent students at the École Polytechnique have been introduced to the profound and beautiful world of global analysis by way of his inspiring lectures.

As is well-known, the Springer Monographs have as their primary goal the publication of research monographs dealing with rather recent advances. However, every now and then, a pedagogical work of great importance and expositional excellence is put forward because of its role as an enabler of current and future research at the highest level. As a specimen of this kind of publication, the book in your hands will have few equals. It is with great pleasure that we offer it within this series.

Edinburgh and Seoul, *Minhyong Kim*
September 2022

Preface

The Calculus of Variations aims to characterise a mathematical object such as a curve, a surface, a function, or more generally a field (in the physicists' sense), which satisfies "*variational*" conditions. In more modern language, this means that this object should be an *extremum of a function* (in this context, one often speaks of *functionals*) defined on the space of objects which are in competition.

Problems arising from the Calculus of Variations have played an important role in the history of mathematics. Substantial contributions to problems of this type were made by, among others, Bernoulli, Euler, Lagrange, Leibniz, Jacobi, Hamilton, Weierstrass, Hilbert, Carathéodory, Morse. This subject also attracts the attention of numerous present-day mathematicians and, thanks to a considerable extension of the methods available, important new developments have appeared recently with spectacular applications, both inside and outside mathematics.

The Calculus of Variations has maintained, and still maintains, a close relationship both with analysis and with geometry. The progress which it inspired has touched the foundations of mathematics as well as the elaboration of new methods. Today, it remains a source of problems of great interest and nourishes a very active branch of the theory of partial differential equations. Many classical variational problems in geometry have recently been the object of intense research activity in theoretical physics (we shall mention them on several occasions in this book).

This course aims to present the basic concepts which allow one to discuss several classical problems of the Calculus of Variations. As well as giving general methods, it also studies several special situations. It is centred around the *search for the extrema of a function defined on a space*. To accomplish this, it is necessary to generalise the notion of a space in two directions: first, to treat effectively the objects which are "varied" (usually functions), one is naturally led to work with *infinite-dimensional spaces* (and there lies the beginning of functional analysis, which has proved to be so powerful for the solution of partial differential equations); second, to find the extrema of the function being studied, one must have available a notion of the derivative on spaces of curves, which are the *configuration spaces* featuring most often in concrete situations, such as in mechanics. It is there that we shall encounter *intrinsic differential geometry* for the first time; this part is often called *differential calculus*. We have deliberately used geometric language because it seems to us the most suitable and most effective for treating the problems we have in mind, whence the title *Variational Calculus* of this book.

Another idea which appears here repeatedly is this: relate the *global* form of a space to the properties of the functions defined on it, which physicists like to call *observables* (a terminology which we have retained). The interest which this approach excites today, both inside and outside mathematics, has led to the introduction of a special term to denote it. One speaks of "*Global Analysis*", or in German "*Variationsrechnung im Großen*" (the old terminology in English was "*Calculus of*

Variations in the Large"). For this reason, we were tempted to add to the title Variational Calculus the adjective *global*. We have decided against this, however, since in the limited space we have available, it is difficult to go very far in this direction. In fact, a more profound study would necessitate excursions into more advanced areas of mathematics such as algebraic topology and differential topology. Moreover, the most spectacular applications make extensive use of results from the theory of partial differential equations.

In many places another theme appears: the study of the *symmetries* of a system, which Felix Klein, in his Erlangen programme of 1872, showed is identical to the search for properties which are invariant under the action of a group. These "*continuous groups of transformations*" (the action of which is the mathematical formulation of symmetry) have taken up, thanks to the work of Sophus Lie, a privileged place in the Calculus of Variations. Moreover, the presence of symmetries in a variational problem is accompanied, according to a theorem of Emmy Noether, by that of *conserved quantities*, which often facilitate the solution of a problem.

The climax of the book occurs in the chapters devoted to the *Euler–Lagrange equations* and *Hamilton's equations*. These equations are very far-reaching (one is seriously tempted to say "*universal*"), and underlie numerous theories in diverse branches of physics, mechanics and economics. The multiplicity of situations where they are used successfully testifies to the power of the concepts elaborated by mathematicians for solving these problems. This shows how completely wrong was Felix Klein's judgement on this subject, as expressed in his presentation of the development of mathematics in the XIXth century. "*Trotz der unzweifelhaften Schönheit dieses Gebietes möchte ich jedoch vor einem einseitigen Studium warnen..... In der Tat kann der Physiker wenig, der Ingenieur gar nichts von diesen Theorien für seinen Aufgaben brauchen. Sie sind sozusagen ein Schema mit leeren Fächern, in welche die bunte Welt der Erscheinungen erst eingeordnet werden muß, um sie sinnvoll erscheinen zu lassen*".[2] In the edition of these lectures edited in 1926 by Richard Courant and Otto Eduard Neugebauer under the title "*Vorlesungen über die Entwicklung der Mathematik im 19. Jahrhundert*" (volume XXIV of the Springer series *Grundlehren der mathematischen Wissenschaften in Einzeldarstellungen*), they inserted the following footnote: "*Wir haben diese, durch die Entwicklung der letzten Jahren widerlegt Bemerkungen stehen lassen, da sie zu der gerade von Klein oft gennanten Erscheinung einen Betrag liefern, wie scheinbar rein mathematische Theorien auch für die Nachbarwissenscheften unvermutet von größter Bedeutung werden.*"[3] This historical reference, in which the actors are among the most eminent mathematicians, emphasises *the great difficulty involved in judging* a priori *the applicability of a mathematical result*.

---

[2] "*Despite the undoubted beauty of this field, however, I would caution the reader.... In reality, physicists can extract only very little from these theories for use in their own work, and engineers almost nothing. They constitute a kind of scheme with empty spaces, which only takes on meaning when the gaps are filled from the varied world of natural phenomena.*"

[3] "*We have retained these remarks, which are refuted by recent developments, because they are an example of a phenomenon often evoked by Klein himself, namely, that mathematical theories which may seem completely pure can turn out to be extremely relevant to other sciences.*"

Today, the Euler–Lagrange equations form the conceptual framework for all stud-ies, based on energy methods, of the *mechanics of systems defined by a finite number of parameters*. The dual methods related to Hamilton's equations are now used in practice in numerous questions of *automation*, especially in problems arising from *spatial engineering* and from *econometrics*. They also serve as the point of depar-ture for *quantum mechanics*, which cannot be developed starting from the Newtonian view of mechanics, and as the setting for the theory of fields in physics. They illustrate the power of the notion of *duality*, once it has been suitably extended to this setting. Note, however, that the practical study of these questions often cannot be divorced from that of the numerical approximation of their solutions. These considerations lead to the introduction of the idea, which can sometimes be forgotten, that *in real life (and in industry), there are problems of a mathematical nature whose solution requires the use of among the most recent mathematical results, and which give rise to new research which can be carried out only by teams of mathematicians devel-oping the most modern theories*. In this process, the modelling of the phenomena which one is studying plays, of course, an essential role.

In a book of this level, it is difficult to include the newest results. We have, above all, tried to put the recent evolution of the subject in perspective, and to bring out a truth which is unfortunately often concealed: "*mathematics is, today more than ever, a living science open to interaction with other sciences*". In the last forty years, its development has undergone an unprecedented acceleration, with the result that there are probably more professional mathematicians alive today than have existed before since the birth of humanity. Such an explosion makes the selection of basic material more difficult than ever, since this expanding activity goes hand-in-hand with a very rapid diversification accompanied by a multitude of the most unexpected cross-fertilisations between sub-disciplines of mathematics.

The prerequisites for this book are those acquired by most students in the first two years of an undergraduate course. We regularly appeal to the fundamental results of linear algebra. An appendix collects together the basic ideas and concepts of general topology.

The subtlety of the knowledge contained in this book may not be as surprising to the reader as the *change of style* which it exhibits compared to previous courses. This change is, in part, inherent in the fact that this is a more advanced course; but one can find another explanation in the general evolution of styles of exposition which reflects the manner in which knowledge is developing: to a period of linear devel-opment of mathematics centred around the structure of great theories corresponds a "one-dimensional" style of presentation; to the recent explosion of sub-disciplines and multiple interactions corresponds a more personal synthesis. From these consid-erations follows, in particular, the great importance of the bibliography, since a book on mathematics offers, as well as the presentation (correct, if possible!) of theorems, a vision of the connections between concepts. In the opinion of this author, a real understanding of the fundamental ideas can only develop out of the diversity of these approaches.

Palaiseau, 4 August 1987   *Jean-Pierre Bourguignon*

Preface to the English Edition

This book is the English translation of the notes of a course delivered more than 30 years ago, several years in a row, to the entire student cohort at École Polytechnique (more than 400 students each year). Because of the special context of the teaching at the school, the level of the course is between undergraduate and graduate. Indeed, students there are selected on the basis of their rather high level of competence in mathematics. They are exposed to lectures in different sciences and also some in the social sciences and humanities. The lectures were delivered in a large lecture hall on the basis of the notes. They were complemented by exercise sessions, hence the inclusion of a number of exercises in the notes that could be used during these sessions – without being an obligation since some teachers preferred to introduce their own.

Because of the time passed, a number of adjustments of dates mentioned in the text were necessary and have been made. This edition also provided an opportunity to be more systematic about the introduction of the short bibliographical notes of scientists whose names appear in the text.

I am very grateful to Professor Andrew Pressley, who did a very careful and faithful job translating the notes. He completed his work 30 years ago and I thank him also for his patience. We owe the existence of this edition to the persistence of the mathematical editors at Springer, now Springer Nature. They managed to lead this project to its fruition as a volume in the 'Springer Monographs in Mathematics' series. I am thankful to them for that. I also owe a lot to the students and my colleagues at École Polytechnique for improving the text and spotting typos and places where it needed to be clarified, if not plainly corrected. Any inaccuracies that remain are my sole responsibility.

I dedicate this book to all my mathematics teachers Mr Lemaître, Mr Thovert, Mr Martin, Mr Gontard, Mr Choquet and Mr Berger, who exposed me to different ways of looking at mathematics and who, unfortunately, have all left us.

Bures-sur-Yvette,
21 July 2022

Jean-Pierre Bourguignon

To the Reader

These lecture notes comprise 11 chapters which are *naturally decomposed into three parts*, each being better approached with a different mindset.

The first, entitled *"The Analytic Setting"*, covers Chapters I, II and III. Its purpose is to amplify the students' knowledge of analysis. It is a kind of warm-up.

The second, entitled *"The Geometric Setting"*, covers Chapters IV, V, VI and VII. It introduces another approach, involving new concepts. It requires some reflection, and the reader must practice a number of exercises to become comfortable with the new language.

The third, entitled *"The Calculus of Variations"*, covers Chapters VIII, IX, X and XI. This is the real goal of the course. It contains many applications coming from different fields, and it is this variety that gives the presented theorems their value. One must be perseverant enough to embrace all facets of the theory. In this part, solving a number of exercises drawn from a wide range of different topics is compulsory.

Some typographical remarks about the text: Chapters are numbered using Roman numerals while references to the appendix use Arab numerals. Within each chapter, the numbering is strictly linear to facilitate internal cross-references; sections are ordered alphabetically to visually organise the text... and the table of contents. At the beginning of each section, one finds a short résumé of its content.

We use two sizes of characters: the *normal size* is that of the largest part of the text; a *smaller size* is used for some complementary material for readers interested in extensions of the notions presented. We also use the latter for proofs of theorems, propositions and lemmas. This distinction has two advantages: it gives some rhythm to the text, avoiding an excessively uniform appearance. It also suggests something more substantial, namely that, during a first reading, proofs can be skipped.[4]

The text contains a number of *exercises*. One of the objectives is to remind readers that it falls on them to control their understanding while progressing in their study.

---

[4] This gives me an opportunity to recall that a mathematical text should ideally be read three times: the *first* time reading only definitions and statements, even if one gets the feeling of having lost track at some stage; the *second* time forming a precise idea about each statement, still avoiding reading proofs; and the *third* one coming finally to a complete and thorough reading of the whole text.

A warning though: the exercises are of a variable difficulty, but the reward of having tried seriously to solve an exercise is almost as great as that of resolving it. Exercises appearing in smaller characters are likely to be more difficult and are reserved for motivated readers.

Each chapter ends with short *historical notes*. These are rather incentives to learn more than complete stories. When the name of a scientist appears in the text, a short *bibliographical reference* is provided as a footnote. The objective is in particular to remind readers that *mathematics is a living science produced by human beings* in which the birth of a new concept takes efforts by many who, sometimes, can get bogged down in dead-ends.

At the end of the book, indexes are provided, one for *notation* and the other for *terminology*.

Last but not least, a short *bibliography*, deliberately selective and multilingual, aims at encouraging readers to explore other viewpoints on the theme of this course and also to discover some further developments.

Bures-sur-Yvette,   *Jean-Pierre Bourguignon*
21 July 2022

Contents

Part I THE ANALYTIC SETTING

Part II THE GEOMETRIC SETTING

Part III THE CALCULUS OF VARIATIONS

Part I
THE ANALYTIC SETTING

"*Considérée sous ce point de vue, l'Analyse mathématique est aussi étendue que la Nature elle-même... Son attribut principal est la clarté ; elle n'a pas de signes pour exprimer des notions confuses.*"

J. FOURIER,
in *Théorie analytique de la chaleur.*

Chapter I
A First Generalisation of the Notion of Space: Spaces of Infinite Dimension

The object of the first three chapters is to provide the analytic framework for the development of Variational Calculus. We present a number of basic ideas of functional analysis in a language which is deliberately geometric. To avoid making these chapters too lengthy, the basic definitions of topology used in the study of metric spaces are collected in an appendix which should be regarded as a glossary rather than as a course on this subject. For a more systematic introduction to topology, the interested reader may refer to (Choquet 1984) or (Schwartz 1970).

In this first chapter, the emphasis is on the differences which exist between vector spaces of finite dimension and those of infinite dimension. The algebraic properties of finite-dimensional vector spaces are part of the prerequisites, and are assumed known.

Despite the introductory character of this chapter, it includes several non-trivial results. This refutes once again the, unfortunately widespread, idea that mathematics is merely a language.

After a brief presentation of some examples of infinite-dimensional spaces in Sect. A, the basic properties of normed spaces are reviewed in Sect. B.

The concept of a compact space presented in Sect. C is quite general, but is particularly important in metric spaces, a context to which we shall almost always restrict ourselves. We review the fundamental properties of maps (or functions) defined on compact spaces. We also give several criteria for a space to be compact.

Thanks to the theorem of Riesz presented in Sect. D, this concept provides a simple criterion to distinguish spaces of finite dimension from those of infinite dimension.

J.-P. Bourguignon, *Variational Calculus*, Springer Monographs in Mathematics,
https://doi.org/10.1007/978-3-031-18307-2_1

A. A First Encounter with Infinite-Dimensional Vector Spaces

In this section the simplest examples of infinite-dimensional vector spaces are reviewed, namely spaces of sequences and spaces of functions defined on domains in finite-dimensional vector spaces.

I.1. If $\mathbb{K}$ is one of the fields $\mathbb{R}$ or $\mathbb{C}$ and n a natural number, the space $\mathbb{K}^n$ of n-tuples $(x^1, \cdots, x^n)$ of elements of $\mathbb{K}$ serves as the model for all finite-dimensional vector spaces over the field $\mathbb{K}$. This space, which we shall call *numerical space*, possesses a natural basis $(e_i)_{1 \le i \le n}$: by definition, the n-tuple e_i has a 1 in the ith place and zeros elsewhere.

This example immediately suggests the idea of considering infinite sequences of elements of $\mathbb{K}$ indexed by the set $\mathbb{N}$ of natural numbers. These *numerical sequences* form a space which we denote by $\mathbb{K}^{\mathbb{N}}$. To avoid confusion with an n-tuple of elements of $\mathbb{K}$, it will be useful to *denote* by (x) the sequence $(x_n)_{n \in \mathbb{N}}$.

I.2. One problem presents itself immediately: since the allowed linear combinations in this vector space are all finite sums (we are in the realm of algebra!) the generalisation of the natural basis mentioned above does not generate $\mathbb{K}^{\mathbb{N}}$, but only the subspace $\mathbb{K}_0^{\mathbb{N}}$ formed by the sequences all of whose terms are zero outside a certain range. To have an algebraic basis of $\mathbb{K}^{\mathbb{N}}$, it is necessary to have *many more* basis vectors (it is necessary to take vectors corresponding to sequences which have an infinite number of non-zero terms; such vectors form an uncountable space).

I.3. It is particularly instructive to consider other subspaces of $\mathbb{K}^{\mathbb{N}}$ which can be introduced by using some simple properties of sequences and series. For example:

  i) the space $\mathbb{K}c_0$ (or more simply c_0) is by definition *the space of sequences of elements of $\mathbb{K}$ which tend to 0*;
 ii) the space $\mathbb{K}l^p$ (or more simply l^p) is formed by *the sequences (x) such that the series $\sum_n |x^n|^p$ converges*;
iii) we denote finally by $\mathbb{K}l^\infty$ (or more simply l^∞) *the space of bounded sequences*.

I.4. It is, of course, possible to carry out this construction with infinite sets other than $\mathbb{N}$, possibly uncountable. These lead to isomorphic spaces only if the index sets have the same cardinal.

I.5. From our point of view, it is natural to consider spaces of functions which take their values in a numerical space (we then speak of numerical functions and *scalar functions*, if they take their values in $\mathbb{K}$).

Note that, in spaces of scalar functions, it is not only possible to add functions but also to multiply them. This gives spaces of scalar functions an *algebra* structure.

We shall be particularly interested in functions defined on domains Ω in a numerical space $\mathbb{K}^n$. This emphasises the fact that the numerical space does not really interest us as a vector space, but rather as an *affine space*, i.e., as a space on which the group of translations acts transitively (one can think of this as a vector space in which the origin has been *lost*). An n-tuple will in general be called a *point* and the unique translation which maps a point p to a point q is the n-tuple $\overrightarrow{pq} = q - p$ (considered now as a vector). This justifies the additive notation $p + \vec{v}$ where, although they are of the same nature, p and $\vec{v}$ are viewed as a point and as a vector respectively.

I.6. The space $\mathbb{K}^{\Omega}$ of all functions defined on Ω with values in $\mathbb{K}$ is a vector space of infinite dimension unless Ω is a finite set.

It is rather more interesting to consider the vector subspaces formed by functions having additional properties, such as the vector space $C^0(\Omega)$ of continuous functions on a domain Ω (i.e., the closure of an open set of the space $\mathbb{K}^n$) or the vector space $C^k(\Omega)$ of k-times continuously differentiable functions on Ω (if needed, a definition will be found in Chapter III).[1] The space $C^{\infty}(\Omega)$ of infinitely differentiable functions on Ω will also be important for us.

The case of *curves*, obtained when the domain Ω is an interval $[a, b]$ of the real line, is of course of particular interest (we often use the notation I to denote an interval). The subset formed by the curves γ such that $\gamma(a) = p$ and $\gamma(b) = q$, where p and q are two chosen points, arises in the variational problem of determining shortest paths which will be presented in Chapter IX.

I.7. There are also other spaces of interest such as the space of bounded functions, the *Lipschitz* functions (cf. (I.12)) or more generally the *Hölder[2] continuous* functions (cf. (I.14)) or the *integrable* functions (for us, this will mean in the sense of Riemann, but there are other definitions of the notion of an integrable function such as that of Lebesgue[3] which has become the classical one, as well as several others).

I.8. For all these spaces, it is impractical to use algebraic bases. To work in spaces whose points are functions, it turns out to be essential to use additional structures which allow one to speak of the distance between two functions, and of the length of a function by its distance from the zero function. We turn to these matters now.

B. A Useful Special Case: Normed Spaces

We consider spaces which possess a distance, notably vector spaces in which the distance comes from a norm, since these are a family of spaces which enjoy convenient properties (such as being Hausdorff[4], in finite dimensions, all the norms defined on a vector space are equivalent, as is shown by the fundamental Theorem I.32.

---

[1] These spaces are the natural ones to use when modelling fields appearing in physics or mechanics: a point there defines, at each point of a region of physical space, a scalar or vector quantity.

[2] The German mathematician Otto Ludwig Hölder (1859–1937) contributed to many branches of mathematics, from algebra to the Calculus of Variations, and from group theory to potential theory.

[3] The French mathematician Henri Lebesgue (1875–1941) was the author of numerous works on the analysis of functions of a real variable. His integration theory is still the standard one today.

[4] Felix Hausdorff (1868–1942) was a German mathematician. He is considered to be one of the founders of modern topology. He contributed significantly to set theory and measure theory, and altogether to the development of functional analysis. In 1942, refusing to move to a concentration camp they were sent to as Jews, he and his wife committed suicide.

I.9. We recall that a *distance* d on a set M is a function on $M \times M$ with values in $\mathbb{R}^+$ such that

(D1) d is *symmetric* (i.e., $d(x,y) = d(y,x)$ for all x, $y \in M$);

(D2) the *triangle inequality* holds (i.e., we have $d(x,z) \leq d(x,y) + d(y,z)$ for all x, y, $z \in M$);

(D3) $d(x,y) = 0$ is equivalent to $x = y$.

One then says that (M, d) is a *metric space*.

I.10. It is sometimes convenient to generalise the notion of a distance to that of a *pseudodistance* by requiring that the function d satisfies $d(x,x) = 0$ instead of condition (D3) and by allowing it to take the value $+\infty$.

I.11. In a metric space (M, d), certain subsets play a special role: given a point x of M and a positive real number r, one can consider the open ball centred at x with radius r, i.e., the set $B_r(x) = \{y \mid y \in M,\ d(x,y) < r\}$.

A subset N of a metric space (M, d) is called *bounded* if it is contained in at least one ball (of finite radius !).

It is now possible to define *bounded maps* between metric spaces: these are those which map bounded sets to bounded sets.

Definition I.12. Let f be a map of a metric space (M, d_M) to a metric space (N, d_N). If there exists a non-negative number k such that, for all x, $y \in M$,

$$d_N(f(x), f(y)) \leq k\, d_M(x, y),$$

f is called *Lipschitz*[5] of weight k. If $k < 1$, the map f is called a *contraction*.

Remark I.13. It might be helpful to stress that a Lipschitz function of weight 1 *is not* necessarily a contraction, and to look for an explicit example of such a function.

I.14. For $0 < \alpha < 1$, a map f of a metric space (M, d_M) to a metric space (N, d_N) is called *Hölder* of order α if there exists a non-negative number k such that, for all x, $y \in M$,

$$d_N(f(x), f(y)) \leq k\, (d_M(x, y))^{\alpha}.$$

A Hölder map is necessarily continuous.[6]

I.15. In a metric space (M, d), a *topology* is induced by the distance by declaring that a set U is *open* if, for any point x of U, there exists a number $r > 0$ such that the ball $B_r(x)$ is contained in U. Among all topological spaces, those whose topology can be defined by a distance enjoy pleasant properties.

---

[5] The German mathematician Rudolf Otto Sigismund Lipschitz (1832–1903), professor at the University of Bonn from 1864, was interested in several branches of mathematics, from number theory to differential geometry through the theory of Fourier series. We owe to him a criterion of *flatness* of a space, which was also discovered by Riemann.

[6] This is why one often speaks of *Hölder continuous* maps as the Hölder condition strengthens continuity.

> *Throughout this course, we shall restrict our considerations to*
> *metric spaces.*[7]

I.16. Of all the properties enjoyed by metric spaces, we shall need only one (which will be used frequently in the sequel).

Proposition I.17. *Every metric space is Hausdorff, i.e., any two distinct points have disjoint neighbourhoods.*

Proof. Let x and y be two distinct points of a metric space (M, d_M) whose distance apart $d_M(x, y) = r$ is thus $\neq 0$. The open balls of radius $\leq r/2$ centred at x and y are disjoint neighbourhoods of x and y. □

I.18. Recall that, *in a Hausdorff space, every convergent sequence has a unique limit.* In fact, if a sequence (x) converges to two distinct points, then all the terms of this sequence, except perhaps for a finite number of them, belong to any given neighbourhood of the two points. This is impossible, since in a Hausdorff space the neighbourhoods can be chosen to be disjoint.

I.19. One should not think that the non-Hausdorff spaces to be excluded from our considerations are fundamentally pathological. In fact, there are a number of geometric situations where such spaces appear naturally. A notable case is that of spaces defined as quotients, which appear for example when one considers the space of trajectories of a differential equation. (Exercise: Give an explicit example of such a situation for a differential equation defined in the plane.)

I.20. In a vector space E, we shall concentrate our attention on distances which arise from a norm.

We recall that a *norm N* is a function (often denoted by $\| \; \|$) defined on E and with values in $\mathbb{R}^+$ with the following three properties:

(N1) N is *positively homogeneous of degree 1*, i.e., for any scalar λ and any vector v in E, $\|\lambda v\| = |\lambda| \|v\|$,

(N2) N is *sub-additive*, i.e., for any vectors v and w in E, we have the inequality
$N(v + w) \leq N(v) + N(w)$,

(N3) N vanishes on the zero vector.

A vector space provided with a norm is called a *normed space* and the associated distance d_N is defined by $d_N(v, w) = N(v - w)$. (We leave to the reader the verification of the fact that d_N is indeed a distance, noting only the relation which exists between the sub-additivity and the triangle inequality.)

Remark I.21. A function defined on E which satisfies only properties (N1) and (N2) is called a *semi-norm*. Such functions will be the object of repeated comments throughout this course.

I.22. There are several natural families of norms. We begin by reviewing some of those which are defined on $\mathbb{K}^n$.

---

[7] One should not conclude, however, that all interesting spaces are metrisable!

For example, to an n-tuple $x = (x^1, \cdots, x^n)$ is associated its *uniform norm* defined by $\|x\|_\infty = \sup_i |x^i|$ (for a justification of the notation $\| \ \|_\infty$, see Exercise I.29). The *standard Euclidean[8] norm* of x is the non-negative real number $\|x\|_2 = (\sum_i (x^i)^2)^{1/2}$; more generally, for $p \geq 1$, the *norm of order p* is $\|x\|_p = (\sum_i |x^i|^p)^{1/p}$.

I.23. The above definitions carry over word for word to spaces of sequences such that the corresponding series converge. Thus, on c_0 and on l^∞, we define the uniform norm by $\|(x)\|_\infty = \sup_i |x^i|$ and on l^p the norm of order p by $\|(x)\|_p = (\sum_i |x^i|^p)^{1/p}$.

I.24. These norms generalise to the *product of a finite number of normed spaces* by replacing the absolute value which, in the above formulas, was used as the norm on the vector space $\mathbb{R}$, by the given norm in each space.

Thus, given two normed spaces (E_1, N_1) and (E_2, N_2), the map which to (v_1, v_2) associates $N(v_1, v_2) = \sqrt{N_1^2(v_1) + N_2^2(v_2)}$ defines a norm on the direct sum space $E = E_1 \oplus E_2$. Similarly for $N' = \max\{N_1, N_2\}$.

I.25. The analogous norms defined on spaces of functions play a very important role in analysis. Many of the advances of modern analysis depend on the use of norms adapted to the problem under consideration: this provides great flexibility.

I.26. We define the *uniform norm* of a bounded function f on a domain Ω by putting $\|f\|_\infty = \sup_{x \in \Omega} |f(x)|$.

On spaces of k-times differentiable functions defined on a *bounded* domain, it is possible, using the uniform norm, to define, as in the case of product spaces, a number of equivalent norms involving the derivatives of order up to k. To avoid complicating the notation, and so as not to anticipate subsequent chapters which treat the derivatives of functions of several variables, we shall restrict ourselves here to functions of *one* real variable, namely *spaces of curves*. For $f \in C^k(I)$, where I is a closed bounded interval in $\mathbb{R}$, we take for example $\|f\|_{C^k} = \sup_{0 \leq i \leq k} |f^{(i)}|_\infty$, where $f^{(i)}$ denotes the derivative of order i of the function f.

I.27. In the case of $C^\infty(\Omega)$, it is natural to look for a definition of closeness of two infinitely differentiable functions involving *all* their derivatives. It is possible to define a distance which involves infinitely many derivatives but it does not arise from a norm (recall that Ω is assumed here to be a closed *bounded* domain): for f_1, $f_2 \in C^\infty(\Omega)$, one can take for example

$$d_\infty(f_1, f_2) = \sum_{i=0}^{\infty} \left(\frac{1}{2}\right)^i \frac{\|f_1 - f_2\|_{C^i}}{1 + \|f_1 - f_2\|_{C^i}}.$$

I.28. The *norm of order p* of a function f defined and integrable on a domain Ω is given by $\|f\|_p = \left(\int_\Omega |f(x)|^p \, dx\right)^{1/p}$. Clearly, for these norms it is necessary to make precise the notion of integral which is being used.

---

[8] Euclid, Greek mathematician of the third century B.C., is the author of the *Elements*, which some say, apart from religious works, has had the greatest influence on the development of human thought. In Antiquity, he was known as "O $\Sigma \tau o \iota \chi \epsilon \omega \tau \eta \zeta$" ("*the author of the Elements*") or as "O $\Gamma \epsilon \omega \mu \acute{\epsilon} \tau \rho \eta \zeta$" ("*the geometer*").

*Throughout this course, the integral used will always be
that of* Riemann.[9]

Exercise I.29. Let f be a continuous real-valued function on a bounded domain I.
Show that $\|f\|_\infty = \lim_{p \to \infty} \|f\|_p$.

I.30. If, on a vector space E, d_N is the distance induced by a norm N, note that, for
two points v and w of E, $d_N(v, w) = \|v - w\|$ is the norm of the translation mapping
v to w. One checks easily that, for this distance, the translations are isometries. For
the topology defined by this distance (whose base of open sets is formed by the open
balls), the translations are thus homeomorphisms. (Exercise: Prove it !) This shows
that for this topology the zero vector, which forms the origin of the vector space,
does not play a distinguished role. This is exactly what we require when working in
an affine space.

Two norms N_1 and N_2 can define the same topology. For this it suffices that every
open ball for one norm contains an open ball for the other, and conversely. This
geometric property translates, in terms of the functions N_1 and N_2, into the existence
of a constant $k > 0$ such that $k^{-1} N_1 \le N_2 \le k N_1$. The norms N_1 and N_2 are then
called *equivalent*.

I.31. We now give a fundamental result which shows that finite-dimensional vector
spaces are particularly simple topologically. Its proof uses the notion of compactness
which is presented a little later in this chapter.[10]

Theorem I.32. *On a finite-dimensional vector space, all semi-norms are continuous
(e.g. with respect to the uniform norm defined by a basis), and all norms equivalent.*

Proof. First of all we show that every semi-norm S on $\mathbb{K}^n$ is continuous. For any point $(x^1, \cdots, x^n)$
we have

$$S\left((x^1, \cdots, x^n)\right) = S\left(\sum_{i=1}^n x^i e_i\right) \le \sum_{i=1}^n |x^i| S(e_i) \le \left(\max_i S(e_i)\right) \sum_{i=1}^n |x^i|.$$

For the topology induced by the distance whose balls are cubes, the function S is thus continuous
at 0. To obtain continuity at an arbitrary point y, it is enough to use the relation which follows
immediately from the triangle inequality $|S(y + x) - S(y)| \le S(x)$.

We now show that all norms on $\mathbb{K}^n$ are equivalent. For this we observe that it suffices to prove
that any norm N is equivalent to the cubic norm N_1. Since the functions N/N_1 and N_1/N are
well-defined and continuous on the boundary of the unit cube, which is a compact subset of $\mathbb{K}^n$
by Proposition I.56 and Corollary I.68, they are bounded there by Proposition I.68. Since these
functions are homogeneous of degree 0 on $\mathbb{K}^n$, they are bounded everywhere.

---

[9] The German mathematician Bernhard Riemann (1826–1866) made decisive contributions to a
great variety of branches of mathematics, from geometry to analysis through number theory and
function theory. His inaugural lecture *"Die Hypothesen, welche der Geometrie zu Grunde liegen"*,
given in 1854, can be considered to be the birth of modern geometry. As was the tradition, this
subject was chosen from among three submitted by the jury of which Gauss was a member.

[10] In a mathematical work it is always a little dangerous to use results which are only established
until later, since one risks introducing vicious circles. We leave it to the reader to check that this is
not the case here.

We now turn to the general case. Choosing a basis of a vector space E of dimension n can be interpreted as defining a linear isomorphism f of E with $\mathbb{K}^n$. Moreover, if S is a semi-norm on E, then $S \circ f^{-1}$ is a semi-norm on $\mathbb{K}^n$. Similarly, if N is a norm on E, then $N \circ f^{-1}$ is a norm on $\mathbb{K}^n$; moreover, the balls defined by these norms correspond bijectively under f. This implies that the result established for $\mathbb{K}^n$ is in fact valid for any finite-dimensional vector space.   □

Corollary I.33. *Any linear map defined on a finite-dimensional space, provided with any norm, taking values in a normed space is continuous.*

Proof. It is enough to note that the composite of a linear map and a norm defines a semi-norm and to apply the preceding theorem.   □

Exercise I.34. Show that, in a finite-dimensional vector space, the balls defined by two different norms are homeomorphic.

I.35. Theorem I.32 and its Corollary I.33 make clear that *for finite-dimensional vector spaces the topologies induced by norms introduce no finer distinctions than the algebraic structure.*

The same is not true in infinite dimensions, for Theorem I.32 has no analogue there.[11] (It is easy to give counterexamples, see Exercise I.36 for example.)

Corollary I.33 does not generalise to infinite dimensions either, and it is advisable to distinguish between those linear maps which are continuous and those which are not.

Exercise I.36. Show that on l^p the norms of order q for $p < q$ are well-defined but inequivalent.

Project I.37. Show that on $C^1([0, 1], \mathbb{K})$ the uniform norm and the norm of order 1 are not equivalent.

Proposition I.38. *On the vector space $L(E, F)$ of continuous linear maps from a normed space $(E, \| \ \|_E)$ to a normed space $(F, \| \ \|_F)$, the function $\| \ \|_{L(E,F)}$ defined by*

$$\|l\|_{L(E,F)} = \sup_{\|v\|_E \leq 1} \|l(v)\|_F$$

is a norm.

Proof. The only part of the proposition which needs to be checked is the fact that, for a continuous linear map l from a normed space E to a normed space F, $\|l\|_{L(E,F)}$ is finite since the inequalities which must be satisfied for $\| \ \|_{L(E,F)}$ to be a norm are then immediate consequences of the same properties for the norms on E and F.

The finiteness of $\|l\|_{L(E,F)}$ is a direct consequence of the continuity of l at 0, for the inverse image under l of an open ball of non-zero radius in F necessarily contains a neighbourhood of 0 in E, and hence a ball of non-zero radius in E.

The map l being linear, this ensures that the image under l of the unit ball of E is contained in a ball of finite radius of F.   □

---

[11] In fact, in many problems of modern analysis the crux is to show that two norms are equivalent, or that a semi-norm is continuous.

In spaces of maps, it is the norm defined in I.38,
called the operator norm, *which will always be used.*

I.39. An immediate extension of the preceding proof establishes the following important equivalence.

Corollary I.40. *For a linear map l from a normed space E to a normed space F, the following four properties are equivalent:*

  i) l is continuous at 0;
  ii) l is continuous at every point;
  iii) $\sup_{\|v\|_E \leq 1} \|l(v)\|_F$ *is finite;*
  iv) l is Lipschitz.

Exercise I.41. Show that, for any continuous map l from a normed space E to a normed space F, $\|l\|_{L(E,F)}$ is the smallest number k such that, for all $v \in E$, $\|l(v)\|_F \leq k \|v\|_E$.

Exercise I.42. Let E, F and G be three normed spaces. For $l_1 \in L(E,F)$ and $l_2 \in L(F,G)$, show that we have $\|l_2 \circ l_1\|_{L(E,G)} \leq \|l_1\|_{L(E,F)} \|l_2\|_{L(F,G)}$.

Deduce that the composition of linear maps is a continuous bilinear map of norm 1 from $L(E,F) \times L(F,G)$ to $L(E,G)$ (for the norm N' on the product, cf. I.24).

Exercise I.43. Let E and F be two normed spaces and l a continuous linear map from E to $L(E,F)$ (it will be convenient to denote the image of a vector v under l by l_v). Show that the map which associates to l the bilinear form $\hat{l}$ on E with values in F defined, for vectors v_1 and v_2 of E by $\hat{l}(v_1, v_2) = l_{v_1}(v_2)$, is an isomorphism of normed spaces between $L(E, L(E,F))$ and the space $L_2(E;F)$ of continuous bilinear forms on E with values in F.

Project I.44. Show that, if norms on E and F are replaced by equivalent norms, the induced norm on $L(E,F)$ is equivalent to the original norm, and hence the induced topology is the same.

C. Compact Spaces

We now embark on the study of a central concept of analysis, which generalises the notion of finiteness to the realm of topology. Thus, this section contains several fundamental results which find numerous applications in very varied parts of mathematics.

Although the subsets of a topological space to be singled out are well understood in the case of a finite-dimensional vector space (see Theorem I.64), it is more difficult to develop an intuitive feeling for them in infinite dimensions, *but this is nevertheless fundamental.*

I.45. Recall that a family of open sets is called a *covering* of a space if every point of the space is contained in at least one open set of the family.

Definition I.46. A topological space is said to be *compact* if it is Hausdorff and if it satisfies the following so-called Borel[12]–Lebesgue property:

(BL) *from every open covering one can extract a finite subcovering.*

I.47. A variant of the preceding definition often called the *finite intersection property* can be stated as follows:

Proposition I.48. *A Hausdorff space is compact if every family of closed sets, of which every finite subfamily has non-empty intersection, itself has non-empty intersection.*

Proof. We show by contradiction that this property is indeed equivalent to Definition I.43. Let $(U_i)_{i \in I}$ be an open covering of a Hausdorff topological space E and denote by F_i the closed complement of the open set U_i. Suppose that the covering $(U_i)_{i \in I}$ has no finite subcovering. This means that every finite subfamily $F_{i_1}, ..., F_{i_p}$ of the family of closed sets $(F_i)_{i \in I}$ has a non-empty intersection consisting of the points which prevent the family $U_{i_1}, ..., U_{i_p}$ from being a covering of E. But $\bigcap_{i \in I} F_i = \emptyset$, hence the contradiction.

Conversely, if we start with a family of closed sets $(F_i)_{i \in I}$ having the finite intersection property but empty intersection, then the family of open complements is a covering of E with no finite subcovering.   □

Proposition I.49. *In a compact space, every (infinite) sequence has at least one accumulation point.*

Proof. Let $(x_n)_{n \in \mathbb{N}}$ be a sequence of points of a compact space E with no accumulation point. Then every point of E has a neighbourhood, hence an open neighbourhood, which contains only a finite number of points of the sequence. We can thus construct in this way an open covering of E. Since E is assumed to be compact, this covering admits a finite subcovering, which implies that the sequence has only a finite number of elements, a contradiction.   □

Corollary I.50. *Every closed subspace of a compact space is compact.*

Proof. We note first of all that every closed subspace of a Hausdorff space is itself a Hausdorff space.

The corollary is then immediate from the finite intersection property, since the closed subsets of a closed subspace are all closed subsets of the space.   □

Exercise I.51. Show that every compact subset of a Hausdorff topological space is closed.

---

[12] Émile Borel (1871–1956), French mathematician and politician, was one of the founders of measure theory. He also contributed to the study of summability of divergent series, a subject which is of interest today to theoretical physicists in the study of perturbation expansions in quantum mechanics and in field theory. He received the first Gold Medal of the Centre National de la Recherche Scientifique.

I.52. Compact sets have many interesting properties. We do not wish to go into too much detail about this, particularly since we want to work mainly in metric spaces for which other formulations of these properties are more effective.

We mention here only one property whose novelty is that it deals with the direct image of a space under a map, as opposed to what generally does in topology where one more often considers the inverse images to express properties of maps.

Proposition I.53. *The image of a compact space under a continuous map taking its values in a Hausdorff space is compact.*

Proof. Let f be a continuous map from a compact space E to a Hausdorff space F. Since F is Hausdorff, so is $f(E)$. Now, if $(U_i)_{i \in I}$ is an open covering of $f(E)$, the $f^{-1}(U_i)$ form an open covering of E, from which it is possible to extract a finite subcovering which we denote by $(f^{-1}(U_i))_{i \in J}$. Since we have $f(f^{-1}(U_i)) = U_i$, the $(U_i)_{i \in J}$ form a finite subcovering of the original covering. □

Remark I.54. Note that *it is not true* that the inverse image of a compact subspace under a continuous map is necessarily compact. (Exercise: Give an example of such a situation.)

A map which preserves compact sets under taking inverse images is called *proper*.

Corollary I.55. *Every continuous bijection from a compact space to a Hausdorff space, in particular to a metric space, is a homeomorphism.*

Proof. Let f be a bijection from a compact space E to a space F. As f is assumed to be continuous, it remains to prove that f^{-1} is continuous, for example that, for any closed (hence compact) subset A of E, its image $f(A)$ is closed in F. By Proposition I.53, $f(A)$ is compact, hence closed. □

Theorem I.56. *Any finite product of compact spaces is compact.*

Proof. For simplicity we give the proof for the product of two spaces, but the case of a finite number of spaces clearly reduces to this (1.5 gives a review of the product topology on a product space).

Let $E = E_1 \times E_2$ be a product of compact spaces. We show that E is Hausdorff. In fact, if $x = (x_1, x_2)$ and $y = (y_1, y_2)$ are two distinct points of E, they differ in one of their coordinates, say the first. The neighbourhoods $U = U_1 \times E_2$ of x and $V = V_1 \times E_2$ of y, where U_1 and V_1 are open sets separating x_1 and y_1, are obviously disjoint.

Now let $(U_i)_{i \in I}$ be an open covering of E. If $x = (x_1, x_2)$ is a point of E, there exists an open set U_{i_x} such that $x \in U_{i_x}$. Now there exists a product neighbourhood $V^x = V_1^x \times V_2^x$ of x which is contained in U_{i_x}. Since the subspace $E_{x_1} = \{x_1\} \times E_2$ of E is homeomorphic to E_2, it is compact. The open sets $(V^x)_{x \in E_{x_1}}$ form an open covering of E_{x_1} from which one can extract a finite subcovering $(V^{x^j})_{1 \le j \le l}$. This picks out l open neighbourhoods of x_1 (which is the common projection of all the x^j in E_1), namely the factors in E_1 of the product neighbourhoods V^{x^j}. Let V^{x_1} be their intersection, which is then a neighbourhood of x_1. Then $V^{x_1} \times E_2 \subset \bigcup_{1 \le j \le l} V^{x^j}$. The open sets $(V^{x_1})_{x_1 \in E_1}$ form a covering of E from which one can extract a finite subcovering which picks out a finite number of points $(x^k)_{1 \le k \le m}$ of E_1. The open sets of the original covering which we have associated to the points $(x_1^k)_{1 \le k \le m}$ cover E since

$$E_1 \times E_2 \subset \bigcup_{1 \le k \le m} V^{x_1^k} \times E_2 \subset \bigcup_{\substack{1 \le j \le l \\ 1 \le k \le m}} U_{i_{(x_1^k, x_2^j)}} \subset E_1 \times E_2 \, .$$

From the original covering we have thus extracted a finite subcovering. □

I.57. A theorem of Tychonov[13] asserts that *any product of compact spaces is compact*. Its proof requires a serious excursion into the *theory of cardinals* and the introduction of very interesting objects called *ultrafilters* which are very useful in topology and logic.

I.58. We shall now be interested in properties of compact metric spaces. Before taking up this subject, it is useful to have available a fundamental lemma.

Lemma I.59. (Lebesgue's Uniformity Lemma) *For any open covering $(U_i)_{i \in I}$ of a metric space E, there exists a number $\rho > 0$, called a Lebesgue number of the covering, such that, for any point x of E, the ball $B_\rho(x)$ is contained in at least one of the open sets U_i, unless there exists in E an infinite sequence of points having no convergent subsequence.*

Proof. Suppose that for any number $\rho > 0$, there exists a point x_ρ such that the ball $B_\rho(x_\rho)$ is not contained in any of the open sets U_i. By taking successively $\rho = 1, 1/2, \cdots, 1/n$, we obtain a sequence of points which we denote by $(x_n)_{n \in \mathbb{N}}$.

For the sequence thus formed, there are two possibilities: either no subsequence of it converges and we are in the second case in the statement of the lemma, or at least one subsequence converges.

If a subsequence of the sequence of the x_n converges to a point x_∞, we shall obtain a contradiction. In fact, the point x_∞ belongs to at least one open set U_{i_0} of the covering. It is thus possible to find a number ρ such that the open ball $B_\rho(x_\infty)$ is contained in U_{i_0}. Since the sequence $(x_n)_{n \in \mathbb{N}}$ converges to x_∞, for n sufficiently large the ball of centre x_n and radius $1/n$ is contained in $B_\rho(x_\infty)$, and hence in U_{i_0}, whence the stated conclusion.   □

I.60. We can now introduce a characteristic property of compact metric spaces which generalises that of Bolzano[14]–Weierstrass[15] for sets of real numbers.

Theorem I.61. *A metric space is compact if and only if every infinite sequence of points of the space contains a convergent subsequence.*

Proof. Let E be a compact metric space. Let $(a_n)_{n \in \mathbb{N}}$ be an infinite sequence of points of E. In order for the sequence $(a_n)_{n \in \mathbb{N}}$ to contain a subsequence which converges to a point x, it is necessary and sufficient that every neighbourhood of x contains infinitely many points of the sequence. If we assume that every point of E has a neighbourhood which contains only a finite number of points of the sequence, we obtain a contradiction from I.49.

Conversely, suppose that in a metric, hence Hausdorff, space E, from every infinite sequence one can extract a convergent subsequence. We show that the compactness of E follows from Lebesgue's Uniformity Lemma I.59. In fact, let $(U_i)_{i \in I}$ be an open covering of E. By the Lebesgue lemma, we can find a number ρ such that, for any point x of E, the ball $B_\rho(x)$ is contained in at least one U_i. Let x_1 be a point of E. If $B_\rho(x_1)$ does not cover E, we can find a point x_2 such that $d(x_1, x_2) \geq \rho$. In general, suppose points $x_1, x_2, ..., x_i$ are defined whose mutual distances apart are at least ρ. We can continue the construction as long as the union of the $B_\rho(x_i)$ does not cover E completely. If the construction could be continued indefinitely, we would be able to construct an infinite sequence of points whose mutual distances apart are bounded below by $\rho \neq 0$, and hence a sequence no subsequence of which can converge.   □

---

[13] Andrei Nikolayevich Tychonov (1906–1993) was a Russian mathematician who contributed to functional analysis, mathematical physics, and to the solution of ill-posed problems.

[14] Bernhard Bolzano (1781–1848) was a Czech philosopher and mathematician of Italian origin.

[15] The German mathematician Karl Theodor Wilhelm Weierstrass (1815–1897) was one of the founders of the modern theory of functions, completing in particular the work of Abel and Jacobi on elliptic functions. He had a decisive influence on mathematics at the end of the XIXth century.

I.62. Recall that a continuous map f from a metric space E to a metric space F is said to be *uniformly continuous* if, given $\epsilon > 0$, there exists $\delta > 0$ such that, for any points x, y of E satisfying $d_E(x, y) < \delta$, $d_F(f(x), f(y)) < \epsilon$.

Theorem I.63. *Every continuous map from a compact metric space to a metric space is uniformly continuous.*

Proof. For any $x \in E$, put $U_x = f^{-1}(B_\epsilon(f(x)))$, which is an open neighbourhood of x since f is continuous. The open sets U_x clearly form an open covering of E. Let ρ be a Lebesgue number of this covering. Since E is assumed to be compact, every ball of radius ρ in E is contained in at least one of the U_x, which means that, for any y and z in such a ball, $d_F(f(y), f(z)) < \epsilon$. □

D. Looking For Compact Sets

In view of the important role played by the compact subsets of a topological space, it is crucial to obtain criteria for recognising these sets. This leads us to give a very clear distinction between (normed) spaces of finite dimension and those of infinite dimension.

We begin with a classical but fundamental theorem.

Theorem I.64. *The compact subsets of a finite-dimensional vector space are the closed bounded sets.*

Proof. Let E be a vector space of finite dimension n. We show first of all that every compact subset K of E is closed and bounded. By Lebesgue's Uniformity Lemma I.59 and Theorem I.61, K can be covered by a finite number of balls, and hence is bounded. Moreover, the same Theorem I.61 ensures that every convergent sequence of elements of K has a limit there, and hence that K is closed.

We now show that every closed bounded subset B of E is compact. We note first that E is Hausdorff. Since a cube of $\mathbb{K}^n$, being a product of compact intervals, is compact by Proposition I.56, we see on taking a basis of E that the balls for the uniform norm associated to this basis, being the images of compact subsets of a Hausdorff space, are compact by Proposition I.53. All the norms on E being equivalent by Theorem I.32, the bounded set B is contained in at least one ball for the cubic norm. Since B is also closed, it is compact by I.50. □

Corollary I.65. *In a normed space, every finite-dimensional subspace is closed.*

Proof. Let F be a finite-dimensional subspace of a normed space E. Suppose that there is a point x_∞ in the closure of F but not in F. We can find a sequence $(x_n)_{n \in \mathbb{N}}$ of points of $B_{2\|x_\infty\|} \cap F$ which converges to x_∞. Since the points of this sequence belong to a bounded closed subset of F, by Theorem I.64 we can extract from the sequence a subsequence which converges to a point of $B_{2\|x_\infty\|} \cap F$, which is thus different from x_∞. By the uniqueness of the limit, this is a contradiction. □

Exercise I.66. Exhibit subspaces of c_0 and of $C^0(\Omega)$ which are not closed.

I.67. The Calculus of Variations is especially concerned with maxima and minima of functions. Consequently, it is of fundamental importance to have criteria for showing that such points exist.

Corollary I.68 is a result of this type in the case of compact spaces, and it explains the central role which compactness theorems play in the Calculus of Variations.

Corollary I.68 *Every continuous numerical function on a compact space is bounded. If it has values in* ℝ*, it attains its upper and lower bounds.*

Proof. The image of the space under the numerical function is a compact subset of $\mathbb{K}^n$ by Proposition I.53, and hence is closed and bounded by Proposition I.64.

If the function is real-valued, then its upper and lower bounds are among the accumulation points of the image. Since we have seen that the image is closed, these bounds belong to the image, which means precisely that the function attains its bounds. □

Exercise I.69. Show that the restriction of a linear form (or a quadratic form) defined on a finite-dimensional vector space to the unit ball attains its maximum and minimum on the ball.

I.70. The result of Exercise I.69 does not generalise to infinite dimensions, as Exercise I.71 shows. This proves that the unit ball in an infinite-dimensional space is not necessarily compact. Theorem I.77 asserts that this is, in fact, never the case.

Exercise I.71. Let K be a compact connected subset of $\mathbb{R}^n$. Show that, if $(t_n)_{n \in \mathbb{N}}$ denotes a sequence of points which is everywhere dense in K, the linear form l defined for f in $C^0(K, \mathbb{R})$ (provided with the uniform norm) by $l(f) = \sum_n (\frac{1}{2})^n f(t_n)$ is continuous. Show that l does not attain its upper bound on the unit ball.

I.72. Among the subsets of a normed space (E, N_E) it is natural to consider the spheres and (closed) balls associated to the norm N_E. Theorem I.64 asserts that these subsets are compact if the space is finite-dimensional. Since every point of these spaces has a compact neighbourhood, these spaces are said to be *locally compact*.

This is the first time we have met the idea that one can *localise* notions which, up to now, we have considered globally on the space. On numerous occasions throughout this course, we shall consider properties of a space which hold only *in a neighbourhood of each point*.

In the case of spaces which are locally compact but not compact, Corollary I.68 has an interesting generalisation which is often useful in the Calculus of Variations when the function one is interested in behaves simply at infinity.

Exercise I.73. Let f be a continuous numerical function defined on a locally compact space E. Show that, if for any given real number a it is possible to find a compact subset K of E such that, for all x outside K, $a < f(x)$ then f is bounded below and attains its lower bound.

I.74. The result stated in Exercise I.73 has a spectacular application in the form of a proof of the fundamental theorem of algebra, due to d'Alembert.[16]

Theorem I.75. *Every non-constant polynomial of a complex variable has at least one zero.*

Proof. (Sketch) Let P be a non-constant polynomial with complex coefficients. The real-valued function $z \mapsto f(z) = |P(z)|$ is continuous on $\mathbb{C}$. Moreover, one sees that $f(z) \to \infty$ as $z \to \infty$ by extracting as a factor, for $z \neq 0$, the highest power of z which occurs. The function f satisfies the hypotheses of the proposition mentioned above and thus attains its lower bound at a point z_0. We show that this is necessarily zero. For this it suffices to expand the polynomial about z_0 and to show that, if $f(z_0)$ is not zero, one obtains a contradiction by considering the behaviour of the function f along a suitably chosen line segment passing through z_0.   □

I.76. The property of local compactness actually characterises finite-dimensional spaces among normed spaces as the following theorem of Riesz[17] shows.

Riesz Theorem I.77. *A normed space is locally compact if and only if it is finite-dimensional.*

Proof. It only remains to prove the 'only if' part. Let E be a locally compact normed space and denote by U a compact neighbourhood of the origin which we can assume to be a closed ball (since every closed subset of a compact set is compact). Since dilations are homeomorphisms, we can thus assume that the closed unit ball B_1 is compact.

Consider now the covering of B_1 by the open balls of radius $\frac{1}{2}$. Since B_1 is compact, we can cover B_1 by a finite number of balls of radius $\frac{1}{2}$, whose centres we denote by v_i ($1 \leq i \leq k$). Let F be the vector space generated by these points. It remains to show that we necessarily have $E = F$.

Suppose this is not so. Then it is possible to find a vector v in E which is not in F. Since F is closed by Corollary I.65, there exists $\epsilon > 0$ such that the ball $B_\epsilon(v)$ does not meet F, but that $B_{2\epsilon}(v) \cap F \neq \emptyset$. Let $w \in B_{2\epsilon}(v) \cap F$. Introduce the unit vector $z = (v - w)/\|v - w\|$. By definition of the vectors v_i, there exists i such that $\|z - v_i\| \leq 1/2$. By construction v can be written $v = w + \|v - w\| v_i + \|v - w\|(z - v_i)$. Since $w + \|v - w\| v_i \in F$, we thus have $\|v - w\| \|z - v_i\| > \epsilon$; but by hypothesis $\|z - v_i\| \leq 1/2$, hence $\|v - w\| > 2\epsilon$, a contradiction.   □

Exercise I.78. Show that any normed space in which a sphere is compact is finite-dimensional.

I.79. In spaces of functions, as already said, it is important to have criteria for determining whether a space is compact in a suitable topology. This problem has many solutions which often bring into play chains of spaces such as spaces of *Hölder continuous* functions, or *Sobolev*[18] *spaces* which we shall not have time to consider in this course.

---

[16] Jean Le Rond d'Alembert (1717–1783) was a French mathematician, physicist and philosopher who contributed greatly with Denis Diderot to the *Encyclopédie* (with the subtitle *Dictionnaire raisonné des sciences, des arts et des métiers)*. One of his mathematical achievements is the formulation of the wave equation describing the propagation of waves.

[17] The Hungarian mathematician Frederic Riesz (1880–1956) was one of the founders of functional analysis. He made notable contributions to the theory of compact operators, and to that of integral equations.

[18] Sergei Lvovich Sobolev (1908–1989) was a Soviet mathematician working in mathematical analysis. He introduced notions now fundamental to solve partial differential equations. Already in 1935, he obtained weak solutions using generalised functions, ancestors of Schwartz's distributions.

As an example, we give Theorem I.81 which is a useful criterion for compactness in spaces of functions. Before stating it, we must introduce the following notion.

Definition I.80 A subset A of a space of maps from a metric space (E, d_E) to a normed space $(F, \| \ \|_F)$ is said to be *equicontinuous at a point x_0* if

$$\forall \epsilon > 0, \ \exists \delta > 0, \ (d_E(x, x_0) < \delta) \Rightarrow (\forall f \in A, \ \|f(x) - f(x_0)\| < \epsilon) \ .$$

The subset A is said to be *equicontinuous* if it is equicontinuous at every point of E.

Theorem I.81. (Ascoli[19]) *In order for a subset A of the space $C^0(M, \mathbb{K})$ of continuous functions defined on a compact metric space M with values in $\mathbb{K}$ to have compact closure, it is necessary and sufficient that A takes its values in a bounded subset of $\mathbb{K}$ and that A is equicontinuous.*

I.82. For the proof, we refer for example to (Dieudonné 1968). It uses a classical construction of analysis, that of diagonal sequences extracted from sequences of sequences.

From this result one can deduce a theorem which is the prototype of the compactness theorems used in functional analysis, and we leave it as an exercise.

Exercise I.83. Show that, if K denotes a compact subset of $\mathbb{K}^n$, the unit ball of $C^1(K)$ has compact closure in $C^0(K)$.

Project I.84. Show that if K denotes a compact subset of $\mathbb{K}^n$, the unit ball of the space $C^\alpha(K)$ of Hölder continuous functions of order α on K has compact closure in the space $C^\beta(K)$ of Hölder continuous functions of order β when $0 < \beta < \alpha < 1$.

I.85. One of the important themes related to compactness is the study of compact linear operators (an operator f from a normed space E to itself is said to be *compact* if the closure of the image of the unit ball under this operator is compact). This property means roughly that these operators are bounded in a very strong sense. They appear in practice as *integral operators with kernel* of the form $f \mapsto \mathcal{K}(f)$ where $\mathcal{K}(f) = \int_a^b k(x, y) f(y) \, dy$ for a bounded function k on the bounded set $[a, b] \times [a, b]$.

The study of compact operators plays a central role in the *spectral theory of linear operators* in Hilbert[20] spaces, which are defined in Chapter II.

---

[19] The Italian mathematician Giulio Ascoli (1843–1896) made a number of contributions to the theory of functions of a real variable.

[20] The German mathematician David Hilbert (1862–1943) has a place among the greatest mathematicians of all time. He contributed to almost all branches of mathematics, from logic to algebra through analysis and geometry. The 23 problems which he formulated at the International Congress of Mathematicians held in Paris in 1900 have served (and to a lesser extent still serve today) as a reference point for mathematical research. A remarkable school developed around him in Göttingen which played a very important role in the mathematics of that time.

E. Historical Notes

I.86. The study of sets of real numbers played an important role in the birth of topology. The definition of the notion of an accumulation point is due to Weierstrass in about 1860; he used it to prove that every bounded infinite set of real numbers has at least one accumulation point, a result which had been assumed without proof before.

Towards the end of the XIXth century, mathematicians moved away from this rather narrow position and started to consider abstract sets and generalised maps. This did not hinder the development of the calculus of variations, which had already attained a high degree of sophistication. One spoke of *functionals*[21] to denote quantities which depend on objects which are themselves functions. Volterra[22] was certainly one of the first to conceive of a *functional analysis* where the usual arguments about points and vectors are extended to much larger spaces, notably to infinite dimensions (and this is the point which interests us here). This point of view, which today seems commonplace, dates from 1887.

This chapter taught us that, to be able to handle more general objects, a jump in the level of abstraction is necessary. Such a step is frequently productive in mathematics.

I.87. One had to wait until 1906 for Fréchet[23] to introduce the notion of distance on an abstract set. It was only much later (about 1914) that Hausdorff freed himself from the metric notions to introduce what today we call a *topology* by defining the general idea of a neighbourhood.

This period then saw the construction, due mainly to Hilbert, of a true *geometry of infinite-dimensional spaces* with which, for example, he solved certain integral equations. The theory of normed spaces, and above all that of operators on these spaces, was completed in the years around 1930 by Banach.[24] We shall return to it in Chapter II.

I.88. The notion of a *compact* set appeared in 1900 in the work of Hilbert on the existence of certain extrema for variational problems. It was from considerations related to the theory of measures on sets that Borel, and subsequently Lebesgue, developed the modern notion which we have presented in this chapter.

---

[21] Nowadays, this term has fallen somewhat into disuse.

[22] Vito Volterra (1860–1940) was an Italian mathematician and physicist, known for his contributions to the solution of integral equations. He was also interested in mathematical biology.

[23] Maurice Fréchet (1878–1973), French mathematician, studied topology and probability theory. It is to him that we owe the notion of an abstract metric space.

[24] The Polish mathematician Stefan Banach (1892–1945) was one of the founders of modern functional analysis. He created a whole school which contributed to making Poland a country to be reckoned with in the mathematical community.

Chapter II
Banach Spaces and Hilbert Spaces

In this chapter, we present the main families of infinite-dimensional spaces, namely Banach spaces and Hilbert spaces.

The notion of a *Cauchy*[25] *sequence*, known to the reader at least for sequences of numbers, leads naturally to the concept of a *complete metric space* presented in Sect. A. In these spaces (of which compact metric spaces are special cases), one can develop a certain number of general results such as the *fixed point theorem of Picard*[26] which allows one to find the fixed point of a contraction map by successive approximation. When suitably transformed, many of the problems of analysis can be reduced to a form amenable to this theorem.

The category of *Banach spaces*, introduced in Sect. B, can be considered as the cornerstone of functional analysis. They have many interesting closure properties, and contain many of the spaces of functions provided with the topologies introduced in Chapter I.

Particular attention is given to questions of *duality* in Sect. C, since their clarification has constituted important progress in the formalisation of many problems connected with Variational Calculus.

Section D discusses bilinear forms from the viewpoint of duality. Their usefulness will ultimately extend beyond the study of inner products, as we shall also be interested in some special antisymmetric bilinear forms, the *symplectic* forms.

Finally, Sects. E and F are devoted to Hilbert spaces, which are the infinite-dimensional versions of Euclidean or Hermitian spaces (of which they retain many properties). Thus, in such spaces one can speak of *orthogonality* and make use of *orthonormal bases* provided one takes a little more care than in finite dimensions. They provide the mathematical framework in which quantum mechanics can be developed, and also play an important role in many branches of mathematics, especially linear analysis. Some people even consider that the beginner in functional analysis may restrict himself to them. This opinion seems to us less well-founded today since the solution of non-linear problems (whether of physical, geometrical or other origin) is becoming indispensable for a number of mathematicians and engineers.

---

[25] The French mathematician Augustin-Louis Cauchy (1789–1857) made his debut with promising work on polyhedra. From 1814, he obtained important results on definite integrals. His course of algebraic analysis at the École Royale Polytechnique published in 1821 provided a new framework for analysis. He made fundamental contributions to the theory of functions of a complex variable and to the theory of elasticity, the latter leading to the birth of linear algebra. He had a very active public life, cf. the biography written by Bruno Belhoste.

[26] The French mathematician Émile Picard (1856–1941) was professor at the École Centrale from 1894 to 1937. His work deals mainly with analysis and algebraic geometry. We owe to him in particular an extension of Galois theory from algebraic equations to differential equations. Moreover, he was actively involved in the organisation of the scientific activity of his time.

© The Author(s), under exclusive license to Springer Nature Switzerland AG 2022   21
J.-P. Bourguignon, *Variational Calculus*, Springer Monographs in Mathematics,
https://doi.org/10.1007/978-3-031-18307-2_2

A. Cauchy Sequences and Complete Metric Spaces

In metric spaces it is possible to define a notion of a Cauchy sequence which generalises that introduced for sequences of real or complex numbers. It represents a major tool in analysis today.

The study of complete spaces, i.e., those in which all Cauchy sequences are convergent, which is the subject of the present section, has taken a central place in analysis.

Definition II.1 Let (M, d_M) be a metric space. A sequence of points of M is said to be a *Cauchy sequence* if, for any $\epsilon > 0$, there exists an integer n such that, for all $p, q \geq n$, $d_M(x_p, x_q) \leq \epsilon$.

The space (M, d_M) is said to be *complete* if every Cauchy sequence is convergent.

Remark II.2. Although the notion of a Cauchy sequence (and hence that of a complete space) makes explicit use of a norm, it depends on this norm only up to equivalence.

We use this fact implicitly in Proposition II.5.

II.3. Complete spaces enjoy remarkable properties which make *completeness* a central notion in analysis. We give here only a few of them. Their proofs are elementary, and hence details are omitted.

Proposition II.4. *In a complete metric space, closed sets and complete sets are the same.*

Proof. Let C be a complete subset of a metric space M. Any point x of $\overline{C}$ is a limit of a sequence $(x_n)_{n \in \mathbb{N}}$ of points of C lying, for example, in the balls of centre x and radius $1/n$. Being a convergent sequence, the sequence $(x_n)_{n \in \mathbb{N}}$ is Cauchy. It thus converges in C, which is complete, to a point x_∞. By the uniqueness of the limit, we necessarily have $x = x_\infty$.

Conversely, let F be a closed subset of a complete metric space M. If $(x_n)_{n \in \mathbb{N}}$ is a Cauchy sequence of points of F, it converges in M towards a point x_∞ which thus belongs to $\overline{F}$, and this is equal to F since F is closed. Thus, the space F is complete.   □

Proposition II.5. *Every finite product of complete spaces is complete.*

Proof. By induction on the number of factors, it is enough to establish the proposition for a product of two spaces.

Consider the product $M = M_1 \times M_2$ of two metric spaces (M_1, d_1) and (M_2, d_2) provided with the distance

$$d((x^1, x^2), (y^1, y^2)) = \max\{d_1(x^1, y^1), d_2(x^2, y^2)\}$$

(one can also work with the other equivalent distances on the product).

First of all, we note that, because of the definition of the distance on $M_1 \times M_2$, the projections onto M_1 and M_2 of every Cauchy sequence $(x_n)_{n \in \mathbb{N}}$ of M are Cauchy sequences $(x_n^1)_{n \in \mathbb{N}}$ and $(x_n^2)_{n \in \mathbb{N}}$. If the spaces M_1 and M_2 are complete, the sequences of components converge respectively to points x_∞^1 and x_∞^2.

If we put $x_\infty = (x_\infty^1, x_\infty^2)$, then since by definition $d(x_n, x_\infty) = \max\{d(x_n^1, x_\infty^1), d(x_n^2, x_\infty^2)\}$, the sequence $(x_n)_{n \in \mathbb{N}}$ converges to x_∞.   □

Proposition II.6. *Every compact metric space is complete.*

Proof. Let $(x_n)_{n \in \mathbb{N}}$ be a Cauchy sequence in a compact metric space M. We define the sequence of closed sets $F_n = \overline{\{x_n, x_{n+1}, \cdots\}}$. These sets evidently satisfy the finite intersection property (cf. I.48). As the space M is compact, $F_\infty = \bigcap_{n \in \mathbb{N}} F_n$ is non-empty.

We show that F_∞ reduces to a point, which will thus be the only limit point of this sequence, and hence its limit. In fact, if F_∞ contains two distinct points y and z, which consequently belong to all the F_n, then, for any $\epsilon > 0$, there exist arbitrarily large integers n_y and n_z such that $d(y, x_{n_y}) < \epsilon$ and $d(z, x_{n_z}) < \epsilon$. Since the sequence $(x_n)_{n \in \mathbb{N}}$ is Cauchy, for p and q sufficiently large, $d(x_p, x_q) < \epsilon$. It follows from the triangle inequality that $d(y, z) < 3\epsilon$, which finally gives a contradiction. $\quad\square$

Theorem II.7. (Picard's Fixed Point Theorem) *Every contraction map from a complete metric space to itself has a unique fixed point.*

Proof. Let (M, d) be a complete metric space and f a contraction map, which is thus Lipschitz of order $k < 1$, from (M, d) to itself. Let x_0 be a point of M. Put $x_i = f(x_{i-1})$. We thus have

$$d(x_i, x_{i+1}) \le k \, d(x_{i-1}, x_i).$$

From the first n of these inequalities we deduce that

$$d(x_n, x_{n+1}) \le k^n \, d(x_0, x_1),$$

whence

$$d(x_n, x_{n+p}) \le \sum_{i=n}^{n+p-1} d(x_i, x_{i+1}) \le d(x_0, x_1) \sum_{i=n}^{\infty} k^i = \frac{k^n}{(1-k)} d(x_0, x_1).$$

The sequence $(x_n)_{n \in \mathbb{N}}$ is thus a Cauchy sequence. Consequently, in the complete metric space M, it has a limit which we denote by x_∞. By the continuity of f, we have $\lim_{n \to \infty} f(x_n) = f(x_\infty)$. Passing to the limit in the defining relation of the sequence, we obtain $x_\infty = f(x_\infty)$, a relation which means that x_∞ is a fixed point of f.

We now show that x_∞ is the only fixed point of f. Let x be another fixed point of f. Consider $d(x, x_\infty)$. We have

$$d(x, x_\infty) = d(f(x), f(x_\infty)) \le k \, d(x, x_\infty)$$

which implies that $d(x, x_\infty) = 0$, and hence that $x = x_\infty$. $\quad\square$

Project II.8. Give an example of a map f from a complete metric space M to itself such that, for all x and y in M, $d(f(x), f(y)) < d(x, y)$ but which has no fixed point.

Remark II.9. The method used in the preceding proof provides an *effective* procedure for constructing the fixed point since the series with general term $d(x_i, x_{i+1})$ converges faster than a geometric series with ratio k. In particular, it allows one to obtain a good numerical approximation to the fixed point.

Picard's fixed point theorem has numerous applications. In this course, we make use of it to prove the Local Inversion Theorem III.70, for example. It can also be used to establish the existence of solutions of ordinary differential equations (see for example Exercise II.14).

B. A Fundamental Category: Banach Spaces

We introduce Banach spaces, complete normed spaces, and give some of the fundamental examples of such spaces.

*We now restrict ourselves to normed spaces,
which we have introduced in Chapter I.*

II.10. A complete normed space is called a *Banach space*. It is essential to have available examples of Banach spaces and criteria for deciding if a normed space is Banach.

It is well known that $\mathbb{R}$ and $\mathbb{C}$ are complete spaces and, by Proposition II.5, so are $\mathbb{R}^n$ and $\mathbb{C}^n$.

Note that it is not necessary to make explicit what norm is being used, since the notion of a Cauchy sequence is the same for any two equivalent norms. This settles the case of all *finite-dimensional normed spaces*, which *are Banach spaces for any norm*.

Where infinite-dimensional spaces are involved, the question is much more delicate. We start by giving some fundamental examples in the form of propositions or exercises.

Proposition II.11. *If K is a compact subset of $\mathbb{K}^n$, the space $C^0(K)$ provided with the uniform norm is a Banach space.*

Proof. Let $(f_n)_{n \in \mathbb{N}}$ be a Cauchy sequence of functions in $C^0(K)$.

If m is a point of K, by definition of the uniform norm, the sequence $(f_n(m))_{n \in \mathbb{N}}$ is a Cauchy sequence in K, which converges to a limit which we denote by $f(m)$.

Thus, given $\epsilon > 0$, there exists n_0 such that, for all $n, p \geq n_0$, $\|f_n - f_p\| < \epsilon$, in other words, for any point m of K, $|f_n(m) - f_p(m)| < \epsilon$. Leaving n fixed and letting p tend to infinity, we obtain on passing to the limit in the inequality $|f_n(m) - f(m)| \leq \epsilon$ for any point m of K, in other words $\|f_n - f\| \leq \epsilon$ which means that the sequence $(f_n)_{n \in \mathbb{N}}$ converges.

What needs to be verified is that the function f obtained as limit is indeed continuous. For that it suffices to divide ϵ by three and to approximate f within $\epsilon/3$ by some f_n which we already know is continuous. $\square$

Remark II.12. Some comments are in order:

i) In the preceding argument, the numerical space $\mathbb{K}$ in which the functions take their values entered only via the absolute value which is its norm. The result remains true if $\mathbb{K}$ is replaced by any Banach space whatever.

ii) One step in the preceding proof used pointwise convergence of the sequence of functions, often called *simple convergence*, which is implied by uniform convergence. This is a good time to recall that the converse is not true, i.e., *simple convergence alone does not imply uniform convergence.*

Exercise II.13. Show that c_0 is a Banach space (for the uniform norm).

Solved Exercise II.14. Let V be a continuous function defined on $\mathbb{R}$. Show that the
ordinary differential equation

$$(\text{II.15}) \qquad \frac{d^2x}{dt^2} + V(t)\,x(t) = 0 , \quad x(t_0) = x_{t_0} , \quad \frac{dx}{dt}(t_0) = x'_{t_0}$$

has a unique solution defined on the whole of $\mathbb{R}$ which is twice continuously differ-
entiable.

For this, we begin by considering the case where $t_0 = 0$. It is convenient to work
in the Banach space E_τ of continuous real-valued functions on the interval $[0, \tau]$.
We introduce the map h_V from E_τ to itself defined as follows:

$$(h_V(x))(t) = x_0 + x'_0 t - \int_0^t \left(\int_0^s V(u)\,x(u)\,du \right) ds .$$

(The fact that h_V does map E_τ to itself follows immediately from the fact that, V and
x being continuous on the compact interval $[0, \tau]$ by hypothesis, they are bounded
there, thus ensuring the continuity of the function $h_V(x)$.)

We show that a function x is a solution of Eq. (II.15) if and only if it is a fixed point
of h_V. In fact, if the function x (which is *a priori* only continuous) is equal to $h_V(x)$,
then x is of class C^2 on $]0, \tau[$ with the first and second right derivatives at 0 and left
derivatives at τ. Moreover, by an immediate direct calculation, we have the initial
conditions $x(0) = x_0$ and $(dx/dt)(0) = x'_0$. Furthermore, $d^2x/dt^2 + V(t)\,x(t) = 0$.

Conversely, if x is a solution of Eq. (II.15), we can integrate the equation (which
contains only continuous functions) over an interval $[0, s]$ which gives the identity
$(dx/dt)(s) - x'_0 = \int_0^s (d^2x/dt^2)(u)\,du = -\int_0^s V(u)\,x(u)\,du$, and then integrate this
relation again over the interval $[0, t]$, which gives $x = h_V(x)$.

To ensure the existence of a fixed point of h_V, it is enough to study its contraction
properties. For $x, y \in E_\tau$, we have

$$h_V(x) - h_V(y) = \int_0^t \left(\int_0^s V(u)(y(u) - x(u))\,du \right) ds .$$

Consequently,

$$\|h_V(x) - h_V(y)\| \le \frac{1}{2}\tau^2\,\|V\|\,\|x - y\| .$$

The map h_V is thus a contraction whenever $\tau < \sqrt{2/\|V\|}$. Under this hypothesis,
we can apply Picard's Fixed Point Theorem II.7, which not only guarantees the
existence of a fixed point but also its *uniqueness*. By the equivalence which we have
established above, this guarantees the uniqueness of the solution of Eq. (II.15) on
intervals whose length is controlled by $\|V\|$.

We can now establish that this result indeed guarantees the existence of a solution
x of Eq. (II.15) on the whole of $\mathbb{R}$ satisfying the initial conditions $x(t_0) = x_{t_0}$ and
$(dx/dt)(t_0) = x'_{t_0}$.

In fact, we remark first of all that the result established on the interval $[0, \tau]$ applies equally well on an interval $[t_0, \tau]$ provided one replaces 0 by t_0, $x_0't$ in the definition of h_V by $x_{t_0}'(t - t_0)$ and the condition $\tau < \sqrt{2/\|V\|}$ by $\tau - t_0 < \sqrt{2/\sup_{t \in [t_0, \tau]} |V(t)|}$.

Given now an arbitrary interval $[\tau_+, \tau_-]$ around t_0, put

$$v_+ = \sup_{t \in [t_0, \tau_+]} |V(t)|.$$

As we have remarked above, we can solve the differential equation on an interval $[t_0, t_2]$ with initial conditions $x(t_0) = x_{t_0}$ and $(dx/dt)(t_0) = x_{t_0}'$ provided we have $t_2 - t_0 < \sqrt{2/v_+} \leq \sqrt{2/(\sup_{t \in [t_0, t_2]} |V(t)|)}$. We can then solve in the same way on an interval $[t_1, t_3]$ for $t_0 < t_1 < t_2 < t_3$ with initial conditions the values at t_1 of the solution found above and its first derivative provided that $t_3 - t_1 < \sqrt{2/v_+} \leq \sqrt{2/(\sup_{t \in [t_1, t_3]} |V(t)|)}$. To verify that these two solutions coincide on their common domain, it suffices to appeal to the uniqueness of the solution on the interval $[t_1, t_2]$ with the given initial conditions. In a finite number of steps, one can thus construct a solution on the interval $[t_0, \tau_+]$.

To construct the solution on the interval $[\tau_-, t_0]$, we can apply the same method as above after observing that on changing t to $-t$, Eq. (II.15) on an interval $[-\tau, 0]$ is transformed into the equation analogous to (II.15) on the interval $[0, \tau]$ for the potential $\tilde{V}$ defined by $\tilde{V}(t) = V(-t)$ which is again a continuous function.

The solutions obtained on the intervals $[\tau_-, t_0]$ and $[t_0, \tau_+]$ agree at t_0 since the initial conditions ensure that the values of the function agree and their right and left derivatives are equal. The coincidence of their right and left second derivatives follows from the differential equation itself which relates them to the values of the function and of the potential at 0.

We have thus shown that the differential equation (II.15) admits a unique solution with initial conditions x_{t_0} and x_{t_0}' at t_0 on arbitrarily long intervals, and hence finally on the whole of $\mathbb{R}$.

An interesting generalisation of this exercise consists in treating the case where the potential V is only piecewise continuous.

Exercise II.16. Show that, *if K denotes a compact domain in $\mathbb{R}^n$, the space $C^k(K)$ (formed by the maps having partial derivatives up to order k on a neighbourhood of K) is a Banach space for the norm corresponding to uniform convergence of the function and its derivatives up to order k.*

Exercise II.17. Show that *l^p is a Banach space for the norm of order p $(1 \leq p)$.*

Proposition II.18. *Let E and F be two normed spaces. When F is a Banach space, $L(E, F)$ provided with the norm defined in I.38 is a Banach space.*

Proof. Let $(f_n)_{n \in \mathbb{N}}$ be a Cauchy sequence in $L(E, F)$. Given $\epsilon > 0$, we can find an integer n_0 such that, for all $p, q \geq n_0$, $\|f_p - f_q\| < \epsilon$, i.e., for all $v \in E$,

(II.19) $$\|f_p(v) - f_q(v)\| \le \|f_p - f_q\| \, \|v\| \le \epsilon \, \|v\| \, .$$

Thus, for each v in E, the sequence $(f_n(v))_{n \in \mathbb{N}}$ is Cauchy in F, and thus has a limit which we denote by $f(v)$. The new map f thus defined is linear as a simple limit of linear maps. It remains to see that $f \in L(E, F)$. For this, it suffices to pass to the limit in (II.19) by making q tend to infinity. We then obtain $\|f_p(v) - f(v)\| \le \epsilon \, \|v\|$, which implies that f is bounded on the unit ball, and on the other hand that $f_p \to f$ in $L(E, F)$ when $p \to \infty$.   □

Exercise II.20. Show that, *in a Banach space E, any sufficiently small perturbation of the identity (in fact, any endomorphism of E of the form $\mathrm{Id}_E + u$ with $\|u\| < 1$) is invertible.*

Deduce that the set of automorphisms (i.e., invertible endomorphisms) of E is an open subset of $L(E, E)$ and show that, on this open set, the map which to an automorphism associates its inverse is continuous.

Project II.21. Prove the theorem of Banach according to which *every continuous linear isomorphism from one Banach space to another is a homeomorphism.*

II.22. A notion which today plays a very important role in analysis is that of the *completion* of a space. In a given problem, it may turn out that the natural function space in which to work is incomplete. It is then necessary to embed it into a complete space which is as small as possible, namely one of which it is a dense subset. Such a space exists: it is precisely the completion of the space considered.

The construction of this space in the greatest generality is outside the scope of this course, though it is analogous to that of the set $\mathbb{R}$ of real numbers from the set $\mathbb{Q}$ of rational numbers.

However, as an example, we mention that the space of sequences $\mathbb{K}l^p$ is the completion of $\mathbb{K}_0^{\mathbb{N}}$ for the norm of order p.

C. Dual Space and Weak Topology

We re-examine the concept of the dual space of a vector space and related notions such as the transposed map. We discuss the extensions necessary to work in the context of topological vector spaces, in particular in Banach spaces.

II.23. If E is a vector space of finite dimension n over a field $\mathbb{K}$, the space of $\mathbb{K}$-linear forms on E (in the sequel we suppress the reference to $\mathbb{K}$) is also a vector space over $\mathbb{K}$ called the *dual* of E and *denoted* by E^*. (To make the notation more transparent, we shall endeavour to denote elements of E^* by Greek letters.)

The case of $\mathbb{K}^n$, *a priori* the simplest, can however be a source of confusion. In fact, a linear form on the vector space of n-tuples of elements of $\mathbb{K}$ can be identified with a linear combination of the coefficients of the n-tuple, and hence can itself be viewed as the n-tuple of coefficients of the linear combination. Thus we see that the dual of $\mathbb{K}^n$ can be identified naturally with $\mathbb{K}^n$; only the way of writing these n-tuples allows one to finally distinguish them. If an element of $\mathbb{K}^n$ is considered, as one generally does somewhat arbitrarily, as a *column* of n numbers,[27] the dual space appears as the set of *rows* of n numbers necessary to be able to "eat" the column vector.

II.24. As is well known, to any basis (e_i) of a vector space E of dimension n is associated naturally a basis (ϵ^i) of E^*, called the *dual basis*, defined as follows: on a vector v of E which can be expressed as $v = \sum_{j=1}^{n} v^j e_j$, the linear form ϵ^i takes the value $\epsilon^i(v) = v^i$. (This proves in passing that E^* is a vector space of dimension n.)

At this point we introduce the *notation* $\langle \, , \, \rangle_E$ to denote the action of a linear form on a vector of E so that, for example, the definition of the dual basis can be written $v = \sum_i \langle \epsilon^i, v \rangle_E \, e_i$.

Recall that, for three $\mathbb{K}$-vector spaces E, F and G, a map from $E \times F$ to G is said to be *bilinear* if it is linear with respect to each argument, and a *bilinear form* if $G = \mathbb{K}$.

The map $(v, \lambda) \mapsto \langle \lambda, v \rangle_E$ is a bilinear form on $E \times E^*$, often called the *duality bracket* of the space E.

II.25. Moreover, to any linear map l between two vector spaces E and F, it is possible to associate the *transposed map* $^t l$ which maps F^* to E^* and which is defined as follows: for $\mu \in F^*$, $^t l(\mu) = \mu \circ l$, i.e., $^t l(\mu)$ is the linear form on E which, evaluated on a vector v of E, gives $\langle ^t l(\mu), v \rangle_E = \langle \mu, l(v) \rangle_F$. When $E = \mathbb{K}^n$ and $F = \mathbb{K}^m$, it is traditional to represent a linear map l from E to F as the matrix L which has in its ith column the components in the natural basis of $\mathbb{K}^m$ of the image of the ith vector of the natural basis of $\mathbb{K}^n$ (which one often calls its ith *column vector*). To have a convenient way of writing it, taking into account the position of the indices (which will be very useful in the rest of this course), we make the *convention* that L^i_j denotes the element of the matrix L which lies at the intersection of the ith row and the jth column so that, if $v = \sum_{i=1}^{n} v^i e_i$, then

$$l(v) = \sum_{i=1}^{n} v^i l(e_i) = \sum_{i,j=1}^{n,m} L^j_i v^i e_j$$

where (e_i) denotes, according to the range of variation of its index, the natural basis of $\mathbb{K}^n$ or of $\mathbb{K}^m$.

Let us examine the consequences of this convention for the matrix of the transposed map. The dual space of $\mathbb{K}^n$ being identified with $\mathbb{K}^n$, the matrix of the transposed map is precisely the *transposed matrix* $^t L$ of the matrix L of which the element in the jth row and i^{th} column is equal to L^i_j.

---

[27] One could just as well consider them as *row*-vectors, but various traditional notations such as the multiplication of matrices would then be affected. In fact, in Chapter I, we tended to write n-tuples of elements of $\mathbb{K}$ on a single line, contradicting the convention which we have now decided to adopt, but we have not carried out any algebraic operations on them which could be affected by this notation.

II.26. All these considerations apply, of course, *only* in finite dimensions. As we have already mentioned, for the passage to infinite-dimensional vector spaces it is essential to take a fuller account of the technology, and this is what we now turn to.

Definition II.27. The space of continuous linear forms on a topological vector space E is denoted by E' and called the *topological dual space* of E.

Remark II.28. Because of Corollary I.33, the preceding definition generalises the usual definition in finite dimensions.

> *In the context of infinite-dimensional topological vector*
> *spaces, we shall often omit in the sequel the*
> *adjective "topological" when referring to the dual space.*

II.29. One of the important features of the dual space of a topological vector space is that it has good properties when the original space has them. We give several examples of this situation shortly.

It also produces new spaces which can turn out to be very useful. Thus, for the space of continuous functions on a compact space provided with the norm of uniform convergence, the dual is formed by the *Radon*[28] *measures* which can be taken as the basic elements of a theory of integration. We also mention the case of the space of infinitely differentiable functions with compact support on $\mathbb{R}^n$ of which the topological dual (for a suitable topology, which does not arise from a norm) is the space of *distributions* introduced by Laurent Schwartz[29].

II.30. One should not believe that, on an infinite-dimensional space, every linear form, even if its definition is simple, is continuous. Exercise II.31 gives an example of a linear form which is not continuous, but there are plenty of them.

Exercise II.31. Show that, on the space $C^1([0,1], \mathbb{R})$ provided with the norm of uniform convergence, the linear form $f \mapsto f'(0)$ is not continuous.

Proposition II.32. *The dual of a normed space is a Banach space.*

Proof. This is an immediate corollary of Proposition I.38 which defines a norm on E' and of Proposition II.18 since K is a complete metric space. $\square$

---

[28] The Austrian mathematician Johann Radon (1887–1956) was interested above all in the Calculus of Variations, notably in its relations with analysis, geometry and physics. He completed the works of Lebesgue and left under his name a transformation in integral geometry which forms the foundation of the theory of the "scanner".

[29] The French mathematician Laurent Schwartz (1915–2002) was professor at the Université de Paris and at the École Polytechnique from 1959 to 1983 where he founded the Centre de Mathématiques. He contributed to numerous areas of analysis (from harmonic analysis to the theory of partial differential equations) and more recently to probability (notably the theory of martingales). The theory of distributions which he founded in 1945 (and which won him the Fields medal in 1950) has played a decisive role in the success of the methods of functional analysis in the solution of partial differential equations and consequently in very diverse applications of mathematics to the sciences.

Solved Exercise II.33. Show that, for $1 < p$, *the dual of the space* l^p (i.e., the space of sequences whose series of pth powers is convergent) *is the space* l^q *with* $(1/p) + (1/q) = 1$.

One must first make explicit what is meant by the expression "a normed space F *is* the dual of a normed space E". This entails constructing a map which associates to any element of F a continuous bilinear form on E, thus defining an injection of F into E', then showing that this map is an isometry for the given norm on F and the natural norm on E', and finally establishing that it is surjective.

In our case, we first show that any element of l^q naturally defines a continuous linear form on l^p (recall that $1 < p, q$). Let $(x) \in l^q$. To any element $(y) \in l^p$, we associate the real number $\langle (x), (y) \rangle = \sum_{i=0}^{\infty} x^i y^i$ (the notation on the left-hand side rather anticipates the result !). That this number is well-defined follows directly from the *Hölder inequality* (an easy consequence of the convexity of the pth power functions for $1 < p$) since this asserts that, for any two finite sequences of positive real numbers $(a^0, \cdots, a^n)$ and $(b^0, \cdots, b^n)$,

$$(\text{II.34}) \qquad \sum_{i=0}^{n} a^i b^i \leq \left(\sum_{i=0}^{n} |a^i|^q \right)^{\frac{1}{q}} \left(\sum_{i=0}^{n} |b^i|^p \right)^{\frac{1}{p}} .$$

If these finite sequences are taken to be the initial parts of the absolute values of the sequences (x) and (y) that we considered earlier, the left-hand member of Eq. (II.34) is bounded above by $\|(x)\|_q \|(y)\|_p$, and hence by a quantity independent of n. We thus deduce that the series $\sum_i (x^i y^i)$ is absolutely convergent. Moreover, it is clear that the map $(y) \mapsto \langle (x), (y) \rangle$ is linear. Finally, the inequality which we have proved ensures that the norm of the linear form defined on l^p by $(x) \in l^q$ is at most $\|(x)\|_q$.

It remains to show two things: first, that every linear form on l^p is obtained from this construction, and second, that the map which we have constructed defines an isometry from the Banach space l^q to the dual of l^p provided with its natural Banach space norm. Here is a single construction which will do both these things at once.

To show that the norm of $(x) \in l^q$, as a linear form on l^p, is indeed $\|(x)\|_q$, it suffices to exhibit an element of l^p for which the inequality deduced from the Hölder inequality is an equality. For this, consider the sequence $(\tilde{x})$ whose i^{th} term is $\tilde{x}^i = |x^i|^{q-1} \text{sign}\, x^i$ (where $\text{sign}\, a$ denotes the sign of the real number a). We claim that $(\tilde{x}) \in l^p$. In fact,

$$\|(\tilde{x})\|_p = \left(\sum_{i=0}^{\infty} |x^i|^{pq-p} \right)^{\frac{1}{p}} = \|(x)\|_q^{\frac{q}{p}} < \infty .$$

Now, since $q - (q/p) = 1$, evaluating the linear form defined by (x) on $(\tilde{x})$, we obtain

$$\langle (x), (\tilde{x}) \rangle = \sum_{i=0}^{\infty} |x^i|^{q-1} |x^i| = \sum_{i=0}^{\infty} |x^i|^q = \|(x)\|_q^q = \|(\tilde{x})\|_p \|(x)\|_q .$$

The other part follows directly from the same construction. In fact, every linear form λ on l^p gives rise to a sequence (x_λ) by putting $x_\lambda^i = \lambda(e_i)$ (where (e_i) denotes the natural basis of l^p). To show that this sequence belongs to l^q, it suffices to use the continuity of λ as follows. Form the sequence $(\tilde{x}_\lambda)$ and truncate this sequence after the nth term. We clearly obtain an element of l^p. On evaluating λ on it and using the continuity of λ, we obtain

$$\sum_{i=0}^n |x_\lambda^i|^q \leq \|\lambda\| \left(\sum_{i=0}^n |x_\lambda^i|^q \right)^{\frac{1}{p}},$$

and hence finally

$$\left(\sum_{i=0}^n |x_\lambda^i|^q \right)^{\frac{1}{q}} \leq \|\lambda\|$$

for all n. This ensures that the sequence (x_λ) is indeed an element of l^q. Moreover, the linear form on l^p defined by (x_λ) does coincide with λ.

Exercise II.35. Show that *the dual of l^1 is l^∞*.

Project II.36. Show that *the dual of l^∞ is "much larger" than l^1*.

II.37. By II.32, the category of Banach spaces is closed under passage to the dual. But it is possible to use duality in an interesting way if one remembers that the *bidual* (i.e., the dual space of the dual) of a finite-dimensional vector space E can be identified naturally with E itself. We recall the construction.

If x is an element of E, it is possible to associate to it a linear form $\tilde{x}$ on E^* by putting, for $\lambda \in E^*$, $\langle \tilde{x}, \lambda \rangle_{E^*} = \langle \lambda, x \rangle_E$. One can see that this map defines an injective linear map from E to E^{**}. We prove that this map actually sends E into the space of continuous linear forms on E', which itself consists of continuous linear forms. For this it suffices, by Proposition I.38 and Remark I.40, to prove that there exists a number k such that, for all $\lambda \in E'$, $|\langle \tilde{v}, \lambda \rangle_{E'}| \leq k \|\lambda\|_{E'}$. Now the term on the left is equal to $|\langle \lambda, v \rangle_E|$ which, by definition of the norm on E', is bounded above by $\|\lambda\|_{E'} \|v\|_E$. One can thus take $k = \|v\|_E$. What is more, this proves that the injection from E to $(E')'$ is continuous.

II.38. The fact that *a space can be identified with its bidual does not generalise to infinite-dimensional spaces*. Here is a good reason for this: since the dual of a space is automatically a Banach space by Proposition II.32, the bidual of a normed space is a Banach space, and there exists normed spaces which are non complete.

From the preceding discussion, it is clear nevertheless that "*every normed space admits an injection into a Banach space*". However, there exist Banach spaces which can be identified with their bidual by the map defined above. Such spaces are called *reflexive*. Exercise II.33 shows that the spaces l^p are reflexive for $1 < p < \infty$ (and Project II.36 that l^∞ is not).

Project II.39. Show that *the natural map from a Banach space to its bidual is in fact an isometry onto its image*.

II.40. The dual space E' of a normed space E allows one to introduce a new topology on E, called the *weak topology*, which is a good example of a topology defined by a family of semi-norms. One considers E as a space of functions on E'; the weak topology is that of simple convergence of this family of functions.

Its precise definition is as follows: to any linear form λ on E is associated the semi-norm $v \mapsto |\langle \lambda, v \rangle_E|$. For a sequence of points $(x_n)_{n \in \mathbb{N}}$ to converge to 0 in this topology, it is necessary and sufficient that for all $\lambda \in E'$ the sequence $(\langle \lambda, x_n \rangle_E)_{n \in \mathbb{N}}$ converges to 0 in $\mathbb{R}$. The passage to an arbitrary point is made by using the linearity of elements of the dual.

II.41. This topology plays an important role in the analytical techniques of the calculus of variations. It brings in the dual space as a space of test functions, which is another good reason for determining this space precisely.

A fundamental theorem which uses this topology is stated in II.89.

D. Bilinear Forms and Duality

We begin the study of duality by showing that every bilinear form on a vector space defines a map from this space to its dual.

We discuss in some detail the case of scalar products and of symplectic forms.

II.42. A bilinear form is said to be *symmetric* if its value does not change when its arguments are interchanged, and *skew-symmetric* if its value changes sign when its arguments are permuted.

In the case of a vector space F over $\mathbb{C}$, a map s from $F \times F$ to $\mathbb{C}$ is said to be *sesquilinear* if it is *conjugate-linear* with respect to its first argument and *linear* with respect to its second argument (i.e., for α_1, $\alpha_2 \in \mathbb{C}$ and v_1, v_2, $w \in F$, $s(\alpha_1 v_1 + \alpha_2 v_2, w) = \overline{\alpha_1}\, s(v_1, w) + \overline{\alpha_2}\, s(v_2, w))$ where $\bar{z}$ denotes the conjugate of the complex number z. If the sesquilinear form s also satisfies $s(v, w) = \overline{s(w, v)}$, the form s is said to be *Hermitian*[30].

In both cases, the associated homogeneous form q of degree 2 (i.e., its restriction to the diagonal in the product space) is real, and vanishes if the form is skew-symmetric.

Exercise II.43. Show that *every sesquilinear form s can be obtained from its homogeneous form of degree 2, which we denote by q_s, by the so-called polarisation formula*

$$4\, s(v, w) = q_s(v + w) - q_s(v - w) + i\, q_s(w + iv) - i\, q_s(w - iv)\,.$$

Deduce that *s is Hermitian if and only if q_s is real.*

---

[30] after Charles Hermite (1822–1901), a French mathematician who was professor of analysis at the École Polytechnique from 1869 to 1876. He has been one of the dominant figures in the development of the theory of algebraic forms, of the arithmetic theory of quadratic forms and of the theory of elliptic functions. He proved that the number e is transcendental and gave the first solution of the general equation of the fifth degree in terms of elliptic functions.

II.44. Every bilinear form b defined on a vector space E (even if it has no particular symmetry property) can be identified with a map from E to E^* by the following construction: to $v \in E$ one associates the linear form $\tilde{b}(v)$ which maps a vector w to $\langle \tilde{b}(v), w \rangle = b(v, w)$.

One says that the bilinear form is *non-degenerate* if the linear map $\tilde{b}$ is injective, and that b defines a *duality* if $\tilde{b}$ is an isomorphism. If E is finite-dimensional, these two notions are of course equivalent.

It is interesting to consider further the case of forms defined on a *finite-dimensional* vector space E. If we take a basis (e_i) of E, b can be represented by its matrix $B = (b_{ij})$ where $b_{ij} = b(e_i, e_j)$. One then finds that, when one refers E to the basis (e_i) and E^* to the dual basis (ϵ^i), the matrix of the linear map $\tilde{b}$ is again the matrix B (in fact, the elements of this matrix are obtained by evaluating the dual basis of (ϵ^i), which is precisely (e_i) under the natural isomorphism between E and its bidual E^{**}, on $\tilde{b}(e_j)$, in other words the values $\langle \tilde{b}(e_i), e_j \rangle = b_{ij}$).

We can now give a simple criterion for a bilinear form on a finite-dimensional space to be non-degenerate.

Proposition II.45. *For a bilinear form defined on a finite-dimensional space to be non-degenerate it is necessary and sufficient that its matrix in an arbitrary basis has non-zero determinant.*

Proof. Since the representation of $\tilde{b}$ in the chosen bases of E and E^* is the matrix of the bilinear form b in the basis of E and since the non-vanishing of the determinant is a necessary and sufficient condition for a linear map to be invertible, the proposition is clear.   □

II.46. To every linear map l from a vector space E to a vector space F provided with a bilinear form b, one can associate a bilinear form l^*b on E by putting, for $v \in E$, $(l^*b)(v, v) = b(l(v), l(v))$ which is called the *inverse image form* of b by l. This inverse image form can also be defined via the map from E to E^* which is associated to it and for which we have

(II.47) $l^{\tilde{*}}b = {}^t l \circ \tilde{b} \circ l$.

A special case of the use of this notion is that of the *restriction*[31] of a bilinear form to a subspace E of its space of definition F for which the injection is *denoted* by i. One usually abuses notation by writing b to denote the bilinear form i^*b on E.

II.48. We now review the two great families of non-degenerate bilinear forms which are of geometric interest: *scalar products* and *symplectic forms*.

II.49. Classically one speaks of a *scalar product* if the bilinear or Hermitian form is symmetric and *positive* (i.e., non-negative for any vector and zero only for the zero vector). By its definition, a scalar product is a non-degenerate form (and thus defines a duality in finite dimensions).

---

[31] It is particularly instructive to understand how the restriction of a non-degenerate bilinear form to a subspace is *not* automatically non-degenerate, contrary to what happens for linear maps. The study of symplectic forms which we consider a little later is quite instructive on this point.

In what follows, we usually adopt the notation
(|) to denote a scalar product.[32]

It is usual to employ the terminology of classical geometry for such forms; thus two vectors v and w such that $(v|w) = 0$ are said to be *orthogonal*. Their sum $v + w$ then satisfies the classical *Pythagoras*[33] *relation* $(v + w|v + w) = (v|v) + (w|w)$.

II.50. On $\mathbb{K}^n$, it is customary to define a scalar product, called the *standard scalar product*, by associating to $((x^1, \cdots, x^n), (y^1, \cdots, y^n))$ the scalar $\sum_{i=1}^n \overline{x^i} y^i$ (where the conjugation is taken to be equal to the identity when $\mathbb{K} = \mathbb{R}$).

If now a space of finite dimension n over $\mathbb{K}$, say F, is provided with a scalar product $(|)_F$, it is well-known that there are always bases (f_i) in which the scalar product has the simple expression

$$
\text{(II.51)} \qquad\qquad (f_i|f_j)_F = \begin{cases} 0 & \text{if } i \neq j; \\ 1 & \text{if } i = j. \end{cases}
$$

Such bases are said to be *orthonormal* if $\mathbb{K} = \mathbb{R}$ and *unitary* if $\mathbb{K} = \mathbb{C}$.

If we interpret (as we have done already) a basis of F as an isomorphism f of $\mathbb{K}^n$ with F (defined by $f(e_i) = f_i$, where (e_i) denotes as usual the canonical basis of $\mathbb{K}^n$), we see that, by extending relations II.51 bilinearly, f satisfies, for all x and y in $\mathbb{K}^n$, $(f(x)|f(y))_F = (x|y)$. The basis f thus defined is an *isometry* between $\mathbb{K}^n$ with the standard scalar product and F with the scalar product $(|)_F$.

Proposition II.52. *Every scalar product on a vector space E satisfies the following fundamental properties:*

i) the Cauchy–Schwarz[34] inequality holds, namely, for all v, $w \in E$,

$$
|(v|w)|^2 \leq (v|v)(w|w);
$$

ii) the map $v \mapsto (v|v)^{1/2}$ is a norm.

---

[32] It is rather the variant $\langle | \rangle$ of this notation which is preferred by physicists. The left-hand vector is often called a "*bra*" and denoted by $\langle v|$ and the right-hand vector a "*ket*" and denoted by $|v\rangle$. Note that the convention of taking the scalar product to be conjugate-linear in the left member is consistent with writing *elements of a matrix L* in the form $\langle v|L|w\rangle$, dear to physicists, but that numerous authors take the opposite convention.

[33] Pythagoras (570–495 B.C.) was a Greek philosopher, who developed a very influential school in the South of Italy. His teachings embraced philosophy, politics, mathematics and music. Several major discoveries, including Pythagoras' Theorem and the list of the five regular solids, are attributed to him but the reality of his personal contribution remains debated. In the philosophy and the theory of music he and his followers advocated rational numbers play a fundamental role.

[34] Karl Hermann Amandus Schwarz (1843–1921), a German mathematician who had great geometric intuition, solidly established the uniformisation theorem for simply-connected domains in the complex plane. He solved the problem of determining the minimal surfaces bounded by simple contours and stated a symmetrisation principle for such surfaces.

Remark II.53. The proof of the preceding proposition is sufficiently elementary and classical to be left to the reader. Nevertheless, we would like to comment on the arguments used in the proof of the Cauchy–Schwarz inequality, since when presented in a certain way, they are typical of those used in this course.

The key point is of course the positivity of the quadratic form associated to the bilinear form. Given two non-zero vectors v and w, one considers the family of vectors $v + tw$ ($t \in \mathbb{R}$) which they generate. The argument most often presented uses the special fact that a polynomial of second degree has no root precisely when its discriminant is non-positive, and this discriminant is precisely the quantity which appears in the Cauchy–Schwarz inequality. We are going to obtain the same result by using an argument which is a little more general (and still very elementary).

The function $t \mapsto \phi(t) = (v + tw|v + tw) = t^2(w|w) + 2t\,(v|w) + (v|v)$ attains its minimum when $t = -(v|w)/(w|w)$, as one sees by setting its derivative equal to zero. At its unique extremum, the value of the function ϕ is equal to $(w|w) - (v|w)^2/(v|v)$, precisely the expression whose positivity gives the Cauchy–Schwarz inequality.

This search for critical points of partial functions is a method which is currently used in the Calculus of Variations. Moreover, from the study of the critical values of the function under consideration, it is often possible to extract remarkable identities.

Exercise II.54. Show that, *in a real vector space E, every norm $\| \ \|$ coming from a scalar product satisfies the parallelogram identity, i.e., for any triple of vectors u, v and w,*

$$(\text{II.55}) \qquad \|u + v\|^2 + \|u - v\|^2 = 2\,\|u\|^2 + 2\,\|v\|^2 .$$

Deduce that, *in any real affine space $\mathcal{E}$ modelled on a vector space E provided with a norm $\| \ \|$, the median identity is satisfied, i.e., given three points A, B and $C \in \mathcal{E}$, then*

$$(\text{II.56}) \qquad 4\,\|AI\|^2 = 2\,\|AB\|^2 + 2\,\|AC\|^2 - \|BC\|^2 ,$$

if I denotes the middle of the segment BC.

Solved Exercise II.57. Show that *the parallelogram identity (II.55) characterises, among the norms on a vector space, those which come from a scalar product.*

For this, we must reconstruct from the norm itself the bilinear form which gives rise to it. This is done by imitating the operation of *polarisation*. In fact, for $u, v \in E$, put

$$f(u, v) = \|u + v\|^2 - \|u - v\|^2 .$$

We must show that $f(u, v)$ depends linearly on u and on v. Since, by its definition, f is symmetric in its two arguments, it suffices to prove linearity with respect to one of them. We show first of all that $f(2u, v) = 2 f(u, v)$. This is a direct consequence of the parallelogram identity since

$$f(2u, v) = \|2u + v\|^2 - \|2u - v\|^2$$

$$= \|2u + v\|^2 + \|v\|^2 - \|2u - v\|^2 - \|v\|^2$$

(II.58)

$$= 2\|u + v\|^2 + 2\|u\|^2 - 2\|u - v\|^2 - 2\|u\|^2$$

$$= 2 f(u, v) .$$

By a simple change of variable, we thus have $f(\frac{1}{2}u, v) = \frac{1}{2}f(u, v)$.

Now we show that f is additive with respect to its first argument. In fact, for all u, v and $w \in E$,

$$2\left(f(u, w) + f(v, w)\right) = 2\|u + w\|^2 + 2\|v + w\|^2 - 2\|u - w\|^2 - 2\|v - w\|^2$$

$$= \|u + v + 2w\|^2 + \|u - v\|^2$$

$$ - \|u + v - 2w\|^2 - \|u - v\|^2$$

$$= f(u + v, 2w) ,$$

whence the desired result by applying Formula (II.58) to the second argument.

Thanks to the additivity property, and to a repeated application of Formula (II.58) to allow positive and negative powers of 2, we thus obtain that, for all dyadic numbers α, $f(\alpha u, v) = \alpha f(u, v)$. Since f is continuous by construction and since the dyadic numbers are dense in $\mathbb{R}$, we have finally proved that f is linear.

II.59. If b is a scalar product on F, the inverse image bilinear form $l^* b$ under a linear map l from E to F is again a scalar product on E if and only if l is injective since it is clear that in general the new form constructed remains non-negative and symmetric or Hermitian. In particular, this is the case when one considers the restriction of a scalar product to a subspace.

Thanks to Formula (II.47), one easily finds the matrix characterisation of the isometries of the standard metric on $\mathbb{K}^n$: in fact, the natural identification of $\mathbb{K}^n$ with its dual $\mathbb{K}^n$ coincides with the map induced by the standard scalar product if $\mathbb{K} = \mathbb{R}$ and with its complex conjugate if $\mathbb{K} = \mathbb{C}$ so that l is an isometry of $\mathbb{K}^n$ for its standard metric precisely when, for $\mathbb{K} = \mathbb{R}$, the matrix ${}^t L L$ is the identity, in other words, when L is *orthogonal*, and, for $\mathbb{K} = \mathbb{C}$, when the matrix ${}^t \bar{L} L$ is the identity, in other words, when L is *unitary*.

II.60. It is also important (though unfortunately less usual) to consider the *skew-symmetric* forms defined on a vector space.

A bilinear form ω is said to be a *symplectic[35] form* if it is skew-symmetric and non-degenerate.

An example of a universally defined symplectic form appears on the space $E \times E^*$ and is directly related to the duality bracket: for v, $w \in E$, and λ, $\mu \in E^*$, one puts $\omega_E((v, \lambda), (w, \mu)) = \langle \mu, v \rangle_E - \langle \lambda, w \rangle_E$. The bilinear form ω_E is skew-symmetric by construction. It is also non-degenerate since, if the right-hand side is zero for all

---

[35] This term, which is a Greek version of the word *"complex"*, was introduced by Hermann Weyl in 1939 in *"The Theory of Classical Groups"*, a book which has become a great classic.

(w, μ), this means, by taking first $\mu = 0$, that $\lambda = 0$ (since then $\langle \lambda, w \rangle$ is zero for every vector w) and then by taking $w = 0$, that $v = 0$ (since every linear form vanishes on it).[36] In this context, it is actually more convenient to use additive notations (rather than the notation of pairs), i.e. to work in $E \oplus E^*$ (which allows us to add elements of E and its dual E^*, an operation which seems surprising at first but which is perfectly legitimate in $E \oplus E^*$). This changes nothing since $E \times E^*$ and $E \oplus E^*$ are naturally isomorphic.

If E is a space of dimension n and if we take a basis (e_i) of E and the dual basis (ϵ^i), the form ω_E (which we now consider as defined on $E \oplus E^*$) takes a very simple form. In fact, we have

$$(\text{II}.61) \qquad \omega_E(e_i, e_j) = 0, \quad \omega_E(e_i, \epsilon^j) = \delta_{ij}, \quad \omega_E(\epsilon^i, \epsilon^j) = 0,$$

where δ_{ij} denotes the *Kronecker*[37] symbol which is equal to 0 if $i \neq j$ and 1 if $i = j$. In other words, in the basis $(e_1, \cdots, e_n, \epsilon^1, \cdots, \epsilon^n)$ of $E \oplus E^*$, the matrix of the symplectic form ω_E is very "sparse": it is of the form

$$\begin{pmatrix} 0 & I_n \\ -I_n & 0 \end{pmatrix}$$

i.e., it contains only zeros except on the main diagonals of the $n \times n$ off-diagonal matrices of its decomposition into $n \times n$ blocks. This form is called *canonical*.

Solved Exercise II.62. Show that, *if ω is a symplectic form defined on a vector space F of dimension n, then*

  i) *its dimension is even, say $n = 2m$,*
 ii) *there exists at least one subspace E of dimension m of F to which the restriction of the form ω is the zero form,*
iii) *there exists an isomorphism l from F to $E \oplus E^*$ such that $l^*(\omega_E) = \omega$.*

This result can be proved in various ways. We give a proof which is elementary except at one point (though this point is an excellent example of a fruitful digression).

The fact that symplectic forms exist only on spaces of even dimension is a direct consequence of Proposition II.45 which implies that, if Ω is the matrix of the symplectic form in a basis (f_i) of F (i.e., that whose matrix elements are the $\omega(f_i, f_j)$), then $\det \Omega \neq 0$. Since ω is skew-symmetric, Ω is a skew-symmetric matrix, which implies that $\det(^t\Omega) = \det(-\Omega) = (-1)^n \det \Omega$, and hence that n is necessarily even.

To prove the other properties, we consider once more the matrix Ω of the symplectic form in a basis (f_i) of F. Since Ω is real and skew-symmetric, the matrix $i\Omega$ is Hermitian (i.e., $i\Omega$ is equal to the transpose of its complex-conjugate, which we call its *transconjugate*). It is thus diagonalisable and its eigenvalues are real, which

---

[36] Note that, in an infinite-dimensional space, this point requires the use of Zorn's lemma.

[37] The German mathematician Leopold Kronecker (1823–1891) was mainly interested in the theory of algebraic equations and the theory of numbers. He was insistent on demanding that all mathematics should be founded on the integers.

implies that the eigenvalues of Ω are purely imaginary and occur in pairs of opposite sign (since the characteristic polynomial of Ω has real coefficients).

To finish the proof, it is nevertheless necessary to consider the space in which the eigenvectors are to be found. For this, one must *complexify* the space F to a space $F_{\mathbb{C}}$ (which we should describe as the *tensor product* over $\mathbb{R}$ of F with $\mathbb{C}$ viewed as a vector space of dimension 2 over $\mathbb{R}$) of which (f_i) is still a basis (the vectors of $F_{\mathbb{C}}$ now have complex components with respect to this basis). The space $F_{\mathbb{C}}$ has a well-defined conjugation automorphism γ, which sends a vector $v = \sum_{i=1}^{n} v^i f_i$ to the vector $\gamma(v) = \sum_{i=1}^{n} \bar{v}^i f_i$. The map γ is intrinsically defined as it is the anti-linear involution of $F_{\mathbb{C}}$ whose fixed points are exactly those vectors whose components in the basis (f_i) are real (i.e., the subspace F of $F_{\mathbb{C}}$), and for which the vectors whose sign is changed are those whose components in the basis (f_i) are purely imaginary, i.e., the image of F under multiplication by i in $F_{\mathbb{C}}$.

Since the matrix Ω has real coefficients, the linear map on $F_{\mathbb{C}}$ defined by the matrix $i\,\Omega$ with respect to the basis (f_i) anticommutes with γ. It follows that the eigenspaces of $i\,\Omega$ corresponding to two (necessarily non-zero) complex conjugate eigenvalues λ and $-\lambda$, say F_λ and $F_{-\lambda}$, are interchanged by γ. Note that $F_\lambda \oplus F_{-\lambda}$ is the complexification of a real space: in fact, if (ϕ_j) is a basis of F_λ, then $(\gamma(\phi_j))$ is a basis of $F_{-\lambda}$, whence the fact that $(\phi_j + \gamma(\phi_j), i(\phi_j - \gamma(\phi_j)))$ is a basis consisting of real vectors of the complex space $F_\lambda \oplus F_{-\lambda}$. It is easy to see that in this basis the matrix of the restriction to $F_\lambda \oplus F_{-\lambda}$ of the symplectic form ω is very "sparse": the only non-zero elements occur along the secondary diagonals and, with a suitable choice of the order of these two groups of vectors, are equal to $|\lambda|$ and $-|\lambda|$ respectively. From this point we no longer need the complexification since we are reduced to considering only real objects. If we divide all the vectors in the basis we have constructed by $\sqrt{|\lambda|}$, the matrix of ω in the resulting basis takes the canonical form.

We can thus take for the subspace E on which the symplectic form ω restricts to the zero form the subspace spanned by the half-bases of the invariant real subspaces which we have constructed. It has dimension m by its definition. Denote by $\tilde{E}$ the subspace spanned by the other half-bases. We define a linear map l from F to $E \oplus E^*$ by putting $l(v) = v_E + \tilde{\omega}(v_{\tilde{E}})_{|E}$ where $v = v_E + v_{\tilde{E}}$ denotes the decomposition of v corresponding to $F = E \oplus \tilde{E}$. It is easy to see that the matrix of l in the basis of F which we have constructed and the basis of $E \oplus E^*$ obtained by juxtaposing the basis of E and its dual basis is actually the identity. The linear map l is precisely the map we are looking for, i.e., a map from F to $E \oplus E^*$ which satisfies $l^* \omega_E = \omega$.

Project II.63. Let E be a finite-dimensional real vector space. Show that giving on E two of the three compatible structures, a scalar product, a symplectic form and a complex structure (i.e., an orthogonal endomorphism whose square is minus the identity, like the multiplication by the imaginary unit in a complex vector space) determines a third.

Example II.64. An example of a symplectic form defined on an infinite-dimensional vector space appears naturally in the following context.

Let E be the Banach space formed by the functions of class C^1 defined on the interval $[a, b]$ and vanishing at a and b. For two functions f and g, one then defines $\omega(f, g) = \int_a^b f'(t)\, g(t)\, dt$. Let us show that ω *is a symplectic form on E*.

By performing an integration by parts and using the vanishing of the functions at the endpoints, we obtain that ω is skew-symmetric.

To show that ω is non-degenerate, it suffices to establish that, if f is not identically zero, then there exists a function g such that $\omega(f, g) \neq 0$. If the function f' belongs to E, it will suffice to take $g = f'$ to conclude that the function f of class C^1 is necessarily a constant, and hence 0 because of its values of f at the endpoints of $[a, b]$.

Unfortunately, $f' \notin E$ in general for two independent reasons: for $f \in E$, f' is *a priori* only continuous; moreover there is no reason why f' should vanish at a and b. To get round the first difficulty, it suffices to consider a sequence $(g_n)_{n \in \mathbb{N}}$ of functions of class C^1 converging to f' in the uniform norm. This convergence is enough to ensure the convergence of $\omega(f, g_n)$ to $\int_a^b (f'(t))^2 \, dt$. Since each term of the sequence is zero by hypothesis, the limit is also zero and one concludes as before. For the second, it suffices to consider the product of f' (or the elements of the sequence considered above) with an auxiliary family of non-negative functions of class C^1, zero at a and b and equal to 1 on an arbitrary proper sub-interval of $[a, b]$. If $f \not\equiv 0$, the L^2-norm of f' will be non-zero and it will be possible to find a proper sub-interval of $[a, b]$ where the L^2-norm of f' will be greater than any given fraction of its L^2 norm on $[a, b]$, which finally gives a contradiction.

This construction generalises to the case of functions taking values in a finite-dimensional vector space, the product in the integral being replaced by a scalar product on this space.

E. Hilbert Spaces: Fundamental Properties

We give the basic properties of Hilbert spaces, the Banach spaces whose norm is defined by a scalar product. The main theorem is the Projection Theorem II.74, an essential consequence of which is the isomorphism of any Hilbert space with its (topological) dual.

Definition II.65. A pair consisting of a vector space E and a scalar product is called a *pre-Hilbert space*. If E is complete in the norm induced by the scalar product, E is called a *Hilbert space*.

II.66. Being complete for every norm, finite-dimensional vector spaces are Hilbert spaces for every scalar product defined on them. This is another example of the differences between finite and infinite-dimensional spaces.

The practical determination of the smallest Hilbert space containing a given pre-Hilbert space (it is easy to see that it is a question of its *completeness*) is an important problem for many questions in analysis and is at the heart of the theory of Sobolev spaces for example.

Exercise II.67. Deduce from II.57 that *every Banach space of which every finite-dimensional subspace is a Hilbert space is itself a Hilbert space.*

Corollary II.68. *For every vector v in a pre-Hilbert space E, the linear form v^b, which associates to any vector w the scalar $\langle v^b, w \rangle_E = (v|w)$, is continuous.*
  Moreover, the map b from E to E' thus defined is continuous.

Proof. The Cauchy–Schwarz inequality says precisely that the norm of the linear map v^b is $\|v\|$. This means that v^b is continuous, and that the map b is an isometry, and hence continuous. $\square$

Remark II.69. The musical notation b refers to the fact that the components of a vector are written with upper indices (one also speaks of *contravariant* indices), while the components of linear forms are written with lower indices (for these, one speaks of *covariant* indices). The map b defined in Corollary II.68 is thus analogous to the musical operation of a flat, which lowers a note by a semi-tone.
  To be able to define a map $\sharp$ which associates vectors to forms, it is in some sense necessary that there are not *more* elements in the dual than in the space itself, which is not always the case in infinite-dimensional spaces as we have already remarked. This point is the subject of Theorem II.83 a little later in this section.

II.70. From construction II.46, it follows that every subspace F of a pre-Hilbert space E is itself a pre-Hilbert space with the inverse image under the injection of the ambient scalar product. If E is a Hilbert space, it follows from Proposition II.4 that *F is a Hilbert space in the induced scalar product if and only if F is closed.*
  Given a subspace F, it is possible to associate to it another subspace formed by the vectors orthogonal to all the vectors in F, a subspace which is naturally called the *orthogonal complement* of F and denoted by $F^\perp$.

Proposition II.71. *In a pre-Hilbert space, the orthogonal complement $F^\perp$ of an arbitrary vector subspace F is closed.*

Proof. We note first of all that the hyperplane $v^\perp$ formed by the vectors orthogonal to v is closed since it is the set of zeros of the map $w \mapsto (v|w)$ which is continuous by Corollary II.68.
  Since $F^\perp = \bigcap_{v \in F} v^\perp$, this subspace is closed, being the intersection of closed subspaces. $\square$

II.72. We now embark on the first true problem of *Variational Calculus* in this course by generalising to the setting of Hilbert spaces the theorem about the orthogonal projection onto a subspace which is classical in the setting of finite-dimensional spaces. In that case, the discussion of the problem can be conducted purely in terms of algebra. As we have seen on several occasions, the extension to the case of infinite-dimensional spaces will necessitate recourse to analysis.
  Although it is rather simple, the theorem which we now present has a very large field of applications, notably in optimisation theory. A more sophisticated version, which is also very useful, is given in Exercise II.77.

II.73. Given a subspace F of a Hilbert space E, we propose *"to determine whether in F there exists one (or several) point(s) nearest to a given point x of the space"*. This problem of *the shortest path* has a satisfactory solution when F is closed as shown by the projection theorem which we now state.

Projection Theorem II.74. *Let F be a closed subspace of a Hilbert space E which does not consist only of the zero vector.*

   i) *To every point x in E there is a unique nearest point in F, denoted by x_F, which is characterised by the condition $x - x_F \in F^\perp$.*

  ii) *The map $x \mapsto P_F(x) = x_F$ is a continuous linear projection of E of norm 1 called the orthogonal projection onto F.*

 iii) *The space E admits the following orthogonal direct sum decomposition $E = F \oplus F^\perp$.*

Proof. i) Let x be a point in E and put $d = d(x, F) = \inf_{y \in F} \|x - y\|$. By definition of the lower bound, there exists a sequence of points y_i in F such that $\|x - y_i\|$ tends to d as $i \to \infty$.

We show that the sequence $(y_i)_{i \in \mathbb{N}}$ is Cauchy. For this, it suffices to evaluate the quantity $\|y_p - y_q\|^2$ which can be done by using the Median Formula (cf. II.56)

$$(\text{II.75}) \qquad \|y_p - y_q\|^2 = 2\|y_p - x\|^2 + 2\|x - y_q\|^2 - 4\left\|x - \frac{1}{2}(y_p + y_q)\right\|^2.$$

Since $\frac{1}{2}(y_p + y_q) \in F$, $\|x - \frac{1}{2}(y_p + y_q)\|^2 \geq d^2$. Moreover, we know that, by taking p and q sufficiently large, $\|x - y_p\|^2 + \|x - y_q\|^2$ can be made as close as we like to $2\,d^2$. It follows that $\|y_p - y_q\|^2$ is bounded above by a quantity as small as we please. The subspace F being closed by Proposition II.4, the Cauchy sequence $(y_i)_{i \in \mathbb{N}}$ has a limit point y in F. Moreover, $\|x - y\| = \lim_{i \to \infty} \|x - y_i\| = d$.

If there were two distinct points y and y' which realise the distance from x to F, the preceding argument using Eq. (II.75) applied to the sequence $y_{2p} = y$ and $y_{2p+1} = y'$ together with the definition of d shows that the distance from y to y' would be zero, a contradiction.

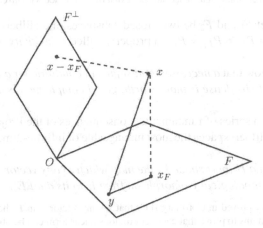

Fig. II.76 Orthogonal projection onto a subspace

We now show that the point z, which we denote by x_F from now on, is characterised by the condition $x - x_F \perp F$. If w is an arbitrary element of F,

$$\|x - (x_F + tw)\|^2 = d^2 - 2t\, \mathcal{Re}((x - x_F)|w) + t^2(w|w) .$$

This function of t is bounded below by d^2 if and only if the coefficient of t is zero, whence the orthogonality (by applying the argument to w and then to iw).

ii) Since $P_F(x) = x$ when x belongs to F, we have $P_F \circ P_F = P_F$ and hence P_F is a projection. The map is linear since $P_F(\lambda x + \mu y)$ is the only vector in F such that $(\lambda x + \mu y) - P_F(\lambda x + \mu y)$ is orthogonal to F and since $\lambda(x - P_F(x))$ and $\mu(y - P_F(y))$ are both orthogonal to F. To show that the map P_F is continuous, it is enough, by Proposition I.38 and the remark which follows it, to show that $\|P_F\| = \sup_{\|x\| \leq 1} \|P_F(x)\| \leq 1$. But this property is a consequence of the identity

$$\|x\|^2 = \|x - P_F(x) + P_F(x)\|^2 = \|x - P_F(x)\|^2 + \|P_F(x)\|^2 \geq \|P_F(x)\|^2$$

which is always satisfied because of the orthogonality and the fact that $P_F(y) = y$ for $y \in F$.

iii) For all $x \in E$, we can write $x = (x - P_F(x)) + P_F(x)$ which ensures that $E = F + F^\perp$ since $x - P_F(x) \in F^\perp$. Moreover, it is clear that $F \cap F^\perp = 0$ by the positivity of the scalar product on non-zero vectors. It then follows from the Pythagoras theorem that the map $x \mapsto (P_F(x), x - P_F(x))$ is an isometry from E to $F \oplus F^\perp$, and hence a topological isomorphism.   □

Exercise II.77. Show that *in a Hilbert space, to every point there is a unique nearest point in any closed convex subset* (i.e., any closed subset A such that, for any two points x and y in A, the line segment $[xy]$ is contained in A).

Remark II.78. Note that in the Projection Theorem II.74, as in the preceding exercise, the fact that the ambient space was complete was only used to ensure that a closed subspace is complete, hence the extension of the projection theorem to complete subspaces of pre-Hilbert spaces. Note also that the projection theorem does not generalise in any way to Banach spaces for the simple reason that in such spaces *it is not true* that every closed subspace admits a closed complement.

Project II.79. Give an example of a closed subspace of a Banach space with no closed complement.

Exercise II.80. Let F_1 and F_2 be two closed subspaces of a Hilbert space such that $F_2 \subset F_1$. Show that $P_{F_2} \circ P_{F_1} = P_{F_2}$, a property called *transitivity of projections*.

Exercise II.81. Show that *a necessary and sufficient condition for a vector subspace of a Hilbert space to be dense is that its orthogonal complement is the zero vector.*

II.82. We now give a series of fundamental consequences of the Projection Theorem II.74 which make Hilbert spaces the most manageable of infinite-dimensional spaces.

Theorem II.83. *In a Hilbert space E, the map which to any vector x associates the linear form $x^\flat$ is a topological isomorphism from E to its dual E'.*

Proof. We have already proved in Corollary II.68 that for any vector x in E the linear form $x^\flat$ is continuous. It thus remains to prove that every continuous linear form can be obtained in this way.

Since $0_{E'}$ is clearly obtained as $0_E^\flat$, it suffices to consider the case of a non-zero continuous linear form λ. The subspace $F = \lambda^{-1}(0)$ is a closed hyperplane in E which with its orthogonal complement $F^\perp$ defines an orthogonal direct sum decomposition of E by part iii) of the Projection

Theorem II.54. The form λ not being identically zero, there exists at least one vector x such that $\langle \lambda, x \rangle_E \neq 0$, and consequently $F^\perp \neq 0$. Let u be a non-zero vector in $F^\perp$. The linear forms λ and u^b have the same kernel and are both non-zero. It is thus possible to find a scalar α such that $\lambda = \alpha \, x^b$, a relation which can be rewritten $\lambda = (\overline{\alpha} \, x)^b$. $\square$

Remark II.84. In the statement of the preceding theorem, the meaning of the term *isomorphism* has been left a little vague from the algebraic point of view, but it becomes clear from the proof that, in the case when the field of scalars is $\mathbb{C}$, this isomorphism is antilinear.

Exercise II.85. State a converse of Theorem II.83.

II.86. We develop briefly the notion of *weak topology* which is very important in the case of Hilbert spaces because of the identification which is possible between the space and its topological dual, and above all because of Theorem II.89 which turns out to be a fundamental tool in Variational Calculus.

II.87. For the weak topology, a sequence $(x_n)_{n \in \mathbb{N}}$ of vectors in a Hilbert space E converges *weakly* to x_∞ if, *for any vector y in E, the sequence of scalars $(x_n | y)_{n \in \mathbb{N}}$ converges to $(x_\infty | y)$*.

Exercise II.88. Prove that a sequence $(x_n)_{n \in \mathbb{N}}$ of points in a Hilbert space converging weakly to x_∞ and satisfying moreover the condition $\lim_n \|x_n\| = \|x_\infty\|$ converges in norm (one also says *strongly*) to x_∞.

Theorem II.89. *The unit ball of a Hilbert space is compact for the weak topology* (for a proof, see pages 401 and 301 of (Schwartz 1970)).

II.90. The weak compactness theorem is very useful for establishing, for example, the existence of an element of a Hilbert space which is the weak limit of a sequence of elements for which a functional tends to its lower bound. This does not completely solve the problem since it is then necessary to show that the functional actually takes the expected value on this element; for this the functional must be sufficiently regular.

Recall that, in contrast to Theorem II.89, Riesz' Theorem I.77 tells us that the unit ball of a normed space is compact for the topology induced by the norm if and only if the space is finite-dimensional. This difference between finite and infinite-dimensional spaces is emphasised once more by the theorem due to Czeslaw Bessaga[38] which asserts that *"the unit sphere of an infinite-dimensional Hilbert space is homeomorphic to the whole space"*.

---

[38] Czeslaw Bessaga (1932–2021) was a Polish mathematician who made several contributions to functional analysis.

F. Hilbert Bases

After the general development in Sect. E, we come here to the fundamental example of a Hilbert space, namely l^2, and we show that this is the model for Hilbert spaces with a countable basis. In this space, the formulas of Euclidean geometry using the components of vectors in orthonormal bases remain valid.

II.91. Being complete for every norm by I.32, and thus in particular for that associated to the standard scalar product, the space $\mathbb{K}^n$ is an example of a Hilbert space, in fact the model of all Hilbert spaces of dimension n by the remark made in II.50.

We now examine spaces of sequences in this context. If we wish to generalise the standard scalar product defined on $\mathbb{K}^n$, for a sequence (x) we put $((x)|(x)) = \sum_n |x^n|^2$ which we again call the *standard scalar product*. To be able to do this, it is necessary that, for the sequence (x), $\sum_n |x^n|^2 < \infty$, and hence that the sequence is at least in l^2. This space turns out to be the prototype of Hilbert spaces as we shall see later. We begin with a basic proposition.

Proposition II.92. *The space l^2 provided with the standard scalar product which to any pair $((x), (y))$ associates the scalar $((x)|(y)) = \sum_n \overline{x^n} y^n$ is complete, and hence is a Hilbert space.*

Proof. The first point is to remark that the standard scalar product on l^2 is indeed a scalar product.

Let $((x)_k)_{k \in \mathbb{N}}$ be a Cauchy sequence in l^2, which means that, if we set $(x)_k = (x_k^n)$, for any $\epsilon > 0$ there exists an integer r such that, for $p, q \geq r, \sum_n \|x_p^n - x_q^n\|^2 < \epsilon$.

A fortiori, this means that, for n fixed, the numerical sequence $(x_i^n)_{i \in \mathbb{N}}$ is Cauchy, and thus has a limit which we can denote by x^n. We denote by (x) the sequence thus defined.

We also deduce from this that, for any integer m, $\sum_{n \leq m} \|x_p^n - x_q^n\|^2 < \epsilon$, p and q being fixed. We obtain $\sum_{n \leq m} \|x_p^n - x^n\|^2 \leq \epsilon$ by passing to the limit in this inequality as $q \longrightarrow \infty$. Since the integer m is arbitrary, by the triangle inequality this implies that the sequence (x) belongs to l^2 and that the sequence $((x)_k)_{k \in \mathbb{N}}$ tends to (x) in l^2.   □

Remark II.93. By Theorem II.83, we know that l^2 is isomorphic to its topological dual. This agrees with the result of Exercise II.33 since $\frac{1}{2} + \frac{1}{2} = 1$!

II.94. We return to the discussion at the beginning of Chapter I which gave a meaning to the decomposition of a vector relative to a basis. We generalise the notion of an algebraic basis and allow the possibility of an infinite number of components.

In a given Hilbert space E, we are led inevitably to the consideration of sequences of vectors of E. In a Banach space, one of the general criteria at our disposal for the convergence of a series is that of *normal* convergence, namely that the (numerical) sequence of norms is required to converge. (Exercise: Prove that this is a convergence criterion.)

In the case of Hilbert spaces, and for particular series, there is another criterion.

Proposition II.95. *A series $\sum_i v_i$ formed by pairwise orthogonal elements v_i of a Hilbert space converges if and only if the numerical series $\sum \|v_i\|^2$ converges. Moreover, in this case the generalised Pythagorean relation $\| \sum_i v_i \|^2 = \sum_i \|v_i\|^2$ holds.*

Proof. Note that the hypothesis that the vectors v_i are pairwise orthogonal implies that the numerical series $\sum_n \|v_n\|^2$ has $\|\sum_{i=1}^k v_i\|^2$ as its series of partial sums by the Pythagorean theorem. If the series $\sum_i v_i$ converges to a vector v, the numerical series $\|\sum_{i=1}^k v_i\|^2$ converges to $\|v\|^2$, and hence so does the series $\sum_i \|v_i\|^2$.

Conversely, we suppose that the numerical series $\sum_n \|v_n\|^2$ converges and show that the sequence $(y_n)_{n\in\mathbb{N}}$ defined by $y_n = \sum_{i=1}^n v_i$ is Cauchy. Since we have $y_p - y_q = \sum_{p+1\le i\le q} v_i$ and since the vectors v_i are pairwise orthogonal, we obtain that $\|y_p - y_q\|^2 = \sum_{p+1\le i\le q} \|v_i\|^2$. The series $\sum_n \|v_n\|^2$ being supposed convergent, this implies that the sequence $(y_n)_{n\in\mathbb{N}}$ is Cauchy, and hence convergent since we are in a Hilbert space.

That the generalised Pythagorean relation holds follows immediately from the fact that the sequences of partial sums $\|\sum_{i\le n} v_i\|^2$ and $\sum_{i\le n} \|v_n\|^2$ coincide. □

II.96. We now turn to the generalisation of the notion of an algebraic basis of a vector space to the context of Hilbert spaces. (Note that most of the definitions which will be given retain a meaning in the case of more general topological vector spaces.)

Given a family of vectors $(f_i)_{i\in I}$ in a Hilbert space E, it is natural to introduce the smallest closed subspace, say F, which the family generates. Since F is closed in E, F is complete, and thus is itself a Hilbert space. This space is called the *Hilbert subspace generated* by the family $(f_i)_{i\in I}$. It can be obtained as the closure of the set of (finite!) linear combinations of vectors of the family. It is often useful to be able to turn this around, i.e., to consider F as given and to ask for families of vectors which generate it. This leads us to the following definition.

Definition II.97. A family of vectors $(f_i)_{i\in I}$ in a Hilbert space E is said to be *total* in E if any $v \in E$ is the limit of a sequence of linear combinations of the $(f_i)_{i\in I}$.

II.98. Another formulation of this definition consists in saying that a family $(f_i)_{i\in I}$ of vectors in a Hilbert space E is total if the Hilbert subspace which it generates is the whole of E.

It follows directly from this definition that the family of sequences $(e_i)_{i\in\mathbb{N}}$ generalising the vectors in the standard basis of $\mathbb{K}^n$ (whose only non-zero element is the ith which is equal to 1) form a total system in l^2.

Theorem II.99. *If (f_i) is a sequence of pairwise orthogonal vectors of norm 1 in a Hilbert space E, then, for all $v \in E$, the vector series $\sum_i (v|f_i) f_i$ converges to $P_F(v)$ (where F denotes the Hilbert subspace of E generated by the f_i) and the Bessel[39] inequality $\|P_F(v)\|^2 = \sum_i |(v|f_i)|^2 \le \|v\|^2$ holds.*

Moreover, if the sequence of vectors $(f_i)_{i\in\mathbb{N}}$ is total in E, then, for all v in E, $v = \sum_i (f_i|v) f_i$, and we have obtained the Parseval[40] identity $\|v\|^2 = \sum_i \|(v|f_i)\|^2$. The sequence of vectors $(f_i)_{i\in\mathbb{N}}$ is then said to be a Hilbert basis *of E.*

---

[39] The German astronomer Friedrich Wilhelm Bessel (1784–1846) had an extended correspondence with Gauss. He contributed to potential theory, and his study of the perturbations of the trajectories of the planets led him to develop the special functions which today bear his name.

[40] Marc-Antoine Parseval des Chênes (1755–1836) is a French mathematician who produced several visionary memoirs around 1800. The identity quoted is also called the *Rayleigh identity* after the English physicist John William Strutt Rayleigh.

Proof. It is natural to introduce the sequence of finite-dimensional subspaces F_n generated by the first n vectors of the sequence, of which by hypothesis they form an orthonormal basis. It then follows from the classical relations expressing a vector in terms of its components relative to an orthonormal basis that, for every vector v in E,

$$P_{F_n}(v) = \sum_{i=1}^{n} (f_i | P_{F_n}(v)) f_i = \sum_{i=1}^{n} (f_i | v) f_i$$

by using the fact that $P_{F_n}(v) - v$ is orthogonal to F_n. Of course, it follows from this relation that

$$(P_{F_n}(v) | P_{F_n}(v)) = \sum_{i=1}^{n} |(f_i | v)|^2 .$$

If we now use the transitivity of projections (see Exercise II.80), we have $P_{F_n}(v) = P_{F_n}(P_F(v))$. Applying part ii) of the Projection Theorem II.74, we thus have $\sum_{i=1}^{n} |(f_i | v)|^2 = \|P_{F_n}(v)\|^2 \leq \|P_F(v)\|^2$ which ensures the convergence of the numerical series $\sum_i |(f_i | v)|^2$, and by Proposition II.95 that of the series $\sum_i (f_i | v) f_i$. The sum w of this series is by construction such that $v - w$ is orthogonal to all the f_i, and hence to F. But of course we know that this is a characteristic property of $P_F(v)$. We thus have $w = P_F(v)$.

The Bessel inequality is then a direct consequence of the generalised Pythagorean relation proved in Proposition II.95 and the inequality given in the preceding paragraph.

For the last part of the theorem, it is enough to observe that in that case $F = E$.   □

Exercise II.100. Give an example of a function defined on the unit ball of a Hilbert space which does not attain its maximum there (one can use the restriction of a quadratic form).

Show by contrast that the restriction to the unit ball of every linear form defined on a Hilbert space does attain its maximum there.

II.101. With the introduction of Hilbert bases, we have thus accomplished what we set out to do, namely to generalise the expression of a vector as the sum of its components relative to a basis. In infinite dimensions, this could only be done by addressing questions of convergence.

Almost by definition, the generalisation of the standard basis is a Hilbert basis of l^2. There are of course many other examples which are important from the point of view of applications, but we cannot go into them in this course.

Let us consider the space of numerical functions defined on a domain Ω whose norm squared is Riemann integrable, so that their norm of order 2 is finite (a space which we would like to call $L^2(\Omega)$). Unfortunately, this space is *not complete* due to the lack of a good convergence theorem for the (Riemann) integrals of a sequence of functions. This gap is filled by the theory of Lebesgue integrals and it is one of the reasons for the final success of this theory over Riemann's.

II.102. With generalised integrals, there are several families of special functions, exponentials or polynomials, which give Hilbert bases of the space of square integrable functions of one variable. The most famous family is given by the exponentials $x \mapsto e^{2i\pi nx}$ which form a Hilbert basis of the space of functions defined on $[0, 1]$.

The theory of *Fourier*[41] *series* consists precisely of the study of the properties of a function which can be deduced from its representation in terms of its components in this basis. It is of great practical importance, and forms the foundation of the branch of mathematics called *harmonic analysis*.

But there are also other families of special functions which have remarkable properties such as the Legendre[42], Hermite or Tchebyshev[43] polynomials. These families play an important role in approximation theory, which has been given new life with the use of computers.

Solved Exercise II.103. We give a beautiful example of an application of the theory of Hilbert bases to the calculation of the sum of the series $\sum_{n=1}^{\infty}(1/n^2)$. One can also consider this as an application of the Projection Theorem II.74.

For this, we introduce the space E of piecewise continuous functions with values in $\mathbb{C}$ defined on $[-\pi, \pi]$, which one provides with the usual scalar product:

$$(\varphi, \psi) \mapsto (\varphi|\psi) = \frac{1}{2\pi} \int_{-\pi}^{\pi} \overline{\varphi(t)} \, \psi(t) \, dt \ .$$

Even though E is only a pre-Hilbert space, as we have remarked in II.78, we shall be able to apply the projection theorem to complete subspaces of E. We are going to consider the subspaces F_n of E generated by the functions $e_k = e^{ikt}$ for $-n \leq k \leq n$. The subspaces F_n of E are finite-dimensional, and thus automatically complete. It follows for example by the theorem of Weierstrass that the subspace F_∞ is dense in E. Let x_n be the orthogonal projection of x onto F_n. We then have $\|x\|^2 = \|x - x_n\|^2 + \|x_n\|^2$ by II.74 i). In particular, since the (e_n) form an orthonormal basis, if we introduce for a vector x its components $a_n(x) = (e_n|x)$, we have that $\sum_{k=-n}^{n} |a_k(x)|^2 = \|x_n\|^2$ is bounded above by $\|x\|^2$, and hence the series $\sum_{n \in \mathbb{Z}} |a_n(x)|^2$ converges.

Let $y_k = \sum_{i=-n(k)}^{n(k)} a_i(y_k) \, e_i$ be a sequence tending to x. By definition one has $\|x - x_{n(k)}\| \leq \|x - y_k\|$ since $x_{n(k)}$ is the point of $F_{n(k)}$ closest to x. It follows that $\|x\|^2 - \sum_{|i| > n(k)} |a_i(x)|^2$, which is equal to $\|x - x_{n(k)}\|^2$, tends to 0 as k tends to infinity. This means that the subsequence $\sum_{i=-n(k)}^{n(k)} |a_i(x)|^2$ converges to $\|x\|^2$. The same is true of the series $\sum_{n \in \mathbb{Z}} |a_n(x)|^2$ which has positive terms.

---

[41] The French mathematician Jean-Baptiste Joseph Fourier (1768–1830) is also known as an egyptologist and administrator. He was professor of analysis at the École Polytechnique from 1796 to 1798. His study of *trigonometric series*, the basis for *Fourier analysis*, has had a considerable impact in the natural sciences, and is often considered as a model of the way in which mathematics can provide a framework for the analysis of experimental phenomena.

[42] Adrien Marie Legendre (1752–1833), French mathematician, author of numerous works in arithmetic and analysis. We owe to him notably the definition of a very important transformation in the calculus of variations relating the Lagrangian and Hamiltonian formulations (see Chapter X of this course).

[43] The Russian mathematician Pafnoutiy Lvovitch Tchebyshev (1821–1894) contributed to constructive function theory and to the study of probability. His works on approximation theory are still classics in the subject.

We now consider the function x defined by

$$x(t) = \begin{cases} -1, & \text{for } t \in [-\pi, 0[\ ; \\ 0, & t = 0\ ; \\ 1, & \text{for } t \in\ ]0, \pi]\ . \end{cases}$$

One then has

$$a_n(x) = (x|e_n) = -\int_{-\pi}^{0} e^{-int}\,dt + \int_{0}^{\pi} e^{-int}\,dt$$

where $a_{2k+1}(x) = -(2i/\pi)(2k+1)^{-1}$ and $a_{2k} = 0$. Consequently,

$$\|x\|^2 = \sum_{n \in \mathbb{Z}} |a_n(x)|^2 = \frac{4}{\pi^2} \sum_{n \in \mathbb{Z}} \frac{1}{(2k+1)^2}\ .$$

By calculating the integral directly, one has $\|x\|^2 = 1$, which gives

$$\sum_{n=0}^{+\infty} \frac{1}{(2k+1)^2} = \frac{\pi^2}{8}\ .$$

On setting $I = \sum_{n=1}^{+\infty} 1/n^2$, one thus has

$$I = \sum_{n=1}^{+\infty} \frac{1}{(2n)^2} + \sum_{n=0}^{+\infty} \frac{1}{(2n+1)^2}$$

so that $I = I/4 + \pi^2/8$ and finally $I = \pi^2/6$.

II.104. It is useful to focus our attention on certain Hilbert spaces which, although they are infinite-dimensional, are still not too big. This is the case for Hilbert spaces which contain a countable dense subset, which are called *separable*.

It is possible to give a classification of such spaces which is classical but spectacular.

Theorem II.105. *Every separable infinite-dimensional Hilbert space is isomorphic to the space l^2.*

Proof. Let E be a separable Hilbert space and $(f_n)_{n \in \mathbb{N}}$ an everywhere dense sequence. We shall obtain the required isomorphism by constructing a Hilbert basis from the sequence $(f_n)_{n \in \mathbb{N}}$. The representation of a vector in terms of its components then gives the isometry with l^2 because of the Parseval identity.

By removing inductively those f_i which are linear combinations of the preceding ones, we may assume that this sequence consists of linearly independent vectors.

To make the sequence orthogonal, it then suffices to define, by means of a second induction, a sequence $(e_n)_{n \in \mathbb{N}}$ by $e_0 = f_0$ and $e_{n+1} = f_{n+1} - P_{F_n}(f_{n+1})$ where F_n denotes of course the subspace of E generated by the first n vectors of the sequence. By its construction, the sequence $(e_n)_{n \in \mathbb{N}}$ consists of pairwise orthogonal vectors, and is total in E because the sequence $(f_n)_{n \in \mathbb{N}}$ is dense in E. By dividing the vectors by their norm, we have constructed an orthonormal basis. $\square$

Remark II.106. The attentive reader will have noticed that the crux of the classification Theorem II.105 is an extension of the procedure which is well-known in the case of finite-dimensional Euclidean or Hermitian spaces, namely the *Gram*[44] *–Schmidt*[45] *orthonormalisation* procedure.

II.107. Because of this theorem, the space l^2 has played a rather special role historically, and in a number of applications, and is sometimes simply called *the Hilbert space*. From a strictly mathematical point of view, it satisfies the perennial desire of mathematicians to *measure the scope of the generalisation of a notion well-known in a more restricted context*, as well as their taste for *classification theorems*.

On the other hand, note that Banach spaces, which are the other theme of this chapter, present an extraordinary diversity which has given birth to a branch of mathematics, the geometry of Banach spaces, which has recently been developed considerably.

II.108. Theorem II.105 has a generalisation to non-separable Hilbert spaces. The fundamental invariant is the cardinal of a Hilbert basis (one shows that this notion makes sense) called the *Hilbert dimension*. For each cardinal I, the model space consists of the families of elements of $\mathbb{K}$ indexed by a set of cardinal I.

G. Historical Notes

II.109. It is Banach who, by making between 1920 and 1930 a profound study of the operators which can be defined on a complete normed space, gave a definitive form to the theory of the spaces which today carry his name.

II.110. It is to David Hilbert that we owe the introduction of the space l^2 and the first developments of the theory of operators defined on such spaces, notably the various notions related to their spectrum which we have not entered into in this course.

The more abstract presentation of Hilbert spaces which we have used appeared only in 1932 with the work of von Neumann[46] which attempted to make rigorous the

---

[44] Jorgen Pedersen Gram (1850–1916), German mathematician, studied the Γ-function and orthogonal systems.

[45] The German mathematician Ehrard Schmidt (1876–1959) developed linear algebra and took part in the founding of the theory of linear equations.

[46] John von Neumann (1903–1957) was a Hungarian-American mathematician, physicist and engineer. He had a very broad coverage of the field of mathematics, pure and applied. He is considered one of the founders of computer science. He also anticipated the possibility of self-replication ahead of the DNA structure discovery. He was a key actor of the Manhattan Project during World War II.

development of the formalism of quantum mechanics by Dirac[47] and Schrödinger[48];
he used tools which were different, but equivalent precisely because of Theorem
II.105. Several of the fundamental theorems about Hilbert spaces, such as the projection theorem or those concerned with the dual, had been obtained much earlier
(around 1910).

---

[47] Paul Adrien Maurice Dirac (1902–1984) was an English theoretical physicist. His creative
contributions make him one of the leading physicists of the XXth century, e.g. his introduction of
spinors in the relativistically invariant formulation of the Schrödinger equation to express the wave
function of fermions. He received the Nobel Prize in 1933 in particular for his *prediction of the
existence of anti-matter*. He is one of the few theoreticians to have received the Prize.

[48] Erwin Rudolf Josef Alexander Schrödinger (1887–1961) was an Austrian-Irish physicist who
contributions to many parts of the discipline. He obtained a number of fundamental results in
quantum theory, in particular the fundamental equation satisfied by a wave function that bears his
name. He was also interested in biology, both at a philosophical and theoretical level. He received
the Nobel Prize in Physics in1933 for his work on quantum mechanics. Opposing the rise of Nazism,
he left Germany that year.

Chapter III
Linearisation and Local Inversion of Differentiable Maps

In this chapter we begin the developments in differential calculus which are necessary for the Calculus of Variations. We always work in real vector spaces. The key idea is that of *approximation by linear maps* (one often speaks of *linearisation*, a neologism which we have used in our title, not without reservations, and whose meaning we shall make explicit a little later).

After having introduced the notion of *directional* derivative, which extends the classical idea of partial derivative in numerical spaces, one can define in a suitable way, the *tangent linear map* of a given map at a point of a normed space, and we do this in Sect. A.

Section B is devoted to establishing the *chain rule* which allows one to compute the tangent linear map of the composite of two maps. We take the opportunity to relate the notations we use to the classical differential notations.

The next fundamental step, which is the object of Sect. C, is the *local inversion theorem*, which provides a simple criterion for deciding when a map may be considered as a change of variables. *Linearisation* consists then in replacing, after a suitable change of variables, a map by its tangent linear map.

In Sect. D, we present succinctly the notions of *higher order* derivatives. The *Schwarz Lemma*, asserting the symmetry of second partial derivatives for sufficiently regular maps, generalises to the case of normed spaces. The only difficulty stems from the inevitable notational complications.

J.-P. Bourguignon, *Variational Calculus*, Springer Monographs in Mathematics,
https://doi.org/10.1007/978-3-031-18307-2_3

A. Differentiable Maps and Their Tangent Linear Maps

Before defining the notion of a differentiable map, we begin with the simpler (but less powerful) notion of a map which admits partial derivatives.

III.1. The notion of the *derivative*, which of course we assume known, is a fundamental tool in the study of variations of functions of a real variable. The accent placed, with practical applications in mind, on the methods of calculating derivatives sometimes obscures the fact that the derived function, when it exists, provides in a neighbourhood of a point a linear approximation to the function.

When one considers functions of several real variables, things become a little complicated from the notational point of view because of the "partial derivatives" which one can define in this case.

It is in the context of normed spaces, which were the subject of the preceding chapters, that the different notions find their proper place. Of the vector space structure of normed spaces we retain only the *translations*, which means that once again we shall in fact be working in the context of affine spaces. To begin with, we are going to outline certain notions which will be central throughout the remainder of this course.

III.2. We begin with an *a priori* naive approach which reduces the problem to the case of functions of *one* real variable.

For this, let f be a map defined on an open subset U of a real topological vector space E and taking values in a topological vector space F (also assumed to be real); if a is a point of U and v a vector in E, we consider the *partial function* $t \mapsto f(a+tv)$ which is defined when $a + tv \in U$, and hence in a neighbourhood of 0 in $\mathbb{R}$. The *partial derivative* of f at a in the direction v is then defined as a derivative of a function of one variable, as the limit in F (if it exists) of $t^{-1}(f(a + tv) - f(a))$ as $t \to 0$. To represent this partial derivative, we introduce the *notation*

$$(\partial_v f)(a) = \lim_{t \to 0} \frac{f(a + tv) - f(a)}{t} .$$

Note that, by changing the time variable, it is clear that, for $\alpha \in \mathbb{R}$, $\partial_{\alpha v} f = \alpha \, \partial_v f$.

III.3. This method of introducing the notion of a derivative by using partial maps is often known as the *directional derivative* and a function which has directional derivatives in all directions is sometimes called *Gâteaux*[49] *-differentiable*. It is rather weak, but there are situations where it is the only one available.

However, it allows one to obtain the following elementary proposition which is very useful in Variational Calculus, as we shall see for example in III.10.

Proposition III.4. *If a numerical function f defined on an open set U in a normed vector space E has a maximum or a minimum at a point a in U, then, for every v in E for which f has a directional derivative, $(\partial_v f)(a) = 0$.*

---

[49] The French mathematician René Gâteaux (1889–1914) was professor at the lycée at Bar-le-Duc before being killed in the First World War.

Proof. For every $v \in E$, the partial function $t \mapsto f(a + tv)$ has a critical point at $t = 0$ and since

$$\frac{d}{dt} f(a + tv)|_{t=0} = (\partial_v f)(a) ,$$

we have the stated result.   □

III.5. Let us relate this to the usual notation for functions defined on an open subset of a vector space of finite dimension n in which a basis (e_i) has been chosen. This basis defines on the space a system of (linear) coordinates (x^i) which are simply the evaluation at a point x of E of the linear functions which are the vectors ϵ^i of the dual basis of the basis (e_i).

It is then classical that the function f can be identified with a function of the n variables $x^1, \ldots, x^n$. A partial function then corresponds to *fixing* all the variables except one. This leads to the *notation* $\partial f / \partial x^i$ which is often used to denote $\partial_{e_i} f$.

Warning III.6. *Careless use of the notation* $\partial f / \partial x^i$ *can lead to serious errors.*[50]

Indeed the notation suggests that the quantity $\partial f / \partial x^i$ depends only on the consideration of the single coordinate x^i, when in fact it depends on the whole system of coordinates (x^j).

If one goes back to the definition, this is not surprising since the derivative which gives the partial derivative is obtained by fixing *all* the coordinates *except* precisely x^i. To illustrate this cautionary tale, we give a simple but instructive example of this phenomenon.

In a space of dimension 2 referred to a basis $(\vec{\imath}_1, \vec{\imath}_2)$, we introduce a new basis by putting $\vec{\jmath}_1 = \vec{\imath}_1 + \vec{\imath}_2$ and $\vec{\jmath}_2 = \vec{\imath}_2$ so that the new coordinates y^1 and y^2 can be expressed in terms of the old ones x^1 and x^2 by the formulas $y^1 = x^1$ and $y^2 = x^2 - x^1$. Consequently, for a differentiable function f, we have with the above notations $\partial f / \partial x^1 = \partial_{\vec{\imath}_1} f$ while $\partial f / \partial y^1 = \partial_{\vec{\jmath}_1} f$, and these are different quantities in general even though $y^1 \equiv x^1$; on the other hand, even though $y^2 \neq x^2$, one has $\partial f / \partial x^2 = \partial_{\vec{\imath}_2} f = \partial_{\vec{\jmath}_2} f = \partial f / \partial y^2$.

Proposition III.7. *For every vector v in a normed space E, the differential operator ∂_v satisfies the Leibniz[51] formula, i.e., for two scalar functions ϕ and ψ which admit partial derivatives at a in the direction v,*

$$\partial_v(\phi\psi)(a) = \phi(a)(\partial_v\psi)(a) + \psi(a)(\partial_v\phi)(a) .$$

Proof. This is a direct consequence of the usual formula for the derivatives of functions of one real variable if we recall the definitions of the two sides as ordinary derivatives along the straight line $t \mapsto a + tv$.   □

---

[50] ...but we shall continue to use it all the same!

[51] Gottfried Wilhelm Leibniz (1646–1716), born in Leipzig, made essential contributions to the sciences and to philosophy as well as to history. In 1676, he founded, independently of Newton, the differential calculus and the theory of motion which he revealed in the *Acta Eruditorum* of Leipzig.

Remark III.8. We thus see the appearance of a new way to consider a vector v in a vector space, i.e., as a linear form (namely $f \mapsto (\partial_v f)(a)$) on the vector space of scalar functions which have partial derivatives in every direction at a point and which satisfies the *Leibniz formula*, a notion which makes sense only because this vector space is also provided with a *product* which is bilinear and which in fact makes it into an *algebra*.

There is a converse to Proposition III.7 which allows one to *identify the vectors in a vector space with the first order differential operators which have the Leibniz property*, and this point of view will be fundamental when we wish, in Chapter VI, to define the notion of a *tangent vector to a configuration space*.

In a vector space of finite dimension n referred to a basis (e_i) whose linear coordinates are denoted by (x^i), this leads us to *denote*[52] the basis vector e_i by $\partial/\partial x^i$, because it can be identified with the differential operator which to a function f associates $(\partial f/\partial x^i)$ (but we shall not go so far as to use the notation $\partial_{\partial/\partial x^i} f$ to denote $\partial f/\partial x^i$!). This notation will above all be used when considering *vector fields*, i.e., vectors whose components in the basis (e_i) depend on the point where they are being considered. E.g., in this formalism, the *radial* vector field X which, at a point v of E, is equal to $X(v) = v$ can be written as $X = \sum_{i=1}^{n} x^i \, \partial/\partial x^i$, where the x^i are viewed here as functions whose value $x^i(v)$ at v is the ith component v^i of v in the basis (e_i).

III.9. We now give an example of the calculation of a directional derivative which is at the heart of the classical theory of Variational Calculus. We shall not be able to exploit it completely owing to the lack of a sufficiently powerful theory of integration.

Example III.10. On the space E of functions defined on a bounded domain K in $\mathbb{R}^n$ and having continuous partial derivatives there (we shall call them C^1 functions from III.60 onwards), consider the numerical function

$$(\text{III.11}) \qquad \mathcal{E}(f) = \frac{1}{2} \int_K \sum_{i=1}^{n} \left(\frac{\partial f}{\partial x^i}(x) \right)^2 dx^1 \ldots dx^n \, ;$$

this is called the *Dirichlet*[53] integral and is interpreted as the energy of deformation of a material medium.

Since $\mathcal{E}$ is quadratic, the calculation of its derivative in the direction of a function F is particularly simple. We obtain

$$(\partial_F \mathcal{E})(f) = \int_K \left\{ \sum_{i=1}^{n} \frac{\partial f}{\partial x^i}(x) \, \frac{\partial F}{\partial x^i}(x) \right\} dx^1 \ldots dx^n \, .$$

---

[52] From now on it is *essential* to be familiar with this notation, which will be used throughout this course.

[53] Gustav Lejeune-Dirichlet (1805–1859), German mathematician, made important contributions to number theory and the study of trigonometric series.

If the function F has compact support contained in the interior of the domain K (and is thus zero on a neighbourhood of the boundary of K) and if the function f is sufficiently differentiable, it is possible to integrate by parts with respect to all the variables x^i and to make all the derivatives operate on the function f. We then obtain

$$(\text{III.12}) \qquad (\partial_F \mathcal{E})(f) = \int_K F\left(-\sum_{i=1}^{n} \frac{\partial^2 f}{(\partial x^i)^2}\right) dx^1 \dots dx^n \ .$$

It is traditional to use the name *Laplace*[54] *operator* or *Laplacian* for the operator Δ which to a function f associates $\Delta f = -\sum_{i=1}^{n} \partial^2 f/(\partial x^i)^2$.

We can reformulate (III.12) by saying that, on the subspace of E consisting of functions vanishing at the boundary, $(\partial_F \mathcal{E})(f)$ *can be expressed as the L^2 scalar product of the infinitesimal variation F and the Laplacian of the function f at which the derivative is calculated. This formulation requires that the function f be sufficiently regular, say that it has continuous second partial derivatives.

Exercise III.13. Generalise Formula (III.12) to the case where the scalar product used in $\mathbb{R}^n$ is no longer the Euclidean scalar product, but a scalar product g with variable coefficients (g^{ij}), so that the numerical function being considered can be written

$$\tilde{\mathcal{E}}(f) = \frac{1}{2} \int_K \left(\sum_{i,j=1}^{n} g^{ij}(x) \frac{\partial f}{\partial x^i} \frac{\partial f}{\partial x^j}\right) dx^1 \dots dx^n \ .$$

III.14. We can now turn to the presentation of a notion which is central for this chapter and for the rest of this course. It will lead us naturally to the notion of a differentiable map.

Definition III.15. Two maps f and g defined on an open subset U of a normed space $(E, \|\ \|_E)$ with values in a normed space $(F, \|\ \|_F)$ are said to be *tangent at a point a* of U if

$$\lim_{v \to 0} \frac{\|f(a+v) - g(a+v)\|_F}{\|v\|_E} = 0 \ .$$

III.16. It follows easily from the triangle inequality that two maps which are both tangent to a third are tangent to each other. The relation of tangency at a point is in fact an *equivalence relation*.

We shall make use of this property in a special (but important) case in Chapter VI.

---

[54] Pierre Simon, *marquis de* Laplace (1749–1827), French astronomer, mathematician and physicist, held the highest positions of his time. He obtained important results in celestial mechanics, probability theory and electrostatics by the use of analysis.

Remark III.17. In order for two maps to be tangent at a point a, they must of course take the same value there. But this is by no means sufficient.

Note for example that two continuous linear maps l_1 and l_2 cannot be tangent at a point, even the origin, without being identical. In fact, in this case the map $l_1 - l_2$ satisfies $\lim_{v \to 0} \|(l_1 - l_2)(v)\|_F / \|v\|_E = 0$, in other words it would have zero norm. This property plays an important role in the sequel.

III.18. To formulate Definition III.15, it is sometimes convenient to use the *o notation of Landau*[55]. Recall that a function of a real variable t defined on a neighbourhood of 0 is *said to be* $o(t)$ if $\lim_{t \to 0} o(t)/t = 0$.

By extension, in the situation of the preceding definition, we say that a function defined on a neighbourhood of 0 in E with values in F is $o(v)$ (v denoting the variable in E) if we have $\lim_{\|v\|_E \to 0} \|o(v)\|_F / \|v\|_E = 0$. Two maps f and g are then tangent at a point a precisely when $f - g$ is $o(v)$.

Definition III.19. Let $(E, \| \ \|_E)$ and $(F, \| \ \|_F)$ be two (real) normed spaces, in which we denote the translation by a vector v by τ_v.

A map f defined on an open subset U of E with values in F is said to be *differentiable at a point a* of U if there exists a continuous linear map from E to F, denoted by $T_a f$, such that $f \circ \tau_a$ and $\tau_{f(a)} \circ T_a f$ are tangent at the origin, in other words such that

$$\lim_{v \to 0} \frac{\|f(a + v) - f(a) - T_a f(v)\|_F}{\|v\|_E} = 0 .$$

The map $T_a f$ is then unique and is called the *tangent linear map* of f at a. One sometimes says the *differential* of f, but we will reserve this term for a special case that we discuss later.

Remark III.20. At first sight, the definition we have arrived at looks rather contrived since it is not clear why translations appear in it. Nevertheless, it generalises directly to the case of the more general spaces which we introduce in Chapter V, since it can be reformulated as saying that f and $\tau_{f(a)} \circ T_a f \circ (\tau_a)^{-1}$ are tangent to each other at the point a, the translations or their inverses serving to reduce the points a and $f(a)$ under consideration to 0_E and 0_F respectively. This means that, up to terms of the second order, we can "replace" the map f by its tangent linear map $T_a f$ in systems of coordinates centred at a and $f(a)$.

The notion of a differentiable map which we have presented is due to Fréchet. The uniqueness of the tangent linear map is a direct consequence of Remark III.17.

III.21. The use of the Landau notation brings out once again the fact that $T_a f$ defines *a first order approximation* to f in a neighbourhood of a. Using it we can write

(III.22) $f(a + v) = f(a) + T_a f(v) + o(v) .$

Clearly, it follows from this that a map differentiable at a is continuous there.

---

[55] The German mathematician Edmund Landau (1877–1938) was interested above all in arithmetic. He should not be confused with the Russian physicist Lev Davidovich Landau (1908–1968), whose course in theoretical physics, written with Lifschitz, is still one of the classics of the subject.

Proposition III.23. *When a map f between two normed spaces E and F is differentiable at a point a, then f has directional derivatives in the direction of every vector v in E, and $(\partial_v f)(a) = (T_a f)(v)$.*

More generally, the restriction of f to a vector subspace E_1 of E is again differentiable and, for $a \in E_1$, $T_a(f_{|E_1}) = (T_a f)_{|E_1}$.

Proof. It is enough to calculate the limit of the variations of f at a corresponding to the variations tv of the variable.

For the second part, it suffices to go back to the definition. □

Corollary III.24. *If f is a differentiable map defined on an open subset U of a finite-dimensional normed space with basis $(e_i)_{1 \leq i \leq n}$, then, at every point a of U and for every vector $v = \sum_{i=1}^{n} v^i e_i$,*

$$(T_a f)(v) = \sum_{i=1}^{n} v^i \, (\partial_{e_i} f)(a) .$$

Proof. This is an immediate application of the formula given in Proposition III.23, and the linearity of the tangent linear map. □

Corollary III.25. *The tangent linear form of a differentiable scalar function at a maximum or minimum is zero.*

Proof. This is simply a matter of combining Propositions III.4 and III.23. □

Exercise III.26. Give an example of a map which has directional derivatives in all directions but which is not differentiable.

Under what conditions can one state a converse to Proposition III.23?

III.27. It might seem strange to use two notations at once (i.e., $\partial_v f$ and $T_a f$) to denote the generalisation of the derivative. They are both acceptable because they reflect different points of view.

In fact, in the notation $(\partial_v f)(a)$ used to denote the directional derivative of a map f in the direction of a vector v at a point a, the emphasis is on the behaviour of the function in the *direction* of the vector v.

On the other hand, the notation $T_a f$ underlines the fact that one associates to f a linear map which is somehow attached to the point a. It is also natural to evaluate $T_a f$ on vectors such as v by writing $(T_a f)(v)$. The expression "*tangent linear map*" has (perhaps) the drawback of being a little long. Nevertheless, it will be with us throughout this course. In fact, we are soon going to consider maps between more general spaces (see Chapter V) and to extend to them the definition we have given.

The terminology *derived map*, which is sometimes used, has the drawback of being a little ambiguous; it rather suggests that the object obtained is of the same nature as that from which it is derived, which is true for real functions of a real variable but not in a more general context.

> *From now on we shall reserve the expression "differential"*
> *to denote the tangent linear map of a map taking*
> *its values in a vector space.*

Exercise III.28. Show that *if, in Definitions III.15 and III. 19, the norms on E and F are replaced by equivalent norms, the notion of a differentiable map remains the same* as does the value of the tangent linear map.

Deduce that *in finite dimensions the notion of a differentiable map depends only on the algebraic structure of the space.*

III.29. In the sequel, a map f being given, we shall be very interested in the map from U to $L(E, F)$ which associates to every point a of U the tangent linear map $T_a f$ at that point. Since $L(E, F)$ is itself a normed space (cf. Proposition I.38), one has the possibility of constructing higher order derivatives, a problem which we shall consider in Sect. D.

Examples III.30. It is time to give examples of differentiable and non-differentiable maps:

i) a *continuous linear map* from a normed space E to a normed space F is differentiable at every point a of E and we have $T_a f = f$ (in fact, for any vector v in E, $f(a + v) = f(a) + f(v)$);

ii) a *continuous bilinear map* b defined on the product of two normed spaces E_1 and E_2 and taking values in a normed space F is differentiable at every point $a = (a_1, a_2)$ of $E_1 \times E_2$ and we have, for every vector $v = v_1 + v_2$ in $E_1 \times E_2$,

$$(\text{III.31}) \qquad\qquad T_a b\,(v) = b\,(v_1, a_2) + b\,(a_1, v_2)$$

(in fact, $b\,(a_1 + v_1, a_2 + v_2) = b\,(a_1, a_2) + b\,(v_1, a_2) + b\,(a_1, v_2) + b\,(v_1, v_2)$ by bilinearity, and it suffices to prove that $b\,(v_1, v_2)$ is $o(v)$ which is a direct consequence of the assumed continuity of b); this formula generalises that which leads to the classical *Leibniz formula* $((fg)' = f'g + fg')$ since the product of numerical functions is a bilinear map;

iii) because of their homogeneity of degree 1, *norms* are never differentiable at the zero vector; on the other hand, the Euclidean norm is differentiable at every non-zero vector, while the norm $\|\ \|_\infty$ on $\mathbb{R}^n$ is differentiable only at the points $(x^1, \cdots, x^n)$ for which there exists no pair of indices (i, j) for which the equalities $\|x^i\| = \|x^j\| = \|x\|_\infty$ hold. (Exercise: Prove it.)

Remark III.32. In III.10 we proved that the Dirichlet integral $\mathcal{E}$ had directional derivatives in the space E of functions having continuous partial derivatives on the domain K. We know that this space has a natural norm which in fact makes it into a Banach space. It is natural to ask whether $\mathcal{E}$ is differentiable in E.

Since $\mathcal{E}$ is a quadratic form, it is enough to check that the map

$$F \mapsto \int_K (\text{Grad} f(x) \mid \text{Grad} F(x))\, dx$$

is continuous for the topology of E, which is clear by bounding the integral by the upper bound of the derivatives of the function.

However, one must be careful since it is not possible to use the formulation used in (III.12) because it makes use of higher derivatives of f.

This scalar product does not make E into a Hilbert space since E *is not* complete for the topology which it defines. This discussion could be fruitfully continued and leads to important notions of analysis, notably to *Sobolev spaces*.

III.33. Of course, the difficulties encountered in Remark III.32 do not appear in finite-dimensional spaces since there is then a unique topology which it is reasonable to work with. In III.65 we give a convenient criterion which applies in finite dimensions.

III.34. The vector space $\mathcal{M}_n$ of $n \times n$ square matrices with real coefficients, or, more geometrically, the spaces $L(E, F)$ of linear maps from a vector space E to a vector space F, together with subspaces of it with geometric significance such as spaces of symmetric, complex or Hermitian matrices, will provide us with many examples of natural maps which are differentiable.

The calculation of the differentials in these examples is not only instructive; some of them will be of great importance in the sequel. We give an initial list in the form of exercises. Another list will be given when the Chain Rule III.46 is available.

III.35. To begin with, we describe explicitly the tangent linear map of the map μ which associates to a pair of matrices (L, M), which are $l \times m$ and $m \times n$ respectively, their product $\mu(L, M) = LM$.

Since we are concerned here with a bilinear map, we can apply Formula (III.31) which in this case gives

$$(III.36) \qquad (T_{(L_0, M_0)}\mu)(L_1, M_1) = L_0 M_1 + L_1 M_0 .$$

Solved Exercise III.37. In the space $\mathcal{M}_n$ of $n \times n$ square matrices, we shall determine the tangent linear map at a matrix L_0 of the map q which associates to a matrix L its square $q(L) = L^2$, and discuss its invertibility.

The map q, being polynomial, is differentiable. Since we can write

$$q(L_0 + H) - q(L_0) = L_0 H + H L_0 + H^2,$$

from $\|H\| < \epsilon$ we obtain the growth estimate

$$\|q(L_0 + H) - q(L_0) - (L_0 H + H L_0)\| \le \epsilon \|H\|,$$

whence the fact that

$$(T_{L_0}q)(H) = L_0 H + H L_0$$

since the right-hand side depends linearly on H (and automatically continuously since $\mathcal{M}_n$ is finite-dimensional).

We now determine at which points L_0 of $\mathcal{M}_n$ the tangent linear map of the map q is invertible.

We show first of all that L_0 must necessarily be invertible. In fact, if L_0 is not invertible, $\ker L_0$ contains at least one non-zero vector v and the image of L_0, which is not the whole space $\mathbb{R}^n$, has a non-zero annihilator, a non-zero element of which we denote by λ.[56] The endomorphism H of rank 1 defined by $H(w) = \lambda(w)\, v$ lies in the kernel of $T_{L_0} q$. In fact, for $w \in \mathbb{R}^n$,

$$(L_0 H + H L_0)(w) = L_0(\lambda(w)\, v) + \lambda(L_0(w))\, v = 0 \, .$$

We show finally that the preceding argument generalises to the case where there is an eigenvalue α of L_0 such that $-\alpha$ is also an eigenvalue. We note first of all that $H \in \ker T_{L_0} q$ if and only if $L_0 H = H(-L_0)$. Since the relation $IH = HI$ can be rewritten $(L_0)^0 H = H(L_0)^0$, we see that the anti-commutation relation between L_0 and H generalises, for every polynomial P, to $P(L_0)H = H(P(-L_0))$. Since L_0 is annihilated by its characteristic polynomial P_{L_0}, we have , for all $H \in \ker T_{L_0} q$, $0 = H P_{L_0}(-L_0)$. If $H \neq 0$, we deduce from this that $P_{L_0}(-L_0)$ is not invertible, so that the spectrum of $-L_0$ intersects that of L_0.

Conversely, if the spectrum of $-L_0$ does not intersect that of L_0, then $P_{L_0}(-L_0)$ is an invertible matrix and the equation $0 = H P_{L_0}(-L_0)$ has only the solution $H = 0$. We have thus proved that $T_L q$ is invertible for all endomorphisms L whose spectrum does not intersect that of $-L$.

The preceding discussion is interesting because of the information one can extract from the invertibility of the tangent linear map by using the Local Inversion Theorem III.70 (see III.83).

Exercise III.38. Show that the *exponential* map, which to a matrix A associates the matrix $\exp A = \sum_i \frac{1}{i!} A^i$, is differentiable and calculate its tangent linear map at the identity.

Solved Exercise III.39. We determine the tangent linear map of the map *denoted* by det defined on the n^2-dimensional vector space $\mathcal{M}_n$ which to a matrix associates its determinant.

We begin by doing the calculation at the identity matrix I. Being a polynomial function of the elements of the matrix which are the linear coordinates on the vector space $\mathcal{M}_n$, the function det is differentiable. We have $(T_I \det)(H) = \lim_{t \to 0} t^{-1}(\det(I + tH) - \det I)$. But, as is well-known, $\det(I + tH)$ is a polynomial in t of which the coefficient of t^k is the elementary symmetric function of order k of the eigenvalues of the matrix H which is obtained by taking the sum of the minors of order k of the matrix H which are situated symmetrically about the diagonal. The term involving t can thus be identified with the *trace* of H (which we *denote* by trace H). We thus have $T_I \det = \text{trace}$.

Before passing to the general case, we are going to suppose first of all that the point A of $\mathcal{M}_n$ at which we are calculating the derivative is an invertible matrix. In this case, we have $\det(A + tH) = \det(A(I + t A^{-1}H))$, and since the determinant is a homomorphism from the group of invertible matrices to the multiplicative group $\mathbb{R}^*$, $\det(A + tH) = (\det A)(\det(I + t A^{-1}H))$. This reduces us to doing the calculation at the identity. We thus find $(T_A \det)(H) = (\text{trace}((\det A) A^{-1}H))$ since the trace is

---

[56] It may be useful to recall at this point that, for a general endomorphism of a vector space, there is no reason for the space to be the direct sum of its kernel and image.

linear. But $(\det A)A^{-1}$ is simply the transpose of the *comatrix* A^*, i.e., the matrix whose element a^{*i}_j is the minor of order $n-1$ of the matrix A obtained by deleting its ith row and j^{th} column. At an invertible matrix A we thus obtain

$$(\text{III}.40) \qquad\qquad (T_A\det)(H) = \text{trace}(^tA^*H) .$$

But this formula contains only polynomial expressions in the elements of the matrix A. We thus have a candidate for the differential of det at every point $A \in M_n$. Since the function det is polynomial, its differential is also a polynomial map with values in the dual of M_n. We know that this map coincides with the formula given in III.40 on the open set of invertible matrices. It thus coincides with it everywhere.

Exercise III.41. Deduce from Exercise III.39 the differential at the identity of the map which to a matrix associates its inverse.

Exercise III.42. Show that *the map* inv *defined on the open set formed by the automorphisms of a Banach space which to an automorphism a associates its inverse a^{-1} is differentiable and has tangent linear map $l \mapsto T_a\text{inv}(l) = -a^{-1} \circ l \circ a^{-1}$.* (One may begin by doing the calculation at the identity and making use of the series expansion of the inverse.)

Project III.43. Is the uniform norm on the Banach space $C^0(\Omega)$ (where Ω denotes an open subset of a numerical space) a differentiable function away from the zero function?

Exercise III.44. Show that the norm of order 1 which is defined on $\mathbb{R}^n$ by $\|(x^1,\cdots,x^n)\|_1 = \sum_i |x^i|$ is differentiable at all n-tuples $(x^1,\cdots,x^n)$ such that $x^i \neq 0$ for $1 \leq i \leq n$.

Project III.45. Show that, in the space l^1 of summable sequences, the norm of order 1 is not differentiable at any sequence.

B. The Chain Rule

The object of this section is the presentation of a technical lemma which generalises the formula for the derivative of composite functions and which is used *very frequently*, hence its importance.

Chain Rule III.46. *Let f be a map defined on an open subset U of a normed space E, taking values in a normed space F, and differentiable at a point a of U. If g is a map defined on an open subset V of F with values in another normed space G and differentiable at $f(a)$, then $g \circ f$ is differentiable at a and*

$$(\text{III}.47) \qquad\qquad T_a(g \circ f) = T_{f(a)}g \circ T_af .$$

Proof. This is particularly simple. To simplify the notation, put $b = f(a)$ and $h = g \circ f$.

Since f is differentiable at a, we know that the function "remainder at a" $\mathcal{R}^a_f$, defined by $\mathcal{R}^a_f(v) = f(a+v) - f(a) - T_a f(v)$, is $o(v)$. Similarly, since g is differentiable at b, we know that the function $\mathcal{R}^b_g(w) = g(b+w) - g(b) - (T_b g)(w)$ is $o(w)$. Moreover, we know that $T_{f(a)} g \circ T_a f$ is a continuous linear map from E to G. For a variation v of a, we consider another auxiliary function $\mathcal{R}^a_h(v) = h(a+v) - h(a) - (T_b g \circ T_a f)(v)$. Since $(T_b g \circ T_a f)(v) = T_b g(T_a f(v))$,
$$\mathcal{R}^a_h(v) = \mathcal{R}^b_g(T_a f(v) + \mathcal{R}^a_f(v)) + (T_b g)(\mathcal{R}^a_f(v)).$$

The theorem is then a direct consequence of the computation of the limit of the composite of two continuous maps.   □

III.48. We mention in the form of corollaries two consequences of this formula which play an important role in the rest of this course and for which the proof is immediate (and thus left to the reader).

Corollary III.49. *Let $f : U \longrightarrow F$ (where U is an open subset of a normed space E) be a map which has an inverse map f^{-1}. If f is differentiable at a point a of E and f^{-1} is differentiable at the point $f(a)$, then $T_{f(a)}(f^{-1}) = (T_a f)^{-1}$. Consequently E and F have the same dimension, possibly infinite.*

III.50. In the context of Hilbert spaces, the second part of Proposition III.23 has a dual version which is often cited.

We restrict ourselves to the case of a scalar-valued function f; at each point the differential of f then defines an element of the dual E' of E which is isomorphic to E by the map $\flat$ (cf. Theorem II.83). It is then traditional to represent the differential at the point a as the scalar product with the *gradient* vector of the function at this point which we *denote* by $\mathrm{Grad} f(a)$.

If E_1 is a closed subspace of a Hilbert space E and f a numerical function differentiable as a function defined on E, at a point a in E_1, then, if P_{E_1} denotes the orthogonal projection of E onto E_1, $\mathrm{Grad}(f|_{E_1})(a) = P_{E_1}(\mathrm{Grad} f(a))$.

III.51. When the source space or the target space is 1-dimensional, it is traditional to adopt somewhat special notations. We give these using the Chain Rule III.46 now.

III.52. If a map γ is a curve defined on an interval I and taking value in a normed space E, we *denote* by $d\gamma/dt$ its *velocity-vector* if it exists (the notation $\dot{\gamma}$ is often used but we shall reserve it for a special purpose in the sequel), in other words, at a point t of the interval I, $d\gamma/dt = (T_t \gamma)(1)$ where 1 is regarded as a vector in $\mathbb{R}$.

In the case where E is of finite dimension n, it is of course possible to reduce to the calculation of derivatives in the usual sense. In fact, by taking a basis (e_i) of E, one represents γ by its components γ^i such that $\gamma(t) = \sum_{i=1}^n \gamma^i(t) e_i$. One then has $d\gamma/dt = \sum_{i=1}^n (d\gamma^i/dt) e_i$.

If the curve γ is followed by a map f into another normed space F, the Chain Rule Formula (III.47) takes the form

$$\frac{d}{dt}(f \circ \gamma)(t) = (T_{\gamma(t)} f)\left(\frac{d\gamma}{dt}\right),$$

which we recognise as a special case of Proposition III.23.

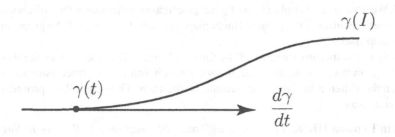

Fig. III.53 A curve and one of its velocity-vectors

III.54. For a scalar function, i.e., a numerical function taking its values in $\mathbb{R}$ (we denote such a function by ϕ to avoid confusion with the general case), which is differentiable at a point a, we *denote* its differential at this point by $d\phi(a)$ rather than by $T_a\phi$; thus, $d\phi(a)$ is an element of the (topological) dual of the space E on which ϕ is defined and, for any vector v in E, we thus have

$$\langle d\phi(a), v \rangle_E = (T_a\phi)(v) .$$

If E is of finite dimension n, from this we recover one of the "magic" formulas of differential calculus. Indeed, if (e_i) is a basis of E with associated coordinates (x^i), since the value of the coordinate function x^i at a point a is obtained simply by evaluating the constant linear form ϵ^i, the ith vector of the dual basis, on the vector $v = \sum_{i=1}^n v^i e_i$, we have

(III.55) $$\langle dx^i(a), v \rangle = \langle \epsilon^i, v \rangle = v^i .$$

Viewing the map ϕ as the composite of the map which to $(x^1, \cdots, x^n)$ associates $x = \sum_{i=1}^n x^i e_i$ and $x \mapsto \phi(x)$, we obtain the celebrated formula

(III.56) $$d\phi = \frac{\partial \phi}{\partial x^1} dx^1 + \cdots + \frac{\partial \phi}{\partial x^n} dx^n$$

defining $d\phi$ as a map from U to E'.

If we return to a more general situation and study the composite of a map f differentiable at a point a with a numerical function ϕ, we have the formula

$$d(\phi \circ f)(a) = d\phi(f(a)) \circ T_a f = {}^t(T_a f)(d\phi(f(a))) ,$$

where ${}^t l$ denotes the transpose of the linear map l (cf. II.25).

If we combine the two situations studied above by considering a curve γ along which we evaluate a numerical function ϕ, we obtain

$$\frac{d}{dt}(\phi \circ \gamma)(t) = (d\phi(\gamma(t)) \circ T_t \gamma)(1) = \langle d\phi(\gamma(t)), \frac{d\gamma}{dt} \rangle .$$

III.57. We now give an application of great practical significance in the 1-dimensional case. From a control of the tangent linear map, the Growth Lemma III.58 gives control of the map itself.

There are numerous variants of the Growth Lemma III.58 known as the *theorem of finite growth* or the *mean value theorem* which aim to give finer estimates or to weaken the differentiability assumptions on the map. They can all be proved in an analogous way.

Growth Lemma III.58. *Let f be a differentiable map on a ball B_r of radius r in a normed space $(E, \| \|_E)$ taking its values in a normed space $(F, \| \|_F)$. If f has differential Tf, then, for any two points a and b in B_r,*

$$\|f(b) - f(a)\|_F \leq \left(\sup_{x \in [a,b]} \|T_x f\|_{L(E,F)} \right) \|b - a\|_E .$$

Proof. Note that there is nothing to prove if $\sup_{x \in [a,b]} \|T_x f\|_{L(E,F)} = +\infty$. We can thus suppose it is finite, say M.

Let a and b be two points of the ball B_r centred at c. Since $\|a - c\|_E < r$ and $\|b - c\|_E < r$, every point x of the line segment $[ab]$, which can be written $x = ta + (1-t)b$ for some $t \in [0,1]$, lies in B_r since $\|x - c\|_E \leq t\|a - c\|_E + (1 - t)\|b - c\|_E < r$.

Consider the partial function $t \mapsto \phi(t) = f(ta + (1 - t)b))$. Since f is differentiable, by the chain rule, ϕ is differentiable and has derivative $d\phi/dt = (T_x f)(b - a)$ where $x = ta + (1-t)b$ so that $\|d\phi/dt\|_F \leq M\|b - a\|_E$.

Given $\epsilon > 0$, consider the subset A_ϵ of $[0,1]$ defined as the set of real numbers s such that, for all t in $[0, s[$, the estimate

$$\|\phi(t) - \phi(0)\|_F \leq t(M\|b - a\|_E + \epsilon)$$

holds. By construction, A_ϵ is a closed interval which contains 0. (In fact, to check that a number a belongs to A_ϵ, one must verify a condition on all the numbers which are strictly less than it.) It suffices to prove that the upper bound of A_ϵ, say u, is actually equal to 1. Suppose $u \neq 1$. By definition of the derivative, we can find an interval $[u, v] \subset [0, 1]$ such that, for all $t \in [u, v[$,

$$\left\|\phi(t) - \phi(u) - (t - u)\frac{d\phi}{dt}\right\|_F \leq \epsilon (t - u) ,$$

whence the estimate

$$\|\phi(t) - \phi(u)\|_F \leq (t - u)\left(\left\|\frac{d\phi}{dt}\right\|_F + \epsilon\right) \leq (t - u)(M\|b - a\|_E + \epsilon) .$$

On using the triangle inequality, we obtain a contradiction to the definition of u, hence we necessarily have $u = 1$.

We have thus proved that

$$\forall \epsilon > 0, \ \|f(a) - f(b)\|_F = \|\phi(1) - \phi(0)\|_F \leq M\|b - a\|_E + \epsilon ,$$

and hence the result stated since ϵ is arbitrary.   □

Remark III.59. At this point, we note that the force of the proof rested on the connectivity of intervals of the real line, the main difficulty residing in the definition of the set which finally turned out to coincide with the maximal set on which the problem is defined. This method is sometimes known as the "*method of continuity*".

Note also that, in the proof of Lemma III.58, the fact that f was defined on a ball entered only via a geometric property of the ball, namely that every line segment whose endpoints are in the ball is entirely contained in it. A set having this property is said to be *convex*.

Such sets (and the functions whose level sets are convex, which are called *quasi-convex*) play an important role in analysis and have given rise to a special branch called *convex analysis*. In this context, it is possible to generalise the differential calculus to maps less regular than differentiable maps.These generalisations have numerous applications, notably in *mathematical economics*, which have greatly stimulated research work in this direction.

C. The Local Inversion Theorem

In this section, we shall generalise Corollary III.49, by obtaining properties of the inverse from stronger assumptions on the map itself.

To do this, we need a supplementary notion, classical in the case of functions defined on finite-dimensional vector spaces, which we introduce first, namely that of a map of class C^1. In passing, we give an important criterion for determining whether a map is of class C^1.

Definition III.60. A map f defined on an open subset U of a normed space E and taking values in a normed space F is said to be *of class* C^1 if

 i) f is differentiable at each point of U,
 ii) its tangent linear map defines a continuous map $a \mapsto T_a f$ from U to $L(E, F)$.

Example III.61. Every continuous linear or bilinear map defined on a normed space with values in a normed space is of class C^1.

C^1-Chain Rule III.62. *When it is defined, the composite of maps of class C^1 is again of class C^1. Similarly, when such notions make sense, the sum and product of two functions of class C^1 is of class C^1.*

Proof. The first part of the proposition follows directly from the Chain Rule III.46 and the continuity of the composition of linear maps.

For the second part, it suffices to compose the C^1 map obtained by juxtaposing in an obvious way the two maps and the sum or product map, which is of class C^1.   □

Project III.63. (*α-Lemma*) Let E, F and G be three normed spaces and f a map of class C^1 from an open subset U of E with compact closure to F. Show that the map α_f which to $g \in C^1(\overline{V}, G)$ (where V is an open subset of F, with compact closure, containing $f(\overline{U})$) associates $g \circ f \in C^1(\overline{U}, G)$ is of class C^1.

Project III.64. (*ω-Lemma*) Let E, F and G be three normed spaces and g a map of class C^1 from an open subset V of F with compact closure to G. Show that the map ω_g which to f in

$C^1(\overline{U}, V)$ (where U is an open subset of E with compact closure) associates $g \circ f$ is a continuous map from $C^1(\overline{U}, V)$ to $C^1(\overline{U}, G)$.

Is the map ω_g differentiable ?

Theorem III.65. *For a map f defined on an open subset U of $\mathbb{R}^m$ with values in $\mathbb{R}^n$ to be of class C^1 on U, it is necessary and sufficient that its Jacobian[57] matrix $(\partial f^j / \partial x^i)$ is a continuous function on U.*

Proof. The 'only if' part is clear, since a map f of class C^1 has a tangent linear map $T_a f$ which depends continuously on a, and thus whose matrix elements in a fixed basis are continuous functions. In the canonical bases of $\mathbb{R}^n$ and $\mathbb{R}^m$, these matrix elements are precisely the partial derivatives $(\partial f^j / \partial x^i)$.

For the 'if' part, we can proceed as follows. The Jacobian matrix defines a linear map which we must prove is the tangent linear map of f. For this, consider a point $q = (q^1, \cdots, q^n)$ near to a point $a = (a^1, \cdots, a^n) \in U$. We want to show that the quantity

$$A = f(q) - f(a) - \sum_{i=1}^{n} (q^i - a^i)(\partial f / \partial x^i)(a)$$

is $o(\|q - a\|)$.

We introduce the $n - 1$ points $q_{(i)}$, $1 \leq i \leq n - 1$, whose first i coordinates are those of q and whose last $n - i$ are those of a (so that by extension $q_{(0)} = a$ and $q_{(n)} = q$). It is easy to see that we can write $A = \sum_{i=1}^{n} (f(q_{(i)}) - f(q_{(i-1)}) - (q^i - a^i)(\partial f / \partial x^i)(a))$. Thus, to establish the required estimate, it suffices to show that, given $\epsilon > 0$, it is possible to find $\eta > 0$ such that, for $\|q - a\| < \eta$, each term $A_{(i)}$ of the preceding sum is less than ϵ. Since the vector $q_{(i)} - q_{(i-1)}$ has only one non-zero component, which is equal to $q^i - a^i$, it is clear that $\|q_{(i)} - q_{(i-1)}\| \leq \|q - a\|$, and hence that $\|q_{(i)} - q_{(i-1)}\| < \eta$. Moreover, each term $A_{(i)}$ can be interpreted as the difference of the values at q^i and a^i of the auxiliary function $g_{(i)}$ of one real variable defined by

$$g_{(i)}(y) = f(q^1, \cdots, q^{i-1}, y, a^{i+1}, \cdots, a^n) - (y - a^i)\, \partial f / \partial x^i(a) .$$

The functions $g_{(i)}$ are of class C^1 by the Chain Rule III.62 for functions of class C^1 and

$$g'_{(i)}(y) = (\partial f / \partial x^i)(q^1, \cdots, q^{i-1}, y, a^{i+1}, \cdots, a^n) - (\partial f / \partial x^i)(a^1, \cdots, a^n).$$

Since the partial derivatives of f are continuous functions by hypothesis, one can find $\eta_i > 0$ such that $\|g'_{(i)}(y)\| < \epsilon$ for $\|y - a^i\| < \eta_i$.

We can thus apply the Growth Lemma III.58 and conclude that

$$\|A_{(i)}\| = \|g_{(i)}(q^i) - g_{(i)}(a^i)\| \leq \epsilon \|q^i - a^i\| \leq \epsilon \|q - a\| ,$$

which enables us to finish the proof by taking $\eta = \sup_{1 \leq i \leq n} \eta_i$.   □

Exercise III.66. For any invertible square matrix A one defines the *adjoint* map ad_A which to a matrix B associates the matrix $A^{-1}BA$ (geometrically it represents the effect of a change of basis). Show that *the map $A \mapsto \mathrm{ad}_A$ is of class C^1 and calculate its tangent linear map.* (One can begin by doing the calculation at the identity.)

---

[57] from the name of Jacob Jacobi (1804–1851), German mathematician, author of numerous works in algebra, arithmetic and analytical mechanics. He contributed in a decisive way to the theory of elliptic functions. He is also the author of a report *"Über die Pariser polytechnische Schule"*.

Exercise III.67. Show that *the map which to a pair of invertible square matrices* (A, B) *associates their commutator* $ABA^{-1}B^{-1}$ *in the group of invertible matrices is of class* C^1.

Exercise III.68. Show that *the map det which to a square matrix associates its determinant is of class* C^1. (One can make use of Exercise III.39.)

III.69. We now come to the key theorem of this chapter. It will be in *constant use* in the rest of this course. It deserves ample reflection.

Paradoxically, although a little long, its proof is elementary. It is a typical example of the method of reduction, in which one puts oneself at all costs in a canonical situation, here that of Picard's Fixed Point Theorem II. 7.

Local Inversion Theorem III.70. *Let* f *be a map, assumed to be of class* C^1, *defined on an open subset* U *of a Banach space* E *taking values in another Banach space* F. *If there exists a point* a *of* E *where the tangent linear map* $T_a f$ *of* f *is an isomorphism from* E *to* F (*i.e., has a continuous inverse*), *then there exists an open neighbourhood* V *of* a *and a neighbourhood* W *of* $f(a)$ *such that:*

i) $f|_V$ *is a continuous bijection on* W,

ii) *the inverse map* $g : W \longrightarrow V$ *is continuous*,

iii) *the map* g *is of class* C^1 *and* $T_{f(a)} g = (T_a f)^{-1}$.

Proof. i) To simplify the notation (and consequently to understand better the essential structure of the proof), it is convenient to reduce to the normalised situation where $a = 0$, $f(a) = 0$ and $T_a f = Id_E$ (and thus $E = F$). For this, we consider the *normalised map* $f_N(v) = (T_a f)^{-1}(f(a+v) - f(a))$ obtained from f by composition with a translation and well-defined linear and bilinear maps.

We show first of all that proving the theorem for f_N is equivalent to proving it for f. In fact, the map f is related to f_N by the formula $f(x) = f(a) + T_a f(f_N(x - a))$, which can be rewritten $f = \tau_{f(a)} \circ T_a f \circ f_N \circ \tau_{-a}$ (recall that τ_v denotes the translation by the vector v); if f_N has an inverse in a neighbourhood of 0, the same will be true of f in a neighbourhood of a and the regularity of the inverse of f then follows from that of the inverse of f_N by applying the chain rule.

By applying the C^1 chain rule, we can find $r_1 > 0$ such that f_N is C^1 on the open ball $B_{r_1}(0)$. Similarly, we can control the distance between f_N and the identity map Id_E. In fact, taking into account that $f_N(0) = 0$ and $T_0(f_N) = Id_E$, and working in a ball of sufficiently small radius r, the map $f_N - Id_E$, which is C^1, has a tangent linear map with norm strictly less than $1/4$ for example. By the Growth Lemma III.58, for all x in $B_r(0)$, we have $\|f_N(x) - x\|_E < (1/2)\|x\|_E$.

We would like to construct an inverse for f_N, in other words to show that, for all y, the equation $f_N(x) = y$ has a unique solution. To prove this, we remark that it suffices to find a fixed point of the auxiliary map $h_y(x) = y + x - f_N(x)$. (To simplify the notation, from now on we shall drop the indices N on f, y on h, and E on the norm, and understand that the balls are centred at the origin). We are going to establish the existence of a fixed point by means of Picard's Theorem II.7 on successive approximations. Thus, we must show that, restricting y if necessary (taking the restriction of g to a smaller ball), h is a contraction map from a ball in E to itself. We note first of all that $\|h(x)\| \leq \|y\| + \|f(x) - x\|$, so that, if x is taken in the closed ball $\overline{B}_r$ and y in $B_{r/2}$, then $\|h(x)\| \leq r$ and h maps the closed ball $\overline{B}_r$ to itself as we wanted. Moreover, since we have $h(x) - h(x') = (f(x) - x) - (f(x') - x')$, by applying the Growth Lemma III.58, we have

(III.71) $$\|h(x) - h(x')\| \leq \frac{1}{2}\|x - x'\|$$

which ensures that the map h is a contraction on this same ball.

We are thus in a position to apply Picard's Fixed Point Theorem: for all y in $B_{r/2}$, there exists a unique point x in B_r such that $f(x) = y$. In other words, on the ball $B_{r/2}$, we have constructed an inverse map of f which we denote by f^{-1}.

ii) We now show that the map f^{-1} is continuous. In fact, it will be more convenient to return to its definition and exhibit a *modulus of continuity*.

From the tautological relation $x = f(x) + x - f(x)$ we obtain

$$\|x - x'\| \le \|f(x) - f(x')\| + \|(f(x) - x) - (f(x') - x')\| .$$

Let y and y' be two points of the ball $B_{r/2}$. Moreover, since the condition $x = f^{-1}(y)$ is equivalent to the condition $y = f(x)$, by III.71 we have $\|y - y'\| \le \|x - x'\|/2$ which finally gives the inequality $\|x - x'\| \le 2 \|f(x) - f(x')\|$, whence

$$\|f^{-1}(y) - f^{-1}(y')\| \le 2 \|y - y'\| .$$

iii) It remains to prove that f^{-1} is of class C^1. From Proposition III.49 for example, it follows that the tangent linear map of f^{-1} at the point y, if it exists, will be $(T_{f^{-1}(y)}f)^{-1}$. Since this map is the composite of the maps $y \mapsto f^{-1}(y)$, $x \mapsto T_x f$ and inv which are continuous by hypothesis or by Exercise II.20, it is continuous, possibly after restricting the ball on which we work to ensure that the inverse makes sense.

We now show that this map is actually the differential of f^{-1}. For this, take two points y and $y' = y + w$ in $B_{r/2}$ whose images in B_r under f^{-1} are denoted by x and $x' = x + v$ respectively. We estimate the distance between f^{-1} and its supposed tangent linear map

$$
\begin{aligned}
& f^{-1}(y + w) - f^{-1}(y) - (T_{f^{-1}(y)}f)^{-1}(w) \\
(\text{III.72}) \quad & = x + v - x - (T_x f)^{-1}(f(x + v) - f(x)) \\
& = (T_x f)^{-1}\left(f(x + v) - f(x) - (T_x f)(v)\right) .
\end{aligned}
$$

Since $x \mapsto (T_x f)^{-1}$ is continuous, by restricting once more if need be the ball in which we are working, there exists a constant C such that $\|(T_x f)^{-1}\| \le C$. Consequently, we obtain that $f^{-1}(y + w) - f^{-1}(y) - (T_{f^{-1}(y)}f)^{-1}(w)$ is $o(w)$ since the last term in Eq. (III.72) is $o(v)$ and since the ratio of the norms satisfies $\|w\|/\|v\| \le 1/2$ as we have shown in ii).   □

Remark III.73. The force of the theorem lies in the fact that, from a regularity hypothesis on the map f on an open set and from the invertibility of its tangent linear map *at a single point*, it is possible to deduce information about f *on a whole neighbourhood of the point*.

By introducing a slightly stronger notion of differentiability, the *strict differential*, it is possible to arrive at an analogous conclusion by making an assumption about f only at the point considered. A map f is said to be *strictly differentiable* at a if, for two increments v and v' in a, there exists a linear map $l \in L(E, F)$ such that $\mathcal{R}^a_{v,v'} = f(a + v) - f(a + v') - l(v - v')$ satisfies

$$\lim_{\sup\{\|v\|_E, \|v'\|_E\} \to 0} \frac{\|\mathcal{R}^a_{v,v'}\|_F}{\|v - v'\|_E} = 0 .$$

Project III.74. Give an example of a differentiable map whose tangent linear map is invertible at a point but which has no local inverse in a neighbourhood of the point.

Corollary III.75. *Let U be an open subset of a Banach space E and f a map of class C^1 defined on U taking values in another Banach space F. If Tf is an isomorphism at every point of U, then f is in particular open, i.e., the image of every open subset of E under f is an open subset of F.*

Proof. Let U_1 be an open subset of E whose image under f we want to study. It suffices to prove that $f(U_1)$ is a neighbourhood of each of its points.

Let $b = f(a) \in f(U_1)$. By the Local Inversion Theorem III.70, there exists an open neighbourhood V of a on which f is a bicontinuous bijection whose inverse we denote by g. Of course, we can assume that $V \subset U_1$. The set $f(V)$ is an open neighbourhood of b since $f(V) = g^{-1}(V)$ and g is continuous. $\qquad\square$

III.76. Note the trivial observation that, in the case of finite-dimensional spaces, the tangent linear map can be an isomorphism between the source and target spaces only if they have the same dimension.

Verifying that the tangent linear map is invertible then reduces to seeing whether its determinant is non-zero, and thus to a simple numerical criterion. We shall make use of this criterion on numerous occasions in the rest of this course, and it may be useful to make it explicit in the case of the numerical spaces $\mathbb{R}^n$ (or what amounts to the same thing, in the case where bases have been chosen in the spaces E and F, assumed to be of the same dimension).

III.77. A map f from $\mathbb{R}^m$ to $\mathbb{R}^n$ can be represented by its components, i.e., by giving n scalar functions f^j, $1 \le j \le n$, such that

$$f(x^1, \cdots, x^m) = \sum_{j=1}^{n} f^j(x^1, \cdots, x^m)\, e_j \,,$$

and the matrix of $T_a f$ in the standard bases of $\mathbb{R}^m$ and $\mathbb{R}^n$ is simply the $m \times n$ matrix of partial derivatives

$$\left(\frac{\partial f^j}{\partial x^i}(a) \right)$$

called the *Jacobian matrix* of the n functions f^1, ..., f^n at a.

In the case which concerns us, we clearly have $m = n$. The determinant of this matrix is called the *Jacobian determinant* (or more briefly the *Jacobian*) of the transformation defined by these functions. A classical notation for the Jacobian is

$$\det\left(\frac{\partial f^j}{\partial x^i}(a^1, \cdots, a^n) \right) = \frac{D(f^1, \cdots, f^n)}{D(x^1, \cdots, x^n)}(a) \,.$$

With these notations, the classical presentation of the local inversion theorem goes as follows: "*if the Jacobian of the transformation defined by the functions f^j is non-zero at $a = (a^1, \cdots, a^n)$, for every x in an open neighbourhood of a, the system of equations*

$$\begin{cases} f^1(x^1, \cdots, x^n) = y^1 \\ \quad \vdots \\ f^n(x^1, \cdots, x^n) = y^n \end{cases}$$

has a unique solution for $(y^1, \cdots, y^n)$ near $b = f(a)$. Moreover, this solution can be written

$$\begin{cases} x^1 = g^1(y^1, \cdots, y^n) \\ \quad \vdots \\ x^n = g^n(y^1, \cdots, y^n) \end{cases}$$

where the functions $g^1, ..., g^n$ are of class C^1 in a neighbourhood of b. The Jacobian matrices $(\partial f^j / \partial x^i)$ and $(\partial g^i / \partial y^j)$ at the corresponding points are inverses of each other."

III.78. In the archaic system of notations where only the variables and not the maps were worthy of mention, one generally wrote the Jacobian matrices as $(\partial y^j / \partial x^i)$ and $(\partial x^i / \partial y^j)$, which is in a certain sense suggestive, but also rather ambiguous, especially if one recalls the Warning III.6.

III.79. In the remainder of this course, we shall make considerable use (especially in the case of finite-dimensional spaces) of maps which satisfy the hypotheses of the local inversion theorem. They deserve a special notation all of their own. We give this now, and make several comments on the new features which it conceals.

By the Local Inversion Theorem III.70, such maps are homeomorphisms, but we want to take account of their additional properties.

Definition III.80. Let f be a map defined on an open subset U of a Banach space E with values in a Banach space F. If f is bijective and if f and f^{-1} are of class C^1, f is said to be *a diffeomorphism of class C^1* (or a C^1-*diffeomorphism*). The map f is said to be a *local diffeomorphism of class C^1* (or a C^1-*local diffeomorphism*) in the neighbourhood of a point a of U if there exist open subsets U' of E and V' of F, which are neighbourhoods of a and $f(a)$ respectively, such that $f_{|U'}$ is a C^1-diffeomorphism from U' onto V'.

III.81. The relation between *diffeomorphisms* and *local diffeomorphisms* is rather subtle. It lies at the origin of a branch of mathematics called *differential topology* which has great vitality and which has brought to light some surprising phenomena, some of which will be described in Chapter V.

It will turn out that a diffeomorphism is a local diffeomorphism in a neighbourhood of each of the points of its domain of definition. But *the converse is far from being automatically true*. To emphasise this difference, we often augment the name "diffeomorphism" by adding the adjective "*global*". Of course, *a bijection which is at each point a local diffeomorphism is a global diffeomorphism*.

As we have said in the introduction, the systematic study of the differences between local and global properties of a space has been one of the principal themes of mathematics in the XXth century.

Solved Exercise III.82. Determine at which points of the plane $\mathbb{R}^2$ the map f : $\mathbb{R}^2 \longrightarrow \mathbb{R}^2$, defined by $f(x, y) = (u(x, y), v(x, y))$ and supposed of class C^1, is a local diffeomorphism.

At a point (x, y) of $\mathbb{R}^2$, the value of the Jacobian matrix of f is

$$\text{Jac}\,(f)(x, y) = \begin{pmatrix} \partial u/\partial x & \partial u/\partial y \\ \partial v/\partial x & \partial v/\partial y \end{pmatrix}.$$

By the Local Inversion Theorem III.70, the map f is a local diffeomorphism precisely at those points (x, y) where the determinant of the Jacobian matrix $Jac(f)$ is non-zero.

As an explicit example whose algebraic interest will become clear eventually, we take: $u(x, y) = x + y$, $v(x, y) = xy$. In this case, the Jacobian matrix is

$$\text{Jac}\,(f)(x, y) = \begin{pmatrix} 1 & 1 \\ y & x \end{pmatrix},$$

whose determinant is equal to $x - y$. We have thus proved that, at every point (x, y) of $\mathbb{R}^2$ outside the main diagonal Δ with equation $y = x$, the map f realises a local diffeomorphism, which means that *in a neighbourhood of a point of $\mathbb{R}^2 - \{\Delta\}$, one can recover in a differentiable way two points from their sum u and product v.* The inverse map of f is of course described by the formula giving the roots of a quadratic equation whose coefficients are precisely the sum and product. The fact that we are working in a neighbourhood of a point whose coordinates are distinct serves to determine which sign should be taken to give the root corresponding to the first coordinate, the other being taken for the second. Along the main diagonal Δ, the map f is only of rank 1.

It is interesting to determine the image of $\mathbb{R}^2$ under f. We note first of all that the image of Δ under f is the parabola with equation $4v = u^2$. Now it is well-known that the sum and product of two real numbers satisfy the inequality $u^2 \geq 4v$ since $(x + y)^2 - 4xy \geq 0$ and that this condition is sufficient for the quadratic equation associated to the given (u, v) to be solvable (over the field of real numbers). We have thus proved that $f(\mathbb{R}^2)$ is the part of the plane $\mathbb{R}^2$ with coordinates (u, v) bounded by the parabola with equation $4v = u^2$ containing the negative v-axis. In particular, the map f is not locally surjective in a neighbourhood of the parabola bounding its image; it has a *fold* there.

One should not think that this phenomenon of non-surjectivity is general as is shown by an interesting extension of this exercise which consists in supposing that f is defined on $\mathbb{C}^2$ so that x and y are taken to be complex numbers. The calculation goes as before even though we have to work with quadruples of real numbers, for the map f is differentiable in the complex sense (one says it is *holomorphic*), i.e., its 4×4 Jacobian matrix has four 2×2 blocks which can be identified with complex numbers. Its determinant vanishes on the main diagonal in $\mathbb{C}^2$, but f continues to be locally surjective in a neighbourhood of the image of this plane. This property comes from the fact that the map from $\mathbb{C}$ to $\mathbb{C}$ which to z associates z^2 is locally surjective in a neighbourhood of 0 even though the differential of the map vanishes at this point (one says that 0 is a *ramification point*).

Solved Exercise III.83. Show that, *on the positive cone S^+ of positive symmetric matrices, one can define a unique square root map $r : S \mapsto S^{\frac{1}{2}}$ which is of class C^1.*

To do this we are going to use the information contained in the solution of Exercise III.37. The tangent linear map of $q : S^+ \longrightarrow S^+$ is invertible since the spectrum of a positive symmetric matrix consists of positive numbers, and is therefore disjoint from its negative. By the Local Inversion Theorem III.70, q is thus a local diffeomorphism in a neighbourhood of each point of S^+.

We now show that q is a global diffeomorphism from S^+ to itself whose inverse is the square root map we want to study. For this, it suffices to prove that q is bijective.

We begin with the surjectivity: let $S \in S^+$. There exists an orthonormal basis which forms a matrix B such that $^t BSB = D$ is a diagonal matrix with positive diagonal coefficients $(\lambda_1, \cdots, \lambda_n)$. If we *denote* by $D^{\frac{1}{2}}$ the diagonal matrix with diagonal coefficients $(\sqrt{\lambda_1}, \cdots, \sqrt{\lambda_n})$, we have $^t BSB = D^{\frac{1}{2}} D^{\frac{1}{2}}$, whence $S = B(D^{\frac{1}{2}})^t BB(D^{\frac{1}{2}})^t B$. If we put $S' = BD^{\frac{1}{2}t}B$, we thus have that $S' \in S^+$ and $S = q(S')$.

We now study the injectivity. Suppose that $S = S'^2 = S''^2$. The matrices S' and S'' preserve the eigenspaces of S. When restricted to the subspace F with associated eigenvalue λ of S, the matrices S'_F and S''_F have square λI_F. They are thus solutions of the equation $(M - \sqrt{\lambda} I_F)(M + \sqrt{\lambda} I_F) = 0$. Since S'_F and S''_F are positive-definite, for both matrices $M + \sqrt{\lambda} I_F$ is a positive, and hence invertible, matrix. They are thus both equal to $\sqrt{\lambda} I_F$, and $S' = S''$.

We have thus proved that q is a (global) diffeomorphism from S^+ to itself. Hence so is its inverse, the map r giving the square root of a positive symmetric matrix, which we were trying to define.

Exercise III.84. How many (not necessarily positive) square roots of a positive $n \times n$ square matrix are there?

III.85. It follows easily from Proposition III.62 that the composite of two C^1-diffeomorphisms is again a C^1-diffeomorphism. A subgroup of the group of homeomorphisms is thus defined, which we can say is formed by the transformations which preserve the "smooth structure" of the space, but we shall return to this point in Chapter V.

III.86. We can now state a simple corollary of Corollary III.49 whose proof consists simply in recalling that isomorphisms of vector spaces exist only between spaces of the *same* dimension. This result brings to light a point of contact between the theory of differentiable maps and that of linear maps.

Corollary III.87. *Two non-empty open subsets of $\mathbb{R}^m$ and $\mathbb{R}^n$ can be diffeomorphic only if $m = n$.*

III.88. In fact, a stronger theorem proved by Brouwer[58] in 1911 asserts that *two non-empty open subsets of* $\mathbb{R}^m$ *and* $\mathbb{R}^n$ *can be homeomorphic only if* $m = n'$ whereas one knows by the work of Cantor[59] that there always exists a bijection between them.

Actually, the problem of defining the *dimension* of a space in purely topological terms played an important role in the development of mathematical research at the end of the XIXth century. Only the case $n = 1$ is simple, since $\mathbb{R}$ minus a point is not connected which is no longer the case for numerical spaces of higher dimension.

The proof of the Brouwer theorem can be viewed as one of the routes into algebraic topology, one of the most active branches of mathematics at the present time, and one whose applications have flourished recently, particularly in the study of non-linear equations.

Solved Exercise III.89. To solve the quadratic system of n equations in n unknowns

$$(\text{III}.90) \qquad \sum_{i_1,i_2=1}^{n} a_{i_1,i_2}^{j} x^{i_1} x^{i_2} + \sum_{i=1}^{n} b_i^j x^i - c^j = 0, \quad 1 \le j \le n$$

which it will be useful to write in the condensed form $^t XAX + BX - C = 0$.

It is not possible, unlike in the $n = 1$ case, to give all the solutions in the form of an explicit formula. By using the fact that, for $C = 0$, we know the trivial solution $X = 0$, we shall be able to find certain solutions when the constant term C is of sufficiently small norm and B is sufficiently regular.

For this, we transform the problem into the study in a neighbourhood of 0 of the image of the map f from $\mathbb{R}^n$ to $\mathbb{R}^n$ defined by $f(X) = {}^t XAX + BX$. We have, of course, $f(0) = 0$ (which means that $X = 0$ is a solution of the equation $^t XAX + BX = 0$). Moreover, $T_0 f = B$, viewed as a linear map from $\mathbb{R}^n$ to $\mathbb{R}^n$, since the term involving A, which is quadratic in X, does not contribute when $X = 0$. If B is an invertible matrix, we are in a situation to which the Local Inversion Theorem III.70 applies, and this ensures that f defines a C^1-diffeomorphism between a neighbourhood U of 0 in the source space and a neighbourhood V of 0 in the target space.

For our equation, this means that, if its second term C is sufficiently small, the inverse diffeomorphism g of f on V expresses X as a function of C of class C^1. In particular, from the calculation of the tangent linear map of g at 0, namely $T_0 g = B^{-1}$, we deduce an expression for the solution X in the limit of the form $X(C) = B^{-1}C + o(C)$.

Remark III.91. Exercise III.89 generalises in two directions, to matrices which are not necessarily square, and to polynomials of higher degrees.

The method used does not, of course, give all the solutions of Eq. (III.90), but the problem of extending a particular solution by perturbing its coefficients often comes up in practice, so the approach presented above is of interest.

---

[58] Luitzen Egbertus Jan Brouwer (1881–1966) was a Dutch mathematician and philosopher. His work had considerable impact in topology, set theory and measure theory. He defended an intuitionistic approach of mathematics in sharp contrast with the formalist approach of Hilbert.

[59] Georg Ferdinand Ludwig Philipp Cantor (1845–1918) was a German mathematician who played a key role in the creation of set theory. He developed the theory of cardinal and ordinal numbers.

D. Derivatives of Higher Order

We define the higher order derivatives of a map (when they exist). After establishing the Schwarz Lemma on the symmetry of the second differential, we give extensions of the fundamental theorems of differential calculus to the case of maps of class C^k.

III.92. To a differentiable map f from a normed space E to a normed space F, we have associated its tangent linear map (or its differential), which is a map from the domain of definition of f to the normed space $L(E, F)$. We can ask for this new map to be differentiable, whence the following definition.

Definition III.93. A map f defined on an open subset u of a normed space E taking its values in a normed space F is said to be *of class* C^2 if its tangent map Tf is of class C^1 as a map from U to $L(E, F)$. The tangent linear map to Tf is called *the second differential* of f.

Remark III.94. It is, of course, possible to be less demanding than we have been in Definition III.93 and to define the second derivative at a point. In our applications of this notion, this short-cut will not cause us any difficulties.

In the situations encountered in the following chapters, it will not be possible to define an intrinsic notion of second derivative, except under the very special conditions we shall meet in Chapter VII.

III.95. We are faced with the problem of finding a suitable notation for the second differential (as well as for the higher order derivatives), particularly as we have decided to use the differential notation ∂ in a restricted way (the most popular notation is $\partial^2 f$). Following the logic of our system of notations, we should write TTf (hum !).

For a map f from $\mathbb{R}^m$ to $\mathbb{R}^n$ represented by its components f^α, the notation of partial derivatives is very convenient for writing second differentials. In fact, to the tangent linear map was associated the matrix $(\partial f^j / \partial x^i)$; to the second differential is associated the three-index quantity $(\partial / \partial x^{i_1} (\partial f^j / \partial x^{i_2}))$ which is traditionally denoted by $(\partial^2 f^j / \partial x^{i_1} \partial x^{i_2})$. We now turn to the question of its precise meaning.

III.96. To understand the meaning of the second differential, we make use of Exercise I.43, which establishes a natural isomorphism between the normed space $L(E, L(E, F))$ and the normed space $L_2(E; F)$ of continuous bilinear forms on E with values in F, which associates to l the form $\hat{l}$ which, on the vectors v_1 and v_2 of E, is equal to $\hat{l}(v_1, v_2) = l(v_1)(v_2)$. Thanks to this isomorphism, one can view the second differential as a bilinear form on E with values in F, a formulation which is very convenient for generalisation to higher order derivatives.

Schwarz Lemma III.97. *The second differential of a map of class C^2 is a symmetric bilinear form.*

In particular, if f is a map of class C^2 defined on $\mathbb{R}^n$ equipped with linear coordinates $(x^1, \cdots, x^n)$, then, for $1 \leq i, j \leq n$,

$$\frac{\partial^2 f}{\partial x^i \partial x^j} = \frac{\partial^2 f}{\partial x^j \partial x^i} \ .$$

Proof. We return to the definition of the second differential as the tangent linear map to its differential.

Let f be a map of class C^2 from an open subset U of a normed space E to a normed space F. Let a be a point of U. For $u \in E$ of sufficiently small norm which we fix for the moment, we introduce the auxiliary map $\Delta_u f$ defined by

$$(\Delta_u f)(v) = f(a + u + v) - f(a + u) - f(a + v) + f(a) - ((T_a Tf)(u))(v) \ .$$

By the Chain Rule III.46, the map $\Delta_u f$ is differentiable and its differential at v is given by

$$T_v(\Delta_u f) = T_{a+u+v} f - T_{a+v} f - (T_a Tf)(u)$$

which, taking into account the linearity of the tangent linear map, can be written

(III.98)
$$T_v(\Delta_u f) = (T_{a+u+v} f - T_a f - (T_a Tf)(u + v))$$
$$- (T_{a+u} f - T_a f - (T_a Tf)(v)) \ .$$

Now, let $\epsilon > 0$. We can find a ball of radius r centred at a such that, for $a + w \in B_{2r}(a)$, $\|T_{a+w} f - T_a f - T_a(Tf)(w)\| \le \epsilon \|w\|$. If $\|u\| < r$ and $\|v\| \le r$, from III.98 we obtain the estimate $\|T_v(\Delta_u f)\| \le 2\epsilon (\|u\| + \|v\|)$. Using the Growth Lemma III.58, we can bound the function $\Delta_u f$ on the line segment $[0, v]$ by $\|\Delta_u f(v)\| \le 2\epsilon (\|u\| + \|v\|)^2$. Consequently, since

$$(T_a Tf(u))(v) - (T_a Tf(v))(u) = (\Delta_v f)(u) - (\Delta_u f)(v) \ ,$$

we obtain finally that the quadratic function $(T_a Tf(u))(v) - (T_a Tf(v))(u)$ is bounded above, for $u, v \in B_r(0)$, by the quadratic function $\epsilon (\|u\| + \|v\|)^2$. The term on the left-hand side being a bilinear function of the pair (u, v), to estimate it, it suffices to work in a ball of arbitrary radius, i.e. to suppose that u and v are of the same norm. On the sphere of radius $r/2$ for example, we find that

$$(T_a Tf(u))(v) - (T_a Tf(v))(u) \le \epsilon r^2 \ .$$

Since ϵ is arbitrary, the second differential is necessarily symmetric. □

Remark III.99. The proof of the Schwarz Lemma also contains a direct definition of the second differential which can be expressed as follows: *a map f has a second differential at a if there exists a bilinear map B such that*

$$\lim_{u,v \to 0} \frac{\|f(a + u + v) - f(a + u) - f(a + v) + f(a) - B(u, v)\|}{(\|u\| + \|v\|)^2} = 0 \ ,$$

and B can be identified with its second differential.

III.100. It is possible to state a *second order Taylor*[60] formula for the map f, namely

$$f(a + v) = f(a) + (T_a f)(v) + \frac{1}{2}(T_a Tf)(v, v) + o_2(v)$$

where the Landau remainder $o_2(v)$ satisfies $\lim_{v \to 0} \|o_2(v)\|/\|v\|^2 = 0$.

---

[60] The name of the English mathematician Brook Taylor (1685–1731) is attached to the series expansion of a function. We also owe to him the first formulation of the *equation of motion of vibrating strings*.

It may be useful to relate this condensed formulation to the partial derivatives in the case when f is defined on $\mathbb{R}^n$ and takes values in $\mathbb{R}^p$. We use the notations of III.77. We have then

$$f^j(x^1, \cdots, x^n) = f^j(a^1, \cdots, a^n) + \sum_{i=1}^{n} \frac{\partial f^j}{\partial x^i}(a)\,(x^i - a^i)$$

$$+ \frac{1}{2} \sum_{i,k=1}^{n} \frac{\partial^2 f^j}{\partial x^i \partial x^k}(a)\,(x^i - a^i)(x^k - a^k) + o_2(\|x - a\|) \,.$$

Examples III.101. i) A continuous linear map is of class C^2, its tangent linear map being constant.

ii) By Formula (III.31), the second differential of a bilinear map is its proper symmetrisation (the Schwarz Lemma again!).

III.102. Since the second differential is again a map with values in a normed space, one can iterate the process of differentiation and define *maps of class* C^k and even *maps of class* C^∞ if they are differentiable to all orders.

It is often important to prove that a map is of class C^∞. Unfortunately, there is no magic method for this. One must proceed one step at a time except for certain special classes of maps such as polynomial maps which are automatically C^∞ because their derivatives are polynomials (in more and more complicated spaces).

Exercise III.103. Show that the map inv (cf. Exercise III.42) is of class C^∞.

III.104. The Schwarz Lemma generalises step by step and has the following expression in the case of kth order derivatives, whose proof we leave as an exercise for the reader.

C^k**-Schwarz Lemma III.105.** *The differential of order k of a map of class C^k defined on an open subset of a normed space E with values in a normed space F can be identified with a k-linear form on E with values in F which is completely symmetric in all its arguments.*

III.106. There exist various formulas which give the higher order differentials of a product or a composite of two maps generalising the Leibniz formula or the Chain Rule III.46. Since we shall not make explicit use of them, we refer the reader for example to page 37 of (Avez 1983), where they are given in sufficient generality.

The fundamental property is of course the following closure property of maps of class C^k, for whose proof we refer again to (Avez 1983).

C^k**-Chain Rule III.107.** *The composite of two maps of class C^k is of class C^k.*

III.108. The Local Inversion theorem III.70 has an extension for maps of class C^k and even C^∞ whose proof we again leave to the reader.

C^k-Local Inversion Theorem III.109. *If, in the statement of the Local Inversion Theorem, the map f is assumed to be of class C^k ($1 \leq k \leq \infty$), then its local inverse is also of class C^k.*

III.110. This result means that there is no ambiguity in speaking of a (local) C^k-diffeomorphism since it ensures that the inverse of a map of class C^k which is a (local) diffeomorphism is also of class C^k. It is thus possible to speak of the *group of diffeomorphisms of class C^k* which, when $k > 1$, preserves smoothness properties of a space more subtle than does the group of C^1-diffeomorphisms (e.g. the notion of curvature of surfaces is preserved by diffeomorphisms of class C^k only for $k \geq 2$).

As far as the global properties of a space are concerned, differential topology, which we mentioned above, establishes, by means of approximation theorems, that there is no difference between those which are preserved by C^1-diffeomorphisms and by C^k-diffeomorphisms or $1 \leq k \leq \infty$.

Project III.111. (C^k ω-**Lemma**) Let E, F and G be three normed spaces and g a map of class C^k from an open subset V of F with compact closure to G. Let U be an open subset of E. Show that the map ω_g, which to f in $C^1(\overline{U}, V)$ associates $g \circ f$, is a map of class C^{k-1} from $C^1(\overline{U}, V)$ to $C^1(\overline{U}, G)$.

E. Historical Notes

III.112. The emergence of the notions of derivatives and differentials played an important role in the birth of analysis in the XVIIIth century. The results have not always had the precision which we try to give them today. The implicit function theorem, for example, (which we choose to present only in the next chapter) appears to have been used without much justification. The first proof of it seems to be due to Cauchy in 1839.

The proof which we give uses a fixed point theorem which was established by Picard towards the end of the XIXth century. It is in Banach's thesis that one finds the generalisation to what we nowadays call... Banach spaces.

III.113. One can ask whether the local inversion theorem continues to be valid for families of spaces more general than Banach spaces. We have seen, for example, that one can define a distance on the space of numerical functions of class C^∞ which is not induced by a norm. This distance, induced by a countable family of semi-norms (which are the norms of the derivatives of order k), is in fact typical of a family of spaces called Fréchet spaces.

Generalisations of the local inversion theorem to these spaces have been obtained only in the second part of the XXth century. They have the generic name of Nash[61] –

---

[61] John Forbes Nash Jr. (1928–2015) was an American mathematician who made fundamental contributions to game theory, differential geometry, and the study of partial differential equations. Nash's work also impacted economics through his concept of Nash equilibrium. He was one of the

Moser[62] theorems. The statement is no longer so simple, for it requires the differential to be invertible in a neighbourhood of the point under consideration. Moreover, one has to control the continuity of this inverse with respect to a family of semi-norms. In fact, it should be regarded more as a method than as a theorem. However, the spectacular geometric applications which it has found justify the excitement which this kind of theorem has generated among mathematicians when it appeared.

Among the famous consequences of such methods is the Kolmogorov[63]–Arnol'd[64]–Moser theorem on the existence of invariant tori stated in the 1950s and completed at the beginning of the 1960s, which is a fundamental result in the theory of conservative dynamical systems.

recipients of the 1994 Nobel Memorial Prize in Economic Sciences. In 2015, he shared the Abel Prize with Louis Nirenberg for his work on nonlinear partial differential equations. He struggled with schizophrenia for a large part of his life, and accepted that his illness be presented in the movie *A Beautiful Mind* to draw attention to the disease.

[62] The German-American mathematician Jürgen Kurt Moser (1928–1999) made major contributions to the study of Hamiltonian dynamical systems and of partial differential equations. He was an accomplished musician.

[63] Andrey Nikolaevitch Kolmogorov (1903–1987) was a Soviet mathematician who contributed to many different fields in mathematics and computer science. He also had a broad knowledge of other sciences, in particular ecology and fluid mechanics. Considered one of the founders of modern probability theory, he proposed an axiomatic approach that is widely accepted to this day.

[64] Vladimir Igorevitch Arnol'd (1937–2010) was a Soviet and Russian mathematician who contributed to several branches of mathematics, including dynamical systems, topology and singularity theory. His interest in symplectic geometry played a key role in its development. He was an exceptional pedagog and his books are models of writing which educated many mathematicians and physicists in the world. He received the Wolf Prize in 2001.

"*La Filosofia è scritta in questo grandissimo libro, che continuamente ci sta aperto innanzi agli occhi (io dico l'Universo) ma non si può intendere, se prima non s'impara a intender la lingua, e conoscer i caratteri, né quali è scritto. Eggli è scritto in lingua matematica, e i caratteri son triangoli, cerchi, ed altre figure Geometriche, senza i quali mezzi è impossibile intenderne umanamenta parola; questi è un aggirarsi vanamente per un oscuro laberinto*"

Galileo GALILEI,
in *Il Saggiatore*.

Chapter IV
Some Applications of Differential Calculus

There are several variants of the local inversion theorem, of which the most well-known is the *implicit function theorem* presented in Sect. A. This form is actually very special, and its proof becomes a little artificial. Other versions of geometric interest can also be interpreted as linearisation theorems.

On the way, we have a brief discussion of the notion of *dimension* which turns out to be so powerful in the setting of linear algebra and which the local inversion theorem allows one to introduce again in the setting of differentiable maps.

Section B contains a presentation of the fundamental notion of a *vector field*. These objects are important in modelling since they allow one to take account of the *dynamics* of a system. They provide the best setting in which to discuss *ordinary differential equations*. In particular, we give the basic theorems on differential equations, such as the *uniqueness theorem* and the *local existence theorem*. Finally, the notion of the *flow* of a vector field is introduced which establishes a correspondence between one-parameter groups of diffeomorphisms and vector fields.

The most important examples of vector fields are presented in Sect. C. These include, in particular, the *linear* and *affine* vector fields, *gradient* vector fields, and *Hamiltonian*[65] vector fields.

Another application of the differential calculus is considered in Sect. D. This concerns the *Poisson*[66] *bracket*, a natural bilinear operation defined on the algebra of functions on the product of a vector space and its dual. This bracket is, in fact, intimately related to Hamiltonian vector fields, and, in particular, allows one to study the conserved quantities of such vector fields.

---

[65] Sir William Rowan Hamilton (1805–1865), Irish mathematician and astronomer, is often considered to be the creator of vector geometry. His formulation of the Calculus of Variations has had a decisive influence on its applications in physics.

[66] Siméon-Denis Poisson (1781–1840), French mathematician and physicist, participated alongside Lagrange and Laplace in the development of mathematical models in physics. His creative activity has been the subject of various appreciations. On the other hand, he played a prominent role in the organisation of the science of his time.

© The Author(s), under exclusive license to Springer Nature Switzerland AG 2022   81
J.-P. Bourguignon, *Variational Calculus*, Springer Monographs in Mathematics,
https://doi.org/10.1007/978-3-031-18307-2_4

A. Geometric Variants of The Local Inversion Theorem

In this section we state the *implicit function theorem* and various other extensions of geometric interest of the local inversion theorem for the case in which the tangent linear map is not invertible.

IV.1. We begin with the presentation of the implicit function theorem which is often given as the central result of this subject rather than the Local Inversion Theorem III.70. However, it forces us to work in a special situation which, historically, has been of great importance but which has restricted the scope of the theorem.

We shall see in Chap. V that it is actually the Local Inversion Theorem III.70 which will be in constant use.

Implicit Function Theorem IV.2. *Let ϕ be a map of class C^1 defined on the product of two open sets U and V of two Banach spaces E and F and taking its values in a third Banach space G.*

Suppose that, at a point (a, b) of $U \times V$, the tangent linear map of the partial map $_a\phi : y \mapsto {}_a\phi(y) = \phi(a, y)$ is an isomorphism of F onto G. Then, there exists an open neighbourhood U' of a $(U' \subset U)$, an open neighbourhood W of $\phi(a, b)$, an open neighbourhood Ω of (a, b) $(\Omega \subset U \times V)$, and a map $\psi : U' \times W \longrightarrow V$ of class C^1 such that the following two conditions are equivalent:

i) *$(x, y) \in \Omega$, $z \in W$ and $\phi(x, y) = z$;*
ii) *$(x, z) \in U' \times W$ and $y = \psi(x, z)$.*

In particular, for all $(x, z) \in U' \times W$, $z = \phi(x, \psi(x, z))$.

Proof. Under our hypotheses, the situation is not symmetric between the source space and the target space. To restore the balance, we introduce an auxiliary map f from $U \times V$ to $E \times G$ defined by $f(x, y) = (x, \phi(x, y))$. By the chain rule III.46, f is of class C^1 and its tangent linear map at (a, b) can be expressed on $u \in E$ and $v \in F$ in terms of those of the partial maps $_a\phi : y \mapsto {}_a\phi(y) = \phi(a, y)$ and $\phi_b : x \mapsto \phi_b(x) = \phi(x, b)$ by the formula

$$(T_{(a,b)}f)(u, v) = (u, T_a(\phi_b)(u) + T_b({}_a\phi)(v)) \ .$$

By hypothesis $T({}_a\phi)$ is an isomorphism from F onto G so that $T_{(a,b)}f$ is itself an isomorphism from $E \times F$ onto $E \times G$, whose inverse isomorphism at $f(a, b)$ can be written explicitly in the form $(u, w) \mapsto (u, T({}_a\phi)^{-1}(w - T(\phi_b)(u)))$. We can thus apply the Local Inversion Theorem III.70 and f is a C^1-diffeomorphism from a neighbourhood of (a, b) onto a neighbourhood of $f(a, b)$. Because of the special form of f, on a possibly smaller product neighbourhood $U' \times W$ of $f(a, b) = (a, \phi(a, b))$, the inverse diffeomorphism f^{-1} can be written $(x, z) \mapsto (x, \psi(x, z))$ for some map ψ defined on $U' \times W$, and this is the map required.   □

IV.3. We describe the finite-dimensional case in detail by using again the notations we have introduced in III.77. For a map ϕ from $\mathbb{R}^m$ to $\mathbb{R}^n$ whose components are denoted by $(\phi^1, \cdots, \phi^n)$ (we suppose here that $m > n$), it is possible to express the condition of invertibility of the partial derivative by saying that *the Jacobian matrix of ϕ contains a non-zero $n \times n$ minor*. This distinguishes an $n \times n$ square submatrix which determines the decomposition of the space $\mathbb{R}^m$ as $\mathbb{R}^n \times \mathbb{R}^{m-n}$ which is referred to in the statement of Theorem IV.2.

We now explain the origin of the terminology "implicit function" which is often used in this context. If, for example, we are given a map ϕ of class C^1 from $\mathbb{R}^2$ with coordinates (x, y) to $\mathbb{R}$, and $c \in \mathbb{R}$, we define the level sets $\phi^{-1}(c)$ of ϕ and we would like to interpret these as curves along which x is a function of y, for example; in this case, $\phi(x, y) = c$ defines x as an *implicit function* of y. The Implicit Function Theorem IV.2 gives a criterion under which this is actually so, namely that $(\partial\phi/\partial y) \neq 0$. For all c, it is then possible to find a function $\psi_{(c)}$ of class C^1 such that we have $\phi(x, \psi_{(c)}(x)) = c$ identically. The theorem has enabled us to make *explicit* the function which was implicit. We could have given an analogous discussion for a map from $\mathbb{R}^m$ to $\mathbb{R}^n$; only the notation would have been more complicated.

Exercise IV.4. Using the notation of Theorem IV.2, show that the tangent linear map of the "implicit function" ψ can be expressed by the formula

$$T_x(\psi_{(b)}) = -(T_{\psi_b(x)}(_x\phi))^{-1} \circ T_x(\phi_{\psi_b(x)}) .$$

Solved Exercise IV.5. Show that *the solution of the differential equation*

$$(\text{IV.6}) \qquad \frac{d^2x}{dt^2} + V(t)x(t) = 0, \quad x(0) = x_0, \quad \frac{dx}{dt}(0) = x_0'$$

studied in II.4 depends in a C^1 *way on the potential V.* We keep the notation of II.14. The map $h : E \times E \longrightarrow E$ (where E denotes the Banach space of continuous functions on an interval $[0, \tau]$), which to (x, V) associates the function $h_V(x)$ defined by

$$(h_V(x))(t) = x_0 + x_0't - \int_0^t \left(\int_0^s V(u)x(u)\,du \right) ds,$$

is of class C^∞, being the sum of a fixed map $t \mapsto x_0 + x_0't$ and a continuous bilinear map.

As we have seen in II.14 that the fixed points of h_V are exactly the solutions of the differential equation IV.6, to characterise the solutions we must consider the implicit relation $g(x, V) \equiv x - h(x, V) = 0$. The map g is evidently of class C^∞.

Suppose then that (x^0, V^0) is a pair satisfying this relation, so that x^0 is then a solution of Eq. (IV.6) for the potential V^0. We have the conditions under which the Implicit Function Theorem IV.2 applies, with the partial map g_{V^0} having an invertible derivative at x^0. Now $(T_{x^0}(g_{V^0}))(y)$ is the difference between the identity map and the linear map L defined for $y \in E$ by $(L(y))(t) = \int_0^t (\int_0^s V^0(u)\,y(u)\,du)ds$, whose norm is bounded above by $\frac{1}{2}\tau^2\|V^0\|$. Provided we have $\tau < \sqrt{2/\|V^0\|}$, it is possible to calculate the inverse of $T_{x^0}g_{V^0}$ as the sum of the convergent series $\sum_{i=0}^{\infty} L^i$. It then follows from the Implicit Function Theorem IV.2 that the implicit relation $g(x, V) = 0$ defines x as a function of class C^1 of V in a neighbourhood of the point V^0. We denote this map by f, so that $g(f(V), V) = 0$ is an identity for V in a neighbourhood of V^0.

To evaluate the tangent linear map of f, we use the Chain Rule III.46 which we apply to the identity $g(f(V), V) = 0$. We obtain $T_{x^0} g_V \circ T_V f + T_V(_{x^0} g) = 0$ (where we have, as usual, denoted by $_x g$ the partial map $_x g(V) = g(x, V)$). Since $T_V(_{x^0} g) = -T_V(_{x^0} h)$, we thus have

(IV.7) $(T_{V^0} f)(v) = (T_{x^0} g_V^0)^{-1}(\partial_v h(x^0, V^0))$.

It is useful to interpret the preceding formula in terms of an inhomogeneous version of the differential equation IV.6. In fact, a solution of the differential equation $(d^2 y/dt^2) + V(t) y(t) = z(t)$ is defined by the implicit relation $y - h_V(y) = \int_0^t (\int_0^s z(u) du) ds$. The map g_V thus sends the solution of this equation with initial conditions x_0 and x_0' to the double primitive of the term on the right-hand side of the equation. Note that the map g_V is not linear because of the presence of the term $x_0 + x_0' t$ in h_V. The inverse of its linearisation at x^0 thus associates to the double primitive of a function z the solution of the inhomogeneous equation with right-hand side z and initial conditions 0 and 0. Moreover, the term $\partial_v(h(x^0, V^0))$ is simply the double primitive of the function $x^0 v$. Formula (IV.7) thus says that the infinitesimal variation of the solution x^0 corresponding to a variation v of V is given by the solution y of the inhomogeneous equation with right-hand side $x^0 v$ and initial conditions $y(0) = 0$ and $(dy/dt)(0) = 0$. Moreover, this is the result we would obtain heuristically by differentiating the equation $(d^2 x_\alpha/dt^2) + (V + \alpha v)(t) x_\alpha(t) = 0$ with respect to α.

From the preceding discussion we can deduce an approximate solution of IV.6 when the potential V is assumed to be uniformly small. We treat the equation as a perturbation of the equation for $V \equiv 0$, of which the solution with initial conditions x_0 and x_0' is clearly $x^0(t) = x_0 + x_0' t$. The approximation given by $(T_0 f)(v)$ is the solution of the inhomogeneous equation with right-hand side $(x_0 + x_0' t) v(t)$ which can be obtained by quadrature

$$((T_0 f)(v))(t) = \int_0^t \left(\int_0^s (x_0 + x_0' t) v(u) \, du \right) ds .$$

IV.8. The conditions which the map in the local inversion theorem must satisfy are very strong. It is natural to ask what can be said when the tangent linear map is not an isomorphism. We now give a theorem which deals with the case where the tangent linear map is *surjective*.

Submersion Theorem IV.9. *Let s be a map of class C^1 from an open subset U of a Banach space E to a Banach space F. If, at a point a of U, $T_a s$ is an isomorphism from a Banach (hence closed) subspace E_1 of E such that $E = \ker T_a s \oplus E_1$ onto F, then there exists a neighbourhood U' of a in E and neighbourhoods V of $s(a)$ in F and $\mathcal{U}_0$ of 0 in $\ker T_a s$ such that the map f from U' to $V \times \mathcal{U}_0$ which to a point x of U' associates the pair $(s(x), \mathrm{pr}_0(x - a))$ (where pr_0 denotes the projection of E onto $\ker T_a s$ parallel to E_1) is a C^1-diffeomorphism.*

Such a map is called a submersion *at a from U to F.*

Proof. We reduce to situation of the Local Inversion Theorem III.70. To this end we introduce the auxiliary map $f : U \longrightarrow F \times \ker T_a s$ which we define by $f(x) = (s(x), \mathrm{pr}_0(x - a))$.

The tangent linear map to f at a is given by $T_a f(u) = T_a s(u) + \mathrm{pr}_0 u$. By hypothesis, $T_a f$ is an isomorphism. We can thus apply the Local Inversion Theorem III.70 to conclude: there exists a neighbourhood U' of a in E and neighbourhoods V of $s(a)$ in F and $\mathcal{U}_0$ of 0 in $\ker T_a s$ such that f is a C^1-diffeomorphism from U' onto $V \times \mathcal{U}_0$. □

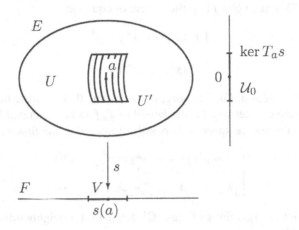

Fig. IV.10 A submersion s at a point a

IV.11. In the case where E is finite-dimensional (or a Hilbert space), the second part of the hypothesis made on s in IV.9 is unnecessary, since we can always take E_1 as the orthogonal complement of the kernel with respect to an arbitrarily chosen inner product (or that defining the Hilbert space structure).

IV.12. The notion of a submersion is particularly useful when the condition on the tangent linear map is satisfied at every point of an open set. In this case one says that the map is a *submersion*.

Corollary IV.13. *Let f be a submersion from an open subset U of a Banach space E to a Banach space F. Then f is an open map.*

Proof. This is an immediate consequence of the Submersion Theorem IV.9. In fact, this asserts that, if s is a submersion at a point a of U, then the image of U in F contains a neighbourhood of $s(a)$. On applying this to every point of an open subset V of the domain U where s is defined, one finds that the image of V under s is a neighbourhood of each of its points, and hence is open. □

Remark IV.14. The geometric content of the Submersion Theorem IV.9 is to give a local description of the space on which the submersion is defined as a product of a fixed neighbourhood of 0 in the kernel of the tangent linear map at a point with a neighbourhood of the image of this point in the tangent space. The source space is thus "fibred"[67] over the target space by copies of an open subset of the Banach space $\ker T_a s$.

---

[67] This is the terminology which mathematicians have chosen to describe such a situation.

We shall give a natural and important example in Chapter IX. We reiterate that the result is obtained from a *pointwise* hypothesis on the derivative.

IV.15. Consider the case of a submersion from $E = \mathbb{R}^m$ to $F = \mathbb{R}^n$. We have necessarily $m \geq n$, since a surjective linear map from $\mathbb{R}^m$ to $\mathbb{R}^n$ only exists in this case.

Following the notation of III.77, if $T_a f$ is surjective, we can find a non-zero $n \times n$ minor of the Jacobian matrix. The Submersion Theorem IV.9 says in particular that, for $(y^1, \cdots, y^n)$ near to $b = f(a)$, the system of equations

$$\begin{cases} f^1(x^1, \cdots, x^m) = y^1 \\ \quad \vdots \\ f^n(x^1, \cdots, x^m) = y^n \end{cases}$$

has at least one solution. Furthermore, if we assume that we have made the linear changes of variable necessary for the kernel of $T_a f$ to be generated by the vectors e_{n+1}, ..., e_m of the source space, it is possible to describe the first n coordinates of the solution as

$$\begin{cases} x^1 = g^1(y^1, \cdots, y^n, x^{n+1}, \cdots, x^m) \\ \quad \vdots \\ x^n = g^n(y^1, \cdots, y^n, x^{n+1}, \cdots, x^m) \end{cases}$$

where g^1, ..., g^n are functions of class C^1 defined on a neighbourhood of a point $(b^1, \cdots, b^n, a^{n+1}, \cdots, a^m)$.

IV.16. The behaviour of this system thus formally resembles that of a system of linear equations, but with two differences: the solution can be given only *locally* near a solution which is already known, and the formulas expressing the solution *are non linear* (only the first order term in the limiting expansion is given by the linearised equation). This process is often given the name "*linearisation*".

It is remarkable that the Submersion Theorem IV.9 provides not only a solution in a neighbourhood of a solution of a neighbouring equation, but also a description of all the neighbouring solutions.

IV.17. The Submersion Theorem IV.9 has interesting applications to the solution of non-linear equations along similar lines.

Suppose, in fact, that a map f from an open subset U of a Banach space F satisfies the hypotheses of the theorem; since the image of f contains a neighbourhood of each of its points it is an open subset of F. If we know *in addition* that this image is closed, then about any point of the image (i.e., an equation of the form $f(x) = y_0$ of which a solution is known), one can *by continuity*, as we have done in III.59, prove that there is a solution of the equation $f(x) = y_1$.

Solved Exercise IV.18. Consider the Banach space E of numerical functions of class C^1 and the Banach space F of continuous functions on the interval $[0, 1]$. We consider the map f from E to F defined by $f(\phi) = \phi' + F(\phi)$ where F is a differentiable function of a real variable which vanishes, together with its derivative, at 0.

It is easy to see that f is of class C^1 and that $T_0 f$ is the map which to a function associates its derivative. By the fundamental lemma of integral calculus, $T_0 f$ is surjective and its kernel consists of the constant functions. This space is finite-dimensional, and thus admits a complement in E, for example the space consisting of the functions whose integral over the interval is zero.

By applying the Submersion Theorem IV.9 we obtain that for any continuous function ψ on $[0, 1]$ sufficiently small in uniform norm, there exists a function ϕ of class C^1 defined on the same interval which is a solution of the equation

$$\phi' + F(\phi) = \psi \ .$$

As a particular map, one can consider the case $F(\phi) = \phi^k$ with $2 \leq k$.

IV.19. In the 1980s many spectacular results have been obtained on the solution of certain non-linear partial differential equations by using information about the linearised equation together with control on the size of the solutions, a procedure known as *"obtaining a priori estimates"*. They justify the importance which is attached nowadays to the operation of linearisation in the context of Banach spaces.[68]

Exercise IV.20. Let S be an $n \times n$ symmetric matrix such that $\det S \neq 0$. Show that the map s defined on Gl_n with values in its subspace SM_n consisting of the symmetric matrices, which to $L \in \mathrm{Gl}_n$ associates $s(L) = {}^t L S L$, is a submersion. Deduce that every symmetric matrix $\tilde{S}$ near S can be reduced to S by a change of basis depending on $\tilde{S}$ in a C^1 fashion.

IV.21. The case where the tangent linear map is *injective* is also of interest. If we stay in the setting of Banach spaces, to be able to state a result it is necessary to assume that the image of the source space under the tangent linear map is closed and admits a closed complement, hypotheses which are, of course, redundant if the target space is finite-dimensional. The map is then called a *(strict) immersion*.

The immersion theorem (whose proof we leave as an exercise) can then be stated.

Theorem IV.22. *Let i be a map of class C^1 from an open subset U of a Banach space E to a Banach space F. Suppose that i is a strict immersion at $a \in U$ and denote by F_1 a complementary subspace to the closed subspace $T_a i(E)$ of F, which we assume exists.*

Then the point $i(a)$ admits a neighbourhood V C^1-diffeomorphic to the product of a neighbourhood U_0 of a in E and a neighbourhood V_1 of 0 in F_1.

IV.23. In finite-dimensional spaces, assuming that the map has *constant rank* in a neighbourhood of a point, it is possible to give another interesting geometric result. Under this hypothesis (which is not a pointwise one), in suitably restricted neighbourhoods, the map can be written as the composite of an immersion and a submersion whence the name *"subimmersion"* given to *maps of constant rank*.

---

[68] ...and the fact that we have decided to use this term as the title of Chapter III.

B. Vector Fields and Ordinary Differential Equations

In this section, we examine the relations between vector fields and certain families of curves, their *integral curves*. Finding these curves amounts to solving the differential equation associated to the vector field. We prove the fundamental theorems, namely the *uniqueness theorem* for the integral curves of a vector field and the *local existence theorem* for integral curves and their dependence on initial conditions.

IV.24. Let U be an open subset of a Banach space E. A *vector field X* on U associates to each point q of U an element $X(q) \in E$ which it will be useful to view as a vector of E *attached to q*.

The *regularity* of a vector field X is that of the map from U to E which defines it.

We concentrate our attention on vector fields viewed as *differential equations*, i.e., *equations whose unknowns are curves which must be determined from a knowledge of their velocity-vectors and initial conditions*.

The primary notion which provides the link between these equations and vector fields is that of an *integral curve*.

Definition IV.25. A curve γ defined on an open interval I of $\mathbb{R}$ and taking its values in an open subset U of a normed space E is an integral curve of a vector field X defined on this open set if, for all t in I, one has (recalling that $(d\gamma/dt)(t) = (T_t\gamma)(1)$)

$$\frac{d\gamma}{dt}(t) = X(\gamma(t)) \ .$$

Remark IV.26. A differential equation in an n-dimensional vector space may look different when it is expressed in coordinates. In fact, one finds that the equation is equivalent to the system of n scalar differential equations

$$\frac{dx^i}{dt} = X^i(x^1, \cdots, x^n) \ ,$$

a form which one finds especially when the intrinsic viewpoint is not mentioned.

Note also that we shall only treat differential equations which only involve the velocity-vector of the curve, equations which are said to be *of first order*.

IV.27. *To solve a differential equation means* (in general) *to find all its integral curves.* Of particular interest is the determination of the integral curve passing through a given point (and, in view of the equation, having at this point a given velocity-vector), a problem which is often referred to a the *Cauchy problem*.

This problem has a solution in general as we shall show later, but we begin by examining the uniqueness of the solution.

Uniqueness Theorem IV.28. *Let X be a vector field of class C^1 defined on an open subset U of a Banach space E. If two integral curves $\gamma : I \longrightarrow U$ and $\eta : J \longrightarrow U$ of X satisfy, for at least one point of $t_0 \in I \cap J$, $\gamma(t_0) = \eta(t_0)$, then $\gamma(t) = \eta(t)$ for all $t \in I \cap J$.*

Proof. Let C be the set where the two integral curves γ and η coincide. Since the curves γ and η are by definition continuous, C is closed in $I \cap J$. We are going to show that C is also open in $I \cap J$.

We will show that C contains a neighbourhood of $t_0 \in C$. Let $a = \gamma(t_0) = \eta(t_0)$. Since X is assumed to be of class C^1, its tangent linear map is bounded in a neighbourhood of a. By applying the Growth Lemma III.58, there exists $\epsilon > 0$ and $A > 0$ such that, for all points q_1 and q_2 of $B_\epsilon(a) \subset U$,

$$\|X(q_1) - X(q_2)\|_E \le A \|q_1 - q_2\|_E .$$

We now consider the map $t \mapsto \gamma(t) - \eta(t)$ defined on a neighbourhood of t_0. This map is of class C^1, and thus, a fortiori, continuous. Consequently, there exists an interval $[t_-, t_+]$ of non-zero length containing t_0 on which $\|\gamma(t) - \eta(t)\| < \epsilon$. Since

$$\frac{d}{dt}(\gamma(t) - \eta(t)) = X(\gamma(t)) - X(\eta(t)) ,$$

if we apply the Growth Lemma III.58 to this map between t_0 and t, with $t \in]t_-, t_+[$, we obtain

$$\|\gamma(t) - \eta(t)\|_E \le |t - t_0| \sup_{s \in [t_0, t]} \|X(\gamma(s)) - X(\eta(s))\|_E$$

which can also be written

$$\|\gamma(t) - \eta(t)\|_E \le A |t - t_0| \sup_{s \in [t_0, t]} \|\gamma(s) - \eta(s)\|_E .$$

Since this inequality is valid for all $t \in [t_-, t_+]$, we deduce, by repeating the argument if necessary on a sub-interval of $[t_-, t_+]$ of non-zero length containing t_0, that unless $\gamma = \eta$ on an open interval containing t_0, the inequality $1 \le A |t - t_0|$ holds, which is absurd since we can take $t - t_0$ as small as we wish. □

IV.29. The Uniqueness Theorem IV.28 shows the importance of the common domain of definition of two integral curves which have a point in common. This leads to the idea of finding integral curves whose domain of definition is as large as possible, and hence to the very useful notion of a *maximal* integral curve.

Definition IV.30. An integral curve $\gamma : I \longrightarrow U$ of a vector field X defined on an open subset U of a Banach space E is said to be *maximal* if, for any integral curve $\eta : J \longrightarrow U$ of X such that, for at least one point $t \in I \cap J$, $\eta(t) = \gamma(t)$, one has $J \subset I$.

Remark IV.31. From the Uniqueness Theorem IV.28 it follows that every integral curve which coincides with a maximal integral curve at some point is, in fact, part of it. This theorem also allows one to *"glue together"* pieces of integral curves and is thus at the heart of the existence theorem for maximal integral curves. However, the Existence Theorem IV.36 will have to wait until we know that at least one integral curve, and hence also a maximal integral curve, passes through every point.

Notice also that the preceding proof uses only the locally *Lipschitz* character of the vector field, though *uniqueness may be lost if the vector field is only continuous.*

Project IV.32. Give an example of a continuous vector field which admits several distinct integral curves passing through the same point.

IV.33. It is useful to introduce here the standard terminology of the *Theory of Differential Equations*, which is often inspired by its applications to the study of the *evolution of systems*, in which the parameter t plays the role of time. Thus, the image in U of an integral curve is called an *orbit* or a *trajectory*.

At a point a, the interval, *denoted* by I_a, on which the maximal integral curve of the differential equation passing through a is defined, is called its *lifetime* at a, and the length of this interval is called its *lifespan*. The partition of U into the orbits of the vector field X is called its *phase portrait*.

IV.34. One of the fundamental ideas introduced by Henri Poincaré[69] was to emphasise the importance of a *qualitative study* of the phase portrait of a differential equation, which forces us to take account of the *global behaviour* of the solutions. We must wait until Chapter VI to take up this idea which leads to the *Theory of Dynamical Systems*, one of the most active (but most difficult) branches of mathematics today. It has important applications notably in the study of the evolution of biological systems, in space travel and in the physics of phase transitions and in many other areas.

IV.35. It is now time to state an existence theorem for the solutions of a differential equation. There are many versions. The most fundamental is the one which asserts the existence of an integral curve with a given initial condition. The most classical version, due to Cauchy–Lipschitz, applies to Lipschitz vector fields. It is obtained by using Picard's method of successive approximations after suitably transforming the problem into the setting of functionals (we have given a special version of it adapted to second order linear ordinary differential equations in Solved Exercise II.14). A more sophisticated version, due to Arzela[70], applies to continuous vector fields.

We give a version which is better suited to the applications we have in mind; it incorporates the functional dependence of the solution on the initial conditions.

Local Existence Theorem IV.36. *Let X be a vector field of class C^1 defined on an open subset U of a Banach space E and a a point of U.*

There exists an open interval I of $\mathbb{R}$ containing 0, an open subset $V \subset U$ containing a and a map Γ of class C^1 from $V \times I$ to U such that, for all $q \in V$, the curve $_q\Gamma : I \longrightarrow U$ defined by $_q\Gamma(t) = \Gamma(q,t)$ is an integral curve of X passing through q, in other words which satisfies the two relations

$$_q\Gamma(0) = q , \qquad \frac{d}{dt} \,_q\Gamma(t)_{|t=t_0} = X(_q\Gamma(t_0)) \,.$$

---

[69] Henri Poincaré (1854–1912) was one of the greatest mathematicians of his time. He was the founder of numerous ideas which have played a central role in mathematical research in the XXth century. Thus, we may cite in particular the introduction of the qualitative study of the theory of ordinary differential equations and his contributions to celestial mechanics, but we should also mention his theory of Fuchsian groups and his work on functions of a complex variable. His summer house was at Lozère, at the corner of the streets known today as François Coppée and... Henri Poincaré, close to the subway station leading to École Polytechnique.

[70] The Italian mathematician Cesare Arzelà (1847–1912) made several contributions to the theory of functions of a real variable.

Proof.[71] To solve the problem we shall transform it so that we can apply the Implicit Function Theorem IV.2. To this end, we now construct a certain auxiliary map.

If $\gamma : J \longrightarrow U$ is an integral curve of X passing through a, for all $t \in J$, we can write

$$\text{(IV.37)} \qquad \gamma(t) = a + \int_0^t X(\gamma(s))\, ds\, ,$$

this form integrating the two constraints which we want to impose on γ.

To reduce to a fixed interval of integration and a fixed initial condition, it is convenient to introduce the map $\delta(q, t, u) = \gamma_q(ut) - q$ for $q \in U$, $u \in [0, 1]$ and $t \in I$, where γ_q denotes the integral curve passing through q. We can then rewrite Eq. (IV.37) in the form

$$\text{(IV.38)} \qquad \delta(q, t, u) = t \int_0^u X\left(\gamma_q(ts)\right)\, ds = t \int_0^u X(q + \delta(q, t, s))\, ds\, .$$

We consider Relation IV.38 as an implicit function relating a bounded continuous map α from $[0, 1]$ to E, a point q in E and a real number t. For this, we introduce the map f from $F \times E \times \mathbb{R}$ (where F denotes the Banach space of continuous maps from $[0, 1]$ to E with the uniform norm) to F defined by

$$f(\alpha, q, t)(u) = \alpha(u) - t \int_0^u X(q + \alpha(s))\, ds\, .$$

It is easy to check that f indeed maps $F \times E \times \mathbb{R}$ to F. Moreover, f is differentiable by the Chain Rule III.46 (recall that, for $u \in [0, 1]$, the map which to $\alpha \in F$ associates $\alpha(u)$ is linear). Furthermore, we have $f(\alpha, q, 0) = \alpha$ so that $f(0, 0, 0) = 0$ and $T_0(f_{(0,0)}) = Id_F$. We can therefore apply the Implicit Function Theorem IV.2. Thus, there exists an open subset $\mathcal{U}$ of E, an open interval I of $\mathbb{R}$ containing 0 and a differentiable map β from $\mathcal{U} \times I$ to F which satisfies $\beta(0, 0) = 0$ and which is a solution of the implicit relation $f(\beta(q, t), q, t) = 0$, i.e., such that

$$\text{(IV.39)} \qquad \beta(q, t)(u) = t \int_0^u X(q + \beta(q, t)(s))\, ds\, .$$

It only remains to go back to the family of integral curves γ, i.e., to show that β can be put in the form $\gamma(tu) - q$, which amounts to showing that its dependence on t and u has a certain homogeneity. For $\lambda \in [0, 1]$, consider the map h_λ from F to F which to α associates $h_\lambda(\alpha) : u \mapsto \alpha(\lambda u)$. It follows directly from the definition of f that $h_\lambda(f(\alpha, q, t)) = f(h_\lambda(\alpha), q, \lambda t)$. The uniqueness of the implicit function defined by f implies that we necessarily have $h_\lambda(\beta(q, t)) = \beta(q, \lambda t)$, in other words $\beta(q, t)(\lambda) = \beta(q, \lambda t)(1)$. Inserting this identity into IV.39, we obtain

$$\beta(q, t)(1) = t \int_0^1 X(q + \beta(q, ts)(1))\, ds = \int_0^t X(q + \beta(q, s)(1))\, ds\, .$$

If we put $\Gamma(q, t) = q + \beta(q, t)(1)$, the map Γ satisfies

$$\Gamma(q, t) = q + \int_0^t X(\Gamma(q, s))\, ds\, ,$$

which is exactly what we want. $\qquad\qquad\square$

IV.40. So far, we have been content to find local solutions of a differential equation, without trying to estimate the lifetime of a maximal solution. This point of view is not sufficient for many purposes.

---

[71] This (particular form of the) proof is due to Joel Robbin (cf. "On the existence theorem for differential equations", *Proc. Amer. Math. Soc.* **19**, 1005–1006 (1968)).

Definition IV.41. When every maximal solution extends to infinity, the vector field is said to be *complete*.

IV.42. There is one trivial case where we are sure to have found the maximal integral curves: *that which corresponds to the zeros of the vector field, since then the integral curves are constant*, i.e., they reduce to a single point.

IV.43. The Local Existence Theorem IV.36 deserves the qualifier "*local*" for two reasons: it is local in *time*, a feature which we have already discussed above; and it is also local in *space*, a restriction which we now try to remove.

IV.44. It is important to consider the *life set* of the differential equation associated to a vector field X defined on an open set U.

This subset of $U \times \mathbb{R}$, which is traditionally denoted by Δ_X, is defined as follows: we put $\Delta_X = \bigcup_{q \in U} \{q\} \times I_q \subset U \times \mathbb{R}$ where I_q denotes the lifetime of the maximal solution passing through the point q. For $(q, t) \in \Delta_X$, we further introduce the notation $\xi(q, t)$ to denote *the point at time t on the maximal integral curve passing through q of the vector field X*.

Theorem IV.45. *The life set Δ_X of a vector field X on an open subset U of a Banach space E is an open neighbourhood of $U \times \{0\}$ in $U \times \mathbb{R}$.*

For $(q, t) \in \Delta_X$, the equality $\xi(\xi(q, t), t') = \xi(q, t + t')$ holds provided one of the two sides is defined; moreover, $\xi(q, 0) = q$.

Proof. We first prove the second part of the theorem. By the definition of ξ, we have $\xi(q, 0) = q$. Now, fix a point (q, t) in Δ_X and put $\tilde{q} = \xi(q, t)$. The two curves $s \mapsto \xi(\tilde{q}, s)$ and $s \mapsto \xi(q, t+s)$ are maximal integral curves of X originating at $\tilde{q}$. They must therefore coincide, and in particular be defined for the same time interval.

The first part is a reformulation of the Local Existence Theorem IV.36 which for any point a gives a uniform interval of existence of the solutions for all points sufficiently close to a.   □

Corollary IV.46. *If the life set Δ_X contains a neighbourhood $U \times]-\epsilon, \epsilon[$ of $U \times \{0\} \subset U \times \mathbb{R}$, for all $t \in] - \epsilon, \epsilon[$, the map $\xi_t : q \mapsto \xi(q, t)$ is a diffeomorphism of U.*

If the vector field X is complete, the map $t \mapsto \xi_t$ satisfies, for all $t \in \mathbb{R}$, the characteristic equation

$$\frac{d}{dt}\xi_t(q) = X(\xi_t(q)) \ .$$

Moreover, $(\xi_t)_{t \in \mathbb{R}}$ is a one-parameter group of diffeomorphisms of U, called the flow of X (i.e., for all t_1 and t_2, we have $\xi_{t_1} \circ \xi_{t_2} = \xi_{t_1 + t_2}$).

Proof. Under the stated hypotheses, the maps ξ_t are well-defined and differentiable by Corollary III.23. Moreover, it follows from Theorem IV.45 that ξ_{-t} is the inverse of ξ_t, so the maps ξ_t are diffeomorphisms of U.

When X is complete, $\Delta_X = U \times \mathbb{R}$ and there is no restriction on t. The homomorphism property is also contained in Theorem IV.45.   □

Remark IV.47. Corollary IV.46 establishes the fundamental relation which exists between vector fields and dynamics, i.e., one-parameter groups of diffeomorphisms of space. This makes it of deep philosophical importance, in view of the central position of dynamics in the natural sciences.

(Trivial) Example IV.48. It is particularly easy to determine the flow of a *constant* vector field. In fact, if X_v is such that $X_v(q) = v$ for every point q in E, then the flow $(\xi_t)_{t \in \mathbb{R}}$ of X_v is simply $\xi_t(q) = q + t\,v$, and thus consists of *translations* by the vector $t\,v$.

The correspondence evoked above associates a *uniform rectilinear motion* to a *constant vector field*.

IV.49. In another direction, to any one-parameter group of diffeomorphisms $(\psi_t)_{t \in \mathbb{R}}$, one can associate its *infinitesimal generator* F defined by $F = (d\psi_t / dt)_{|t=0}$, from which one sees immediately that it is the vector field whose value at the point a is the velocity-vector at $t = 0$ of the curve $t \mapsto \psi_t(a)$. (Exercise: Prove it.)

C. Some Examples of Vector Fields

In this section we develop several important examples of vector fields defined on open subsets of Banach spaces, such as *linear and affine vector fields*, *gradient vector fields* and *Hamiltonian vector fields*.

IV.50. We begin by considering the simplest vector fields which one can define on a vector space, namely the *linear* and *affine* vector fields.

They have a number of advantages, notably that they are defined on the whole space and, at least in the case of finite-dimensional vector spaces, their flow can be described *explicitly*.

Definition IV.51. A vector field X defined on a Banach space E is said to be *linear* if there exists a continuous linear map $A \in L(E, E)$ such that $X(v) = A(v)$ (where $A(v)$ denotes the value of the linear map A at the vector v).

A vector field X defined on a Banach space E is said to be *affine* if there exists a continuous linear map $A \in L(E, E)$ and a vector $w \in E$ such that $X(v) = A(v) + w$.

Example IV.52. Among the linear vector fields, there is one which plays a special role because of the nature of its flow. This is the *Liouville*[72] *vector field*, which is traditionally denoted by D and which is defined on the whole vector space E by $D(v) = v$ for $v \in E$. The linear map associated to it is the identity map of E.

Its flow is formed by the *dilations* with centre the zero vector in E. In fact, the curve $t \mapsto e^t\,v$ passes through v at $t = 0$ and its velocity-vector at time t at the point $e^t\,v$ is the vector $e^t\,v = D(e^t\,v)$.

If E is a vector space of finite dimension n of which (x^i) is a system of linear coordinates associated to a basis (e_i), then D takes the form $D = \sum_i x^i\,(\partial / \partial x^i)$.

---

[72] The French mathematician Joseph Liouville (1809–1882) has a considerable body of work. He contributed notably to the theory of functions of a complex variable, to the theory of numbers and to the theory of algebraic equations. He was also interested in the Calculus of Variations, notably the theory of geodesics.

One can also find its flow (h_t) in this formalism by solving the ordinary differential equation

$$\frac{dx^i}{dt} = x^i(t)$$

which gives of course the dilations $h_t(x^i) = e^t x^i$.

IV.53. The study of an affine vector field $v \mapsto X(v) = A(v) + w$ reduces, if w belongs to the image of A, to that of a linear vector field by changing the origin of the vector space.

IV.54. We shall now consider how to describe the flow of a linear vector field.

Proposition IV.55. *Let X be a linear vector field on a Banach space E with associated linear map A. The flow $(\xi_t)_{t \in \mathbb{R}}$ of X is given by $\xi_t(v) = (\exp tA)(v)$ where $\exp$ denotes the exponential of continuous linear maps defined by $\exp A = \sum_{i=0}^{\infty} (n!)^{-1} A^n$, in other words the solution of the differential equation $dv/dt = A(v(t))$ originating at the vector v_0 is the curve $t \mapsto \exp tA(v_0)$.*

Proof. We note first of all that, for any continuous linear map A, its exponential $\exp A = \sum_{i=0}^{\infty} (n!)^{-1} A^n$ is well-defined since the series is normally convergent and since a linear map is continuous precisely when its operator norm is finite.

Consequently, it is possible to differentiate term-by-term. Thus, we obtain the formula $d(\exp tA)/dt = A \circ \exp tA$ which gives the required result by evaluating on a vector v_0 taken as the initial condition. $\qquad \square$

IV.56. In the case where E is finite-dimensional, it is possible to calculate the flow explicitly by making a suitable choice of basis of E in which the linear map A has a particularly simple form. In view of the way the exponential is calculated, it is appropriate to take A to be *triangular* (upper, for example), which is always possible. In fact, this normal form can be made more canonical by taking the restriction of A to each eigenspace to have the form of a matrix whose only non-zero off-diagonal entries are ones on the diagonal next to the main diagonal.[73]

One can then give the solution quite explicitly. (Exercise: Do it!)

Solved Exercise IV.57. In the plane $\mathbb{R}^2$ with linear coordinates (x, y), determine the flow of the vector field X given by $X(x, y) = (\sqrt{3}x + y)\, \partial/\partial x + (\sqrt{3}\, y - x)\, \partial/\partial y$.

The matrix A of the vector field X in the natural basis of $\mathbb{R}^2$ is thus

$$A = \begin{pmatrix} \sqrt{3} & 1 \\ -1 & \sqrt{3} \end{pmatrix} = \sqrt{3} \begin{pmatrix} 1 & 0 \\ 0 & 1 \end{pmatrix} + \begin{pmatrix} 0 & 1 \\ -1 & 0 \end{pmatrix},$$

of which the exponential is

$$e^{tA} = e^{t\sqrt{3}} \begin{pmatrix} \cos t & \sin t \\ -\sin t & \cos t \end{pmatrix}$$

---

[73] This normal form is often called the *Jordan canonical form.*

(one uses here the fact that the identity matrix commutes with all matrices and hence $e^{t(I+B)} = e^t e^{tB}$). The flow (ξ_t) is thus expressed analytically by $\xi_t(x, y) = (e^{t\sqrt{3}}(x \cos t + y \sin t), e^{t\sqrt{3}}(-x \sin t + y \cos t))$. It thus consists of similarity transformations. One verifies that it is well-defined for all $t \in \mathbb{R}$ (this is what we would expect since the vector field X is linear, and hence necessarily complete).

The orbits of the vector field consist of the point $(0, 0)$ and the logarithmic spirals with polar equations $\rho(t) = \rho_0 e^{t\sqrt{3}}$ and $\theta(t) = \theta_0 - t$ (where (ρ, θ) denotes the polar coordinates of the point with Cartesian coordinates (x, y) in the plane $\mathbb{R}^2$). The vector field X has a single closed orbit, the origin.

IV.58. We now consider other special vector fields defined on a vector space E (or only on an open subset of this space). They are associated to special functions defined on the space, but their definition requires that E has additional structure. In the two cases which we consider, these structures are *bilinear forms*.

IV.59. We begin by assuming that E is a Hilbert space, i.e., that it has a *scalar product g* which we assume is given.

As we have already remarked in Sect. III.52, to any differentiable scalar function ϕ it is possible to associate its *gradient*, which is the vector field whose image under the map from E to E' associated to g is $d\phi$. By the definition of the scalar product, we know that $\partial_{\text{Grad}\phi}\phi = g(\text{Grad }\phi, \text{Grad }\phi) \geq 0$ and zero only if the differential of ϕ vanishes. The value of the function ϕ can thus only increase along the integral curves of Grad ϕ.

Let us analyse in more detail an interesting special case: if E is a vector space of dimension n with a basis (e_i) which defines linear coordinates (x^i) on E, and if the function ϕ whose gradient we are studying is a quadratic function which at the point $q \in E$ with coordinates $(q^1, \cdots, q^n)$ is equal to $\phi(q) = \sum_{i,j=1}^{n} a_{ij} q^i q^j$, then we have Grad $\phi = \sum_{i,j=1}^{n} a_{ij} x^j \partial/\partial x^i$, i.e., the vector field Grad ϕ is linear.

IV.60. In fact, it turns out to be more important for applications to mechanics or physics to consider the case of a vector space E provided with a symplectic form ω. We have seen in Sect. II.60 that the prototype for such spaces is provided by the product of a vector space F and its dual F'.

Definition IV.61. Let ϕ be a differentiable scalar function defined on a symplectic vector space (E, ω). If E is infinite-dimensional, we assume further that the form ω defines a duality between E and E'. The vector field whose image under the duality is $d\phi$ is called the *Hamiltonian vector field* associated to ϕ and is *denoted* by Ω_ϕ.

The differential equation which defines Ω_ϕ is called the *system of Hamilton's equations* associated to ϕ (which is then called the *Hamiltonian function*).

IV.62. We are going to consider in detail the case of $\mathbb{R}^n \times (\mathbb{R}^n)^*$ provided with the standard symplectic form ω_0. We denote by (x^i) the linear coordinates on $\mathbb{R}^n$ and by (X_i) those on $\mathbb{R}^{n*}$ induced by the dual basis of the canonical basis[74] (i.e., for $\lambda \in \mathbb{R}^{n*}$, we have $X_i(\lambda) = \lambda_i$ if $\lambda = \sum_{i=1}^{n} \lambda_i \epsilon^i$).

---

[74] The reasons for these notations will not appear until Chap. X.

Let $\phi : \mathbb{R}^n \times (\mathbb{R}^n)^* \to \mathbb{R}$ be a differentiable scalar function (of the $2n$ variables (x^i, X_i)). The Hamiltonian vector field Ω_ϕ is given by

$$\Omega_\phi = \sum_{i=1}^{n} \left(\frac{\partial \phi}{\partial X_i} \frac{\partial}{\partial x^i} - \frac{\partial \phi}{\partial x^i} \frac{\partial}{\partial X_i} \right)$$

as one easily verifies by using the formula giving ω_0 in the canonical basis (e_i, ϵ^i) of $\mathbb{R}^n \times (\mathbb{R}^n)^*$.

The differential equation on $\mathbb{R}^n \times \mathbb{R}^{n*}$ which defines the vector field Ω_ϕ is traditionally expressed as a differential system. An integral curve η of the vector field Ω_ϕ thus satisfies the equation $d\eta/dt = \Omega_\phi(\eta(t))$, in other words it is the solution of *Hamilton's differential system*

(IV.63)
$$\begin{cases} \dfrac{dx^i}{dt} = \dfrac{\partial \phi}{\partial X_i} \\ \dfrac{dX_i}{dt} = -\dfrac{\partial \phi}{\partial x^i} \end{cases} , \quad 1 \le i \le n .$$

It is generally in this form that Hamilton's equations are written, but opposite sign conventions can also be found in the literature.

IV.64. Differential systems of this type have numerous interesting properties, in particular conservation properties which we consider in Sect. D.

The motion of numerous physical and mechanical systems is determined by solving such systems.

Example IV.65. Consider the function ϕ (suggested by physics) on $\mathbb{R}^n \times (\mathbb{R}^n)^*$ defined by $\phi(x^i, X_i) = (1/2m) \sum_{i=1}^{n} (X_i)^2 + V(x^i)$ (where m is a positive real number and V a differentiable function). The function ϕ represents *the energy of a particle of mass m in a potential V*. Hamilton's system of equations for this Hamiltonian is

$$\begin{cases} \dfrac{dx^i}{dt} = \dfrac{1}{m} X_i \\ \dfrac{dX_i}{dt} = -\dfrac{\partial V}{\partial x^i} \end{cases} .$$

After eliminating the momentum variables X_i, we recover equations of motion for the particle given by Newton[75], namely the classical system $m \, d^2 x^i / dt^2 = -\partial V/\partial x^i$.

In this special case, we see at work one of the virtues of the Hamiltonian approach (whose systematic development is the subject of Chap. X), namely to reduce to first order the system of equations to be solved.

---

[75] Isaac Newton (1642–1727), English physicist and mathematician, proposed in 1687 (three centuries ago!) a very successful model of mechanics and in particular of universal gravitation. Simultaneously with Leibniz, he also introduced differential calculus by a method which was actually called the *calculus of fluxions*. His "*Philosophiae naturalis principia mathematica*", divided into several books, is one of the most outstanding works in the history of humanity.

D. Poisson Brackets and Conserved Quantities

We introduce the *Poisson bracket* operation on the algebra of functions defined on a symplectic vector space, and we show the relation between this operation and the Heisenberg uncertainty relations. We deduce from it a number of conservation properties of Hamiltonian vector fields.

IV.66. There is a naturally defined bilinear map on the algebra of functions on a vector space E provided with a symplectic form ω. This map is of particular interest in the case where the vector space E is the product of a vector space F with its dual space, provided with the symplectic form ω_F.

Definition-Proposition IV.67. *Let ϕ and ζ be two differentiable numerical functions defined on a vector space E provided with a symplectic form ω defining a duality between E and E'. Their* Poisson bracket, *denoted by $\{\phi, \zeta\}$, is the numerical function $-\partial_{\Omega_\phi}\zeta$, which can also be written $\partial_{\Omega_\zeta}\phi$.*

If $E = F \times F^*$ for some space F of dimension n and if ω is the natural symplectic form ω_F defined on $F \times F^*$, then, in terms of the linear coordinates x^i on F defining the dual coordinates X_i on F^*, we have

$$(IV.68) \qquad \{\phi, \zeta\} = \sum_{i=1}^{n} \left(\frac{\partial \phi}{\partial x^i} \frac{\partial \zeta}{\partial X_i} - \frac{\partial \phi}{\partial X_i} \frac{\partial \zeta}{\partial x^i} \right).$$

The Poisson bracket is a skew-symmetric bilinear map.

Proof. Formula (IV.68) is obtained directly from the defining relation of Ω_ϕ and the particularly simple expression for ω_F in the basis of differentials (dx^i, dX_i). The bilinearity of the Poisson bracket also follows from this.

The alternating character of the Poisson bracket is an immediate consequence of

$$\partial_{\Omega_\phi}\zeta = \langle d\zeta, \Omega_\phi \rangle = \omega(\Omega_\phi, \Omega_\zeta) = -\omega(\Omega_\zeta, \Omega_\phi).$$ □

IV.69. We now develop several fundamental properties of the Poisson bracket which are of interest in the study of the evolution of systems.

Proposition IV.70. *Let ϕ be a differentiable numerical function defined on a symplectic vector space (E, ω). Along an integral curve $\tau \mapsto \eta(\tau)$ of the vector field Ω_H defined by a Hamiltonian H, ϕ satisfies the differential equation*

$$d(\phi \circ \eta)/d\tau = -\{H, \phi\} \circ \eta.$$

In particular, ϕ is a constant of the motion defined by the Hamiltonian H if and only if $\{H, \phi\} = 0$.

Proof. This is almost tautological. By the Chain Rule III.48, we have $d\phi/d\tau = \partial_{\Omega_H}\phi$, which by definition is $\{H, \phi\}$.

The second part is immediate. □

Example IV.71. It is interesting to evaluate the Poisson bracket of some functions defined on the model symplectic vector space, namely $\mathbb{R}^{2n}$ with its canonical coordinates (x^i, X_i).

In fact, it follows trivially from Formula (IV.66) that we have

$$\{x^i, x^j\} = 0 \,, \quad \{x^i, X_j\} = \delta_{ij} \,, \quad \{X_i, X_j\} = 0 \,.$$

These relations are a classical version of the *Heisenberg*[76] *uncertainty relations.* To see this, it suffices to interpret the functions X_i as the *momenta*, which is not very surprising if one recalls that, just as linear forms applied to vectors give numbers, so momenta applied to a velocity give an energy. This point is developed at length in Chapters X and XI.

In physics, one often states a *Correspondence Principle* which, *to any differentiable function on* $\mathbb{R}^3$ (a classical observable) *associates a Hermitian operator* acting on the Hilbert space of square-summable complex-valued functions defined on $\mathbb{R}^3$ (a quantum observable) and which *to the Poisson bracket of functions associates the commutator of operators* (up to a factor $h/(2\pi i)$ where h denotes the *Planck*[77] *constant*). Thus, the non-commutativity properties of the quantum observables are related to the differential behaviour of the classical observables to which they are associated.

To the classical observables x^i ($i = 1, 2, 3$), which are differentiable functions on $\mathbb{R}^3 \times (\mathbb{R}^3)^*$, are associated the operators (often denoted by $\hat{x}^i$) of multiplication by the function x^i in the Hilbert space of square-summable functions; to the observables X_j are associated the differential operators $i \, \partial/\partial x^j$. (The presence of the imaginary factor i is essential for this operator to be Hermitian.) The Correspondence Principle can be verified trivially on these particular observables.

Consider now the *components of angular momentum* in $\mathbb{R}^3$ which are given by the functions $\mu_1 = x^2 X_3 - x^3 X_2$, $\mu_2 = x^3 X_1 - x^1 X_3$ and $\mu_3 = x^1 X_2 - x^2 X_1$. We evaluate the Poisson bracket $\{\mu_1, \mu_2\}$. We find

$$(IV.72) \qquad \{\mu_1, \mu_2\} = \sum_{i=1,2,3} \left(\frac{\partial \mu_1}{\partial x^i} \frac{\partial \mu_2}{\partial X_i} - \frac{\partial \mu_1}{\partial X_i} \frac{\partial \mu_2}{\partial x^i} \right) = \mu_3 \,.$$

Similarly, we have $\{\mu_2, \mu_3\} = \mu_1$ and $\{\mu_3, \mu_1\} = \mu_2$.

The differential operators which are associated to μ_1, μ_2 and μ_3, respectively denoted by $\hat{\mu}_1$, $\hat{\mu}_2$ and $\hat{\mu}_3$, are then given by

---

[76] The German theoretical physicist Werner Karl Heisenberg (1901–1976) was one of the creators of quantum mechanics, as recognised by the citation of the Nobel Prize in physics he received in 1932. In 1925 his matrix formulation of the theory was indeed a turning point, and he published the *uncertainty principle* in 1927.

[77] Max Karl Ernst Ludwig Planck (1858–1947) is a German physicist with many major contributions to physics, in particular the introduction of energy quanta. An early achiever – at age 22 he already had successfully passed his Habilitation – it was in struggling with the black-body radiation that he came to the conclusion that energy must be quantised, opening the way to a new mechanics, quantum mechanics. In 1918 he received the Nobel Prize in physics for this.

$$\hat{\mu}_1 = i\,(x^2\,(\partial/\partial x^3) - x^3(\partial/\partial x^2))\,,$$
$$\hat{\mu}_2 = i\,(x^3\,(\partial/\partial x^1) - x^1\,(\partial/\partial x^3))\,,$$
$$\hat{\mu}_3 = i\,(x^1\,(\partial/\partial x^2) - x^2\,(\partial/\partial x^1))\,.$$

In this particular example, one easily verifies the correspondence between the Poisson brackets and the commutator of differential operators, stated as part of the Correspondence Principle.

Exercise IV.73. Show that the Poisson bracket of functions defined on $\mathbb{R}^n \times (\mathbb{R}^n)^*$ satisfies the *Jacobi identity*: for functions ϕ_1, ϕ_2 and ϕ_3,

$$\{\phi_1, \{\phi_2, \phi_3\}\} + \{\phi_2, \{\phi_3, \phi_1\}\} + \{\phi_3, \{\phi_1, \phi_2\}\} = 0\,.$$

Definition IV.74. Let X be a vector field defined on an open subset of a Banach space E. A differentiable numerical function ϕ is called a *first integral* of the vector field X if $\partial_X \phi = 0$, in other words if ϕ is constant along every integral curve of X.

Remark IV.75. i) A vector field does not always admit a non-trivial first integral (i.e., different from a constant function). To have interesting examples, it will be necessary to have available the notion of a vector field in spaces more general than open subsets of vector spaces, spaces which are introduced later in this course.

ii) In the study of dissipative systems, one is sometimes led to search for differentiable functions which behave quite differently from that of first integrals, namely we look for functions which are *strictly monotone along the integral curves*. Such functions are called *Lyapunov*[78] *functions* of the vector field and they play a very important role in the study of particular vector fields such as gradient vector fields (corresponding to some scalar product). In this case also, an arbitrary vector field in general has no Lyapunov function. This is the case for example for a vector field which admits a closed integral curve which does not reduce to a single point (the Lyapunov function would have to increase strictly along the integral curve, something impossible since it is closed).

IV.76. Among the vector fields which are of particular interest to us are the Hamiltonian vector fields defined in Sect. IV.61.

They have the property that they always have a first integral, as is shown in Theorem IV.77.

Theorem IV.77. *The Hamiltonian is constant along the integral curves of the Hamiltonian vector field which it defines.*

Proof. This is immediate. In fact, if ϕ is a differentiable function defined on a symplectic vector space (E, ω), along the integral curve γ of the Hamiltonian vector field Ω_ϕ, we have

$$\frac{d(\phi \circ \gamma)}{dt} = -\partial_{\Omega_\phi}\phi = -\langle d\phi, \Omega_\phi\rangle = -\omega(\Omega_\phi, \Omega_\phi) = 0\,.$$

$\square$

---

[78] The Russian mathematician Aleksandr Mikhailovich Lyapunov (1857–1918) is known for his development of the stability theory of dynamical systems. He also contributed to probability theory.

Remark IV.78. This property is sometimes called *the Law of conservation of energy*, whence the fact that Hamiltonian vector fields are often called *conservative*.

Proposition IV.79. *Let H be a differentiable numerical function defined on a symplectic vector space* (E, ω). *If* ϕ *and* ζ *are two first integrals of the Hamiltonian vector field* Ω_H, *then their Poisson bracket* $\{\phi, \zeta\}$ *is also a first integral of* Ω_H.

Proof. By Proposition IV.68, verifying that $\{\phi, \zeta\}$ is a first integral is equivalent to establishing that $\{H, \{\phi, \zeta\}\} = 0$. This fact is an immediate consequence of the Jacobi identity satisfied by the Poisson bracket (cf, IV.73). In fact, this identity asserts that

$$\{H, \{\phi, \zeta\}\} = \{\{H, \phi\}, \zeta\} + \{\phi, \{H, \zeta\}\}.$$

Now the two terms on the right-hand side of the preceding equation are zero by hypothesis.   □

Solved Exercise IV.80. Show that a Hamiltonian defined on $\mathbb{R}^3 \times (\mathbb{R}^3)^*$ for which two components of the angular momentum are first integrals of its Hamiltonian vector field Ω_H also commutes with the third.

In fact, if a Hamiltonian H defined on $\mathbb{R}^3 \times (\mathbb{R}^3)^*$ is such that Ω_H admits μ_1 and μ_2 as first integrals, then by Proposition IV.79, we have that $\{\mu_1, \mu_2\}$ is again a first integral of Ω_H, but we have remarked in IV.72 that $\{\mu_1, \mu_2\} = \mu_3$.

E. Historical Notes

IV.81. The local existence theorem for solutions of differential equations of class C^1 is due to Cauchy in the 1820s. This was one of the first examples of a rigorous proof using a method of approximation by auxiliary functions (piecewise-linear in fact).

It was in 1866 that Lipschitz gave the proof of this theorem in the case where the vector field is *Lipschitz*, again using a method of successive approximations.

IV.80. It was in his "*Mémoire sur la théorie des variations des éléments des planètes*", which Lagrange[79] published in 1808, that one finds the first discussion of the special geometry which today we call "*symplectic*". In this work, Lagrange gave a linearised version of the equations of motion for the orbital elements of a planet under the effect of a perturbation; by a change of coordinates, this can be reduced to the version of Hamilton's equation we have given. This underlines the very close relationship which exists between the developments contained in this chapter and mechanics, particularly celestial mechanics. Important developments in Hamilton's theory were obtained during the XIXth century, particularly by Jacobi, and we shall present their main themes in Chap. X.

---

[79] Joseph-Louis de Lagrange (1736–1813), astronomer and mathematician, has a considerable body of work which centres around the use of the differential calculus in the study of mechanical systems. He can be considered to be the creator, jointly with Euler, of the calculus of variations, as we shall see in Chapter IX where we present the equations which carry their names. He presided over the commission to establish a system of weights and measures required by the Constituent Assembly of 1790.

Chapter V
A New Generalisation of the Notion of a Space: Configuration Spaces

In this chapter we take up the study of *configuration spaces* which are the natural setting for modelling numerous mechanical and physical systems. They form the global counterpart of local coordinates and are sometimes called "curved spaces". This chapter aims to show that these objects *inevitably* appear whenever one is interested in the global properties of a model of a system. Our treatment does not pretend to be exhaustive, and we avoid giving the most general definitions.

In Sect. A we examine the conditions which systems of local coordinates should satisfy to be able to define the notions useful in modelling. Even though a number of modern theories in physics or mechanics require infinite-dimensional spaces, we restrict ourselves very quickly to finite-dimensional spaces (these generalisations are nevertheless sketched in Chapter VIII).

We develop in Sect. B the fundamental notion of a *differentiable map between configuration spaces*.

The first example, considered in Sect. C, is that of vector spaces where the configuration space point of view leads one to work in *curvilinear coordinates*. Various examples of geometric interest are discussed. In Sect. D we examine the next most fundamental examples, namely *the circles, spheres, tori and projective spaces*. In each case we consider the classical coordinates and also several physical and mechanical systems of which they are the configuration spaces.

Section E is reserved for the particular case of the *group of rotations* since it has a rich geometric structure, and also because it is the configuration space of the *mechanics of a rigid body* moving about a fixed point. The classical parametrisations of Euler[79] and Cayley[80]–Klein[81] allow one to describe a situation which is important both in mathematics and physics, where it is at the heart of the concept of *spin*.

---

[79] The Swiss mathematician Leonhard Euler (1707–1783) has a very considerable body of work which embraces all the mathematics of his time, as well as mechanics and astronomy. He was a (if not *the*) creator of the Calculus of Variations. He was perhaps the most prolific mathematician of all time, in spite of the fact that he ended his life blind. The publication of his collected works is still not complete, but already comprises more than 60 volumes.

[80] Arthur Cayley (1821–1895), British lawyer and mathematician, is the creator of the theory of matrices.

[81] The German mathematician Felix Christian Klein (1849–1925) dominated German mathematics of his time. His Erlangen programme, announced in 1872, in which he identified a geometry with the study of properties which are invariant under a group of automorphisms, sealed the marriage (which has lasted ever since) between geometry and group theory.

© The Author(s), under exclusive license to Springer Nature Switzerland AG 2022   101
J.-P. Bourguignon, *Variational Calculus*, Springer Monographs in Mathematics,
https://doi.org/10.1007/978-3-031-18307-2_5

A. Local Coordinates and Configuration Spaces

We define mathematical objects, configuration spaces, which can be used as *models*[82] of mechanical or physical systems.

V.1. We begin by discussing the modelling of a system. To describe such a system, one wants to carry out *measurements* on it, in other words, to each point of the space which forms the set of configurations of the system it should be possible to associate a collection of numbers, the *measurements of the observables*. The values of these parameters, which we suppose for the moment are finite in number, allow one to fix completely the state of the system (here, the term "state" should not be understood with its restricted meaning of spatial position). It is tempting to consider these quantities as *generalised coordinates*, and to say moreover that two points of the space are close when their coordinates are close.

The most obvious examples of such situations arise from mechanical systems such as the motion of a pendulum or of a system of rigid bodies, but they can arise just as well from the evolution of chemical systems where the parameters are rather the temperature and concentrations of various ions or molecules. We mention in particular two very different examples: that of the model of *space-time* used in physical theories; the space we exist in seems to satisfy the properties we have stated, but there is after all no reason that it should be *globally* as simple as a vector space,[83] a property which is notably at the heart of studies in general relativity; another, more surprising, example concerns the space of *colours* whose study is nowadays of importance in industry because of the appearance of numerous "colourable" composite materials and of the development of computer graphics. The manifold of objective physical colours has an infinity of dimensions but the eye retains only a "projection" of dimension 2. The basic colours in some given intensity can be mixed to give a new colour with a well-defined intensity. The intensities of a given colour are comparable with each other but those of two distinct colours are not. If we interpret the superposition of colours in terms of addition, we are reduced to vector geometry, but since only the ratio of the intensities composing a colour is relevant, we are in the realm of projective geometry. In fact, it is possible to characterise the space of colours as a part of the real projective plane, a space which we describe in V.85.

V.2. We are going to single out a first property of spaces which enables one to label the positions of a system: it should be *"numerisable"*, an expression which is probably dubious from a linguistic point of view; we shall use it temporarily to suggest that points can be labelled by collections of numbers without these numbers being necessarily meaningful parts of the data.

This leads us to give the following partial definition.

---

[82] The notion of a *model* deserves a more profound discussion than we have space for here. It is essential whenever one wishes to discuss the relation between mathematics and other sciences. The interested reader is referred to Chapter I of (Gallavotti 1983) and to the references therein.

[83] Newton already gave arguments, based on energy considerations, to show that this simplistic form of space is not acceptable.

First Part of Definition V.3. *A* configuration space *is, first of all, a Hausdorff topological space, every point of which has a neighbourhood homeomorphic to an open ball of a vector space.*

V.4. The hypothesis in Definition V.3 that the space is Hausdorff can be justified by saying that, in a configuration space, two distinct positions of a system should be perturbable by an arbitrarily small amount without meeting.

We remain deliberately vague about the norm which is taken on the numerical space, since in a concrete situation there is no reason for distinct parameters to correspond to commensurable quantities; in fact, these parameters usually have a *dimension* in the sense usual in physics. Thus, we cannot single out *a priori* any particular norm, such as the Euclidean norm for example.

This is not important since, as we have seen in Exercise I.34, in a finite-dimensional space, to which we restrict ourselves for the moment, the balls for any two norms are homeomorphic.

Remark V.5. It is clear that to check the two rather restrictive properties of configuration spaces we have stated, it is enough to check them in *a neighbourhood of each point.* They are thus the kind of properties we have previously called *local.*

Spaces satisfying the partial Definition V.3 are not traditionally studied by mathematicians. They generally consider spaces which satisfy an additional hypothesis, namely that *"the topological space has a countable number of open sets"* to avoid *problems at infinity.* These more restricted objects are called *topological manifolds.*

Exercise V.6. Consider the topological space M formed by two intersecting lines in the plane, with the induced topology. Is it possible to make M a configuration space?

> *From now on, we reserve the name* "local coordinate system" *for a map which defines a homeomorphism from an open subset of a configuration space onto an open subset of a numerical space homeomorphic to a ball.*

Project V.7. Give an example of a "locally numerisable" space (i.e., one which satisfies the second condition of Definition V.3, that *"every point has a neighbourhood homeomorphic to a ball in a vector space"*) which is not Hausdorff.

V.8. We shall say a few words about *infinite-dimensional* configuration spaces before taking up their study (briefly!) in Chapter VII.

In most current models in physics or mechanics, the objects appearing are *fields*, i.e., of *scalar* or *vector* quantities whose value depends on the point of space-time. Thus, they involve spaces of numerical functions defined on domains in space-time, one of the families of infinite-dimensional spaces on which we have concentrated our attention in the first two chapters.

In more elaborate models (which have actually appeared with great success in theoretical physics in the guise of *gauge theories*), space-time itself can be taken as a finite-dimensional configuration space which needs not be an open subset of a vector space. The configuration spaces which one must then use have a more complex geometric structure which we cannot go into in the context of an introductory course.[84]

V.9. XXth century developments in theoretical physics use extensively the concepts of *fibre space, connection, curvature* and *characteristic classes* elaborated by Christoffel[85], Riemann, Ricci[86], É. Cartan[87], Whitney[88] Chern[89], and many others. We shall say a little about fibre spaces in IX.10.

V.10. It is natural to suppose that *in a configuration space no point is distinguished a priori*. This leads us to assume that the space has a certain *homogeneity*.

In fact, we shall see that the configuration spaces used as models can often be identified with groups (in Sect. E, we treat the case of the orthogonal group as the configuration space of a rigid body moving about a fixed point), or with spaces in which groups can move any point to any other point (one says that these are *homogeneous spaces*, and that the action of the group is *transitive*, see Chapter XI).

---

[84] Note, however, that the objects which theoretical physicists use turn out to be precisely those which geometers have patiently assembled during the XXth century, and which are regarded as the basic concepts of *modern differential geometry*, a remark which is not without interest from the point of view of epistemology.

[85] Elwin Bruno Christoffel (1829–1900), a German mathematician and physicist, contributed to the development of Riemannian geometry through tensor calculus.

[86] The Italian mathematician Gregorio Ricci-Curbastro (1853–1925) developed systematically the tensor calculus and introduced the concept of a covariant derivative, decisive for the success of Riemannian geometry. It was in 1904, i.e. fifty years after Riemann's visionary lecture, that Ricci-Curbastro defined the concept of Ricci curvature which gives the left-hand side of the Einstein equations of general relativity and has proved in recent years a key tool to deform Riemannian metrics.

[87] The French mathematician Élie Joseph Cartan (1869–1951) created many concepts in modern differential geometry, such as that of a *spinor* and that of a *fibre space* (the *"auxiliary variables"* in his terminology). He also contributed to purely *algebraic* topics developing, for example, the representation theory of semi-simple Lie groups, to *analytic* topics (by giving substantial results on the solution of non-linear differential systems), and directly to *geometry* (by establishing Riemannian geometry as a central discipline in geometry). He transformed the *method of moving frames* and the *exterior differential calculus* into powerful geometric tools. His son Henri (hence our keeping his initial when we mention É. Cartan) made very important contributions to algebraic topology and analytic geometry. Moreover, H. Cartan played an important role in the moulding of mathematicians at the École Normale Supérieure.

[88] Hassler Whitney (1907–1989) was one of the founders of the theory of configuration spaces. He also made many contributions to the study of singularities of differentiable maps.

[89] The Chinese-American mathematician Chern Shiing Shen (1911–2004) made fundamental contributions to differential geometry and topology. This makes him one of the most influential mathematician of the XXth century. He received the Wolf Prize in 1883 and the inaugural Shaw Prize in 2002. After graduating from Nankai University (to which he returned after his retirement from the University of Berkeley in the 1990s), he briefly joined Tsinghua University before completing his mathematical education in Hamburg and Paris just before World War II. There, he developed a remarkable understanding and familiarity with the work of É. Cartan. It was during a stay at the Institute for Advanced Study in Princeton (1943–1945) that he developed the theory of *Chern classes*, which continues to strongly impact geometry, topology and theoretical physics. His dedication to developing trustworthy links between the Western and Chinese communities has been decisive.

V.11. We now move on to the next stage. To take account of the evolution of the system, one must have available a means of differentiating, which allows one to calculate the *infinitesimal variations* of certain quantities. To construct this extension of differential calculus, we are forced to introduce a new kind of structure on configuration spaces.

Take, in fact, a local coordinate system $x = (x^1, \cdots, x^n)$ which parametrises a neighbourhood U of a point a of a configuration space. Consider an observable φ defined in a neighbourhood of the position a; let φ^x be its expression in the local coordinates (x^i) defined by $\varphi^x = \varphi \circ x^{-1}$, which is assumed to be differentiable. If we have another local parametrisation $y = (y^j)$ which is also defined in a neighbourhood V of a, we have assumed that it defines a homeomorphism from an open subset V^y of a numerical space onto V and so the transformation $\tau_y^x = y \circ x^{-1}|_{x(W)}$ which passes from the coordinates $x = (x^i)$ to the new coordinates $y = (y^j)$ on their common domain of definition W is by composition also a homeomorphism. (Note that $\tau_x^y = \left(\tau_y^x\right)^{-1}$.)

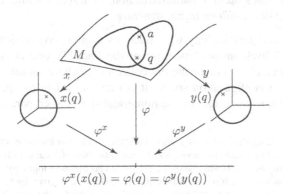

$$\varphi^x(x(q)) = \varphi(q) = \varphi^y(y(q))$$

Fig. V.12 An observable φ expressed in terms of two local coordinate systems x and y

But this *is not sufficient* to ensure that the function $\varphi^y = \varphi^x \circ \tau_x^y$ representing the observable of interest in the new coordinates is differentiable. To ensure this, it is on the other hand enough that *the homeomorphism τ_y^x between open domains of numerical spaces is a local diffeomorphism*, since, by virtue of the Chain Rule III.46, we then have $d(\varphi^y) = d(\varphi^x) \circ T(\tau_x^y)$ and the composite map is then differentiable.

Second Part of Definition V.13. *By a* configuration space *we understand a topological space which, in addition to satisfying the properties stated in V.3, has a coherent* system of local coordinates, *i.e., such that the change of coordinate maps which allow one to pass from one to the other are local diffeomorphisms.*

The domains of definition of these parametrisations are called admissible, *which means that they should form a covering of the configuration space.*

Remark V.14. We emphasise the difference between Definition V.3, which imposes conditions on the space, and Definition V.13, which imposes conditions on an additional structure, of which we are able to make a choice.

Note similarly that, *if the underlying topological space is compact*, it follows from Definition I.46 of a compact space that it is possible to cover the space completely with a finite number of open subsets which are domains of definition of parametrisations, and thus to make it a configuration space the transition maps *need only satisfy a finite number of compatibility conditions*.

Proposition V.15. *An open subset of a configuration space is itself a configuration space for the restrictions of the admissible parametrisations.*

Proof. This follows from the local character of the conditions which are imposed on the parametrisations and from the fact that the restriction of a differentiable map to an open subset is itself differentiable by III.23.   □

V.16. The object we have defined is usually called a *differentiable manifold* rather than a configuration space by mathematicians. They would emphasise this by saying that giving a coherent system of parametrisations on a space defines a structure of a particular kind, called a *differentiable structure*.

V.17. In the classical terminology which derives from a geographical analogy (the word "geometry" itself means "measure of the Earth"!), one often calls a local coordinate system a *chart* and a coherent collection of charts an *atlas*.

At a point of a configuration space, it is often useful to use parametrisations mapping this point to the origin of the numerical space. Such parametrisations are called *centred*.

V.18. To avoid misleading the reader, we mention that the problem of knowing whether, on a given numerisable space, it is possible to choose a coherent collection of local coordinate systems, is a serious one. For a long time, this was one of the great open problems of topology. The first example of a topological manifold which has no differentiable structure was given by Kervaire[90] in 1960. The general problem, known as the *"Hauptvermutung"*, was solved in the early 70s by Kirby[91] and Siebenmann[92].

This aspect of the subject is usually hidden when one parametrises a physical or mechanical system. Nevertheless, it is not superfluous, since recent (and at first sight surprising) results in *differential topology*, which we mention briefly in Sections C and D, show that *even on such simple spaces as vector spaces and spheres, there may be many mutually incompatible choices of collections of coherent systems of local coordinates.* The subtlety lies in the fact that on such a "numerisable" topological space there may be many (incompatible) ways of doing differential calculus, and hence one has the problem of *classifying* the differentiable structures which can be defined on a given topological space, a problem which is still largely open.

---

[90] Michel André Kervaire (1927–2007), a French mathematician who worked in Switzerland, made key contributions to the field of topology, and in particular to differential topology, and also to knot theory.

[91] The American mathematician Robion Cromwell Kirby, born in 1938, is a specialist in low dimensional topology. He has been very involved in developing open access to scientific literature.

[92] Laurent Carl Siebenmann, born in 1939, is a Canadian mathematician who, after graduating from Princeton University under Milnor, moved to the Université Paris-Sud, where he did most of his research. He is a specialist in topology at large.

V.19. In the Second Part of Definition V.13, we have not been precise about the degree of differentiability of the maps which we are considering. This was deliberate, since this simplifies the notation without any serious risk of confusion, the only precaution needed being of course to *keep the same meaning of the word "differentiable" throughout an argument*, whether it be C^1, C^k, $2 \le k < \infty$, or C^∞.

An important result of differential topology asserts in fact that there is no substantial difference, from the mathematical point of view, between configuration spaces in which the changes of local coordinates are of class C^1 or of class C^∞. We shall assume this as a fact.

> *Except where stated otherwise, we shall assume from now on that the maps considered are infinitely differentiable and that the notation M for a configuration space contains implicitly all the objects needed to define it.*

Proposition V.20. *On a connected component (cf. I.6) of a configuration space M, all the parametrisations have the same number of variables, called the* dimension *of M (sometimes one also speaks of the* number of degrees of freedom *of M).*

Proof. Let a be a point of M and suppose that a parametrisation of a neighbourhood of a has m variables. Consider the set A of points of M having the same property.

A chart defined on a neighbourhood of a point being also a chart for all the points of the neighbourhood, A is open.

But A is also closed, for if b is a point of $\overline{A}$, every neighbourhood of b necessarily meets A. Among these neighbourhoods, there is at least one which is the domain of definition of a chart defined at b. This chart is also a system of local coordinates at any point of A. It thus has m variables, for if not, the change of variables transformation between this parametrisation and the parametrisation which makes it a point of A would be a local diffeomorphism between open subsets of numerical spaces of different dimensions, which is impossible by Corollary III.87.

The space A is thus a union of connected components of M. □

V.21. There is a completely elementary operation on configuration spaces, which consists in taking the *product* of two configuration spaces M_1 and M_2. If one takes the underlying topological space to be the product $M_1 \times M_2$, and the family of charts to be the products of the charts of the factors, it is easy to see that we have indeed defined a configuration space, called the *product* of the configuration spaces M_1 and M_2 and denoted of course by $M = M_1 \times M_2$.

B. Differentiable Maps in Local Coordinates

In this section we explain how differential calculus is developed in configuration spaces. We stress above all the use of local coordinates, for which we can state the following basic principle.[93]

---

[93] The word *principle* has tended to disappear from the vocabulary of mathematicians since its meaning is ambiguous, oscillating between "theorem" and "guiding idea". Nevertheless, we use it here to mean that the statement of the Basic Principle V.22 is true, but that its proof is scattered throughout the chapter.

Basic Principle V.22. *In a configuration space, all the formulas and all the statements of differential calculus are expressed in local coordinates in the same way as they are expressed in linear coordinates on vector spaces.*

V.23. We begin with the notion of a differentiable map. To define such a map in Chapter III, we made use of the addition of vectors (or of translations, which amounts to the same thing), a variation of a point being viewed as the addition of a vector whose norm tends to zero. There being no such notion available in a configuration space, we can no longer proceed in this way.

We have recourse to a local coordinate system which, and here lies the difference compared to the way we proceeded in Chapter III, allows us to postpone the definition of the tangent linear map of a differentiable map (which we give in Chapter VI).

Definition-Proposition V.24. *Let M and N be two configuration spaces of dimensions m and n respectively, and let f be a map defined on an open subset U of M and taking its values in N.*

The map f is said to be differentiable *at $a \in U$ if there exist two local charts $x = (x^1, \cdots, x^m)$ at a and $y = (y^1, \cdots, y^n)$ at $b = f(a)$ such that the local expression $f_y^x = y \circ f \circ x^{-1}$ defined on a neighbourhood of $(a^1, \cdots, a^n) \in \mathbb{R}^m$ and with values in $\mathbb{R}^n$ is differentiable at $(a^1, \cdots, a^n)$.*

The local expression of f in any local coordinate system belonging to the collection which defines the configuration space structure is then differentiable at the image point of a.

Proof. This is a trivial consequence of the Chain Rule III.107. It is only writing it down which is (a little) unpleasant.

Let x' and y' be two other local coordinate systems centred at a and at $b = f(a)$ belonging to the families defining M and N respectively. By definition, the change of chart transformations $\tau_{x'}^x$ and $\tau_{y'}^y$ are local diffeomorphisms in neighbourhoods of a and b. Hence, the local expression $f_{y'}^{x'} = y' \circ f \circ (x')^{-1}$ of f in the new systems of local coordinates x' and y' can be written as the composite

$$f_{y'}^{x'} = \tau_{y'}^y \circ f_y^x \circ (\tau_{x'}^x)^{-1},$$

of maps which, by hypothesis, are all differentiable in neighbourhoods of the appropriate points. □

Remark V.25. The notation f_y^x is very useful to indicate the basic reasoning which implies the passage from one system of coordinates to another (and this is why we are going to make use of it). In practice, however, it presents certain dangers since it is not common to use the notations x and y to denote whole local coordinate systems (these letters rather suggest the names of variables). Moreover, one can be tempted to think of the letters x and y as indices belonging to a finite set of values, which is of course not the case here.

V.26. One of course says that *a map f, defined on an open subset U of a configuration space and taking values in another configuration space, is differentiable on U if it is differentiable at every point of U*.

Corollary V.27. *The composite of two differentiable maps between configuration spaces is differentiable.*

Proof. It suffices to take local parametrisations at the points under consideration, to note that the local expression of a composite of maps is the composite of their local expressions, and to use the Chain Rule III.46. □

Corollary V.28. *Let f be a differentiable map defined on a product of configuration spaces. The restriction of f to each partial product of the configuration spaces is differentiable.*

Proof. The statement follows immediately from the definition of a differentiable map expressed in the product parametrisations and from Theorem III.23 on the differentiability of the restriction to a subspace of a differentiable map in the vector space case. □

V.29. When applied to functions with values in $\mathbb{R}^n$ (which we call *numerical* functions), Definition-Proposition V.24 allows one to characterise, in the vector space of numerical functions, the subset formed by the differentiable numerical functions, for which we have the following proposition.

Proposition V.30. *The differentiable numerical functions defined on a configuration space M form a vector subspace of the vector space of numerical functions on M. The differentiable scalar functions form a subalgebra, denoted by $\mathcal{F}M$, of the algebra of the scalar functions defined on M which we call the* algebra of observables.

Proof. This is completely elementary. It is enough to verify that the space of differentiable numerical functions is stable under linear combinations and the space of differentiable scalar functions under multiplication. As this is a matter of checking the properties in a neighbourhood of each point, it suffices to consider them in a local coordinate system where they are well-known. □

From now on, we reserve the word "observable" to mean
"a scalar function which is differentiable on its domain of definition".

V.31. Despite its simplicity, Proposition V.30 opens up the possibility of defining the notion of a differentiable manifold in the following way.

To give a differentiable manifold structure on a topological manifold M, it suffices, in fact, to give a *sheaf* of algebras of numerical functions, which means that to any open subset U of M is associated a subalgebra of the numerical functions defined on U and that these subalgebras satisfy various compatibility conditions under restriction and extension. For more details we refer to (Warner 1983). Note that this definition has the advantage of showing clearly that to give this supplementary structure involves making a choice, an idea which we have already stressed before.

The notion of a sheaf is of central importance in modern developments in *algebraic geometry* and in *analytic geometry*.

V.32. There is another extension of the approach via algebras of observables which is the focus of very active research today. The way we have defined them, these algebras are *commutative* since multiplication of scalars is so. In the programme he stated in his inaugural lecture at the Collège de France in 1985, Alain Connes[94] has been concerned with creating a *non-commutative differential geometry* by considering algebras built out of observables which need not commute.

---

[94] Alain Connes, born in 1947, is a French mathematician, whose work in operator algebras led the way to the establishment of *non-commutative geometry*. For his achievements he received the Fields Medal in 1982. Besides his Chair entitled "Analyse et Géométrie" at the Collège de France, he also held the Léon Motchane Chair at the Institut des Hautes Études Scientifiques (IHÉS) until his retirement in 2017. He received the Gold Medal of CNRS in 2004.

This research intersects the concerns of physicists who, because of the uncertainty relations of quantum mechanics, are also interested in the study of such objects. The complete development of this subject involves very diverse branches of mathematics: algebra and geometry of course, but also analysis and, for certain aspects, probability.

V.33. One problem which naturally suggests itself is: can one adjoin a local coordinate system to a given collection without destroying its coherence?[95]

The following criterion answers this question.

Proposition V.34. *Let M be a configuration space of dimension m. A differentiable map f defined on an open subset U of M with values in $\mathbb{R}^m$ can be adjoined as an admissible chart to M if f is a homeomorphism onto a ball and if, for any point a of U, there exists a local coordinate system x defined at a such that $T_a(f^x)$ is invertible ($f^x = f \circ (x^{-1})$ denotes of course the local expression of f in this chart).*

Proof. Thanks to the Local Inversion Theorem III.70, the condition of invertibility of the differential $T_a(f^x)$ ensures that this map is a local diffeomorphism on a neighbourhood of $x(a)$.

Thus, the only thing left to prove is that the compatibility of the map f with an admissible chart at each point of its domain of definition suffices to ensure its compatibility with every other admissible chart. But the verification of this fact is identical to that which we gave in the proof of Definition-Proposition V.24 which guaranteed that the notion of a differentiable map between configuration spaces is well-defined.   □

V.35. To avoid having to complete the family of charts, one often appeals to the collection of *all* admissible charts which is possible to define on a given configuration space. To verify that it forms a new admissible system (often called a *complete atlas*), one has to appeal to Zorn's[96] lemma.

V.36. At this point, we can give a meaning to the notion of a *local diffeomorphism between configuration spaces* by saying that such a map f should have a local expression f_y^x, in every local coordinate system x in the source space meeting its domain of definition and every system y in the target space, which is itself a local diffeomorphism. Note that this *does not imply* that f is a global diffeomorphism, in particular that it is a bijection. We give two important examples related to this issue in Sect D and Sect E.

Note that, with this definition, two configuration spaces of the same dimension are *always* locally diffeomorphic to each other since among the admissible charts of such a space are by definition diffeomorphisms of an open subset of the space with a ball in $\mathbb{R}^n$.

V.37. In our treatment of modelling, there are two situations which are frequently encountered in concrete problems which we have so far neglected: these are the cases involving "*boundaries*" and the cases involving "*constraints*".

---

[95] We have seen that this coherence was indispensable for giving meaning to the notion of a differentiable map.

[96] Max A. Zorn (1906–1993) was interested in the properties of cardinals, and also in differential calculus.

We say that the configuration space of a system has a *boundary* if the generalised coordinates are restricted so as not to exceed certain values, *although they may attain these values*. For example, this is the case for a particle in a box, which is bound to move in a space that has *boundaries*, "*corners*" and, if there are more than three dimensions, "*hypercorners*".

Unfortunately, it is difficult to give an elementary mathematical treatment of such situations, since the notion of a differentiable function defined on a closed set is rather delicate. In one variable, one gets round the difficulty by talking about the right and left derivatives at the ends of an interval. In higher dimensions, things are clearly more complicated. Nevertheless, we shall come back to this question when we talk about critical points of functions in Chapter VII.

When the evolution of a system is restricted by additional conditions, we say that *constraints* are imposed on the system. If its positions are parametrised by a configuration space, the presence of such constraints leads one to ask whether it is possible to define a new, smaller configuration space for the system. When they are sufficiently regular (in a sense which we make precise in VII.12), it is in fact possible to define a new configuration space for the system, taking account of the constraints.

C. Vector Spaces in Curvilinear Coordinates

We begin with the most elementary examples of configuration spaces, open subsets of vector spaces.

V.38. To make a configuration space out of a finite-dimensional vector space E (we should rather speak of an *affine space* since in this paragraph the origin never plays a special role) or more generally out of open subsets of affine spaces, we must describe the admissible systems of local coordinates: we take simply the *linear* coordinates which, in this special case, are actually global parametrisations.

The coherence of this family is trivial to check, the change of coordinate transformations being linear, and hence differentiable as we want.[97]

V. 39. As examples of physical or mechanical systems whose configuration spaces are a vector space, we mention the systems formed by particles (possibly just one) which in the classical case are identified with *material points*. Some of these points might be abstractions as is the case, for example, when one studies the motion of the *centre of gravity* of a system.

V.40. Other parametrisations, which have a central or axial symmetry, are also used in vector spaces.

For example, this is the case with *polar coordinates* in the plane which are well suited to the study of systems invariant under plane rotations, for example a particle moving under a central force.

---

[97] As Molière's character Monsieur Jourdain did in using *prose* when speaking, in developing differential calculus in Chapters III and IV, we were working in *configuration spaces* without knowing it.

Classically, to any point (x, y) in $\mathbb{R}^2 - \{0\}$, one associates the pair of numbers $(r, \theta) \in \mathbb{R}^{**} \times [0, 2\pi[$ defined by $x = r \cos \theta$ and $y = r \sin \theta$, so that $\tau_p^c(x, y) = (\sqrt{x^2 + y^2}, \text{Arctan}(y/x))$ (where the letters p and c signify "*polar*" and "*Cartesian*" coordinates respectively).

These systems of coordinates are admissible by Proposition V.34, since the Jacobian of the change of coordinates is

$$\frac{D(r, \theta)}{D(x, y)} = \left(\frac{D(x, y)}{D(r, \theta)} \right)^{-1} = \left| \begin{matrix} \cos \theta & -r \sin \theta \\ \sin \theta & r \cos \theta \end{matrix} \right|^{-1} = \frac{1}{r} \neq 0 \quad .$$

(for the notations, see III.77), except for a subtle point which we now make explicit. The interval $[0, 2\pi[$ in which θ takes its values is not an open interval, a condition which, however, is necessary for polar coordinates to give a local diffeomorphism at each point of the configuration space.

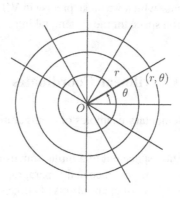

Fig. V.41 Polar coordinates (r, θ) in the plane minus the origin

We can circumvent this difficulty by excluding from the domain of the polar coordinates all the points of $\mathbb{R}^2$ such that $x > 0$ and $y = 0$. It is then necessary to take another polar coordinate system, this time excluding another half-line so that $\mathbb{R}^2 - \{0\}$ is completely covered.

The subtlety lies in the fact that the circle is itself a configuration space, as we shall make explicit in Sect. D.

V.42. A generalisation of this system of local coordinates which applies in $\mathbb{R}^3$ consists of the *cylindrical coordinates* which to the point with Cartesian coordinates (x, y, z) associates the triple (r, θ, z), where the coordinates (r, θ) are the polar coordinates in the plane $z = 0$.

We leave it to the reader to check that this parametrisation is admissible in $\mathbb{R}^3$ minus the half-plane $x > 0$, $y = 0$.

This chart is suited to the study of systems invariant under translation in some direction and under rotation in a perpendicular plane.

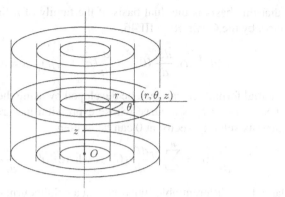

Fig. V.43 Cylindrical coordinates (r, θ, z) in three-dimensional space minus the vertical axis

V.44. In vector spaces, there exist of course other admissible systems of local coordinates which may be suited to the study of particular systems. We return in Sect. D to one of them: the *spherical coordinates*.

In fact, it is traditional to call all these local charts *curvilinear coordinates*. This refers to the fact that, by fixing all the parameters except one, we obtain *curves* which play the role of coordinate axes. We note that they are transformed into these axes by the local diffeomorphism which is the chart. Better, by using such coordinates, it is possible to reproduce the formulas of differential calculus, as we now show.

V.45. Suppose that $x = (x^1, \cdots, x^n)$ is a system of curvilinear coordinates in $\mathbb{R}^n$, and that a is a point in its domain of definition, having coordinates $(a^1, \cdots, a^n)$ in this system.

If, for $1 \leq i \leq n$, ξ_i are the n coordinate curves passing through a, whose curvilinear coordinates are given by $\xi_i(t) = (a^1, \cdots, a^i + t, \cdots, a^n)$, it is easy to see that their velocity-vectors $(d\xi_i/dt)(0)$ at 0 generate the vector space $\mathbb{R}^n$ (since the tangent linear map is invertible).

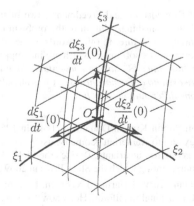

Fig. V.46 A system of curvilinear coordinates

Note that this basis is the dual basis of the family of n linear forms $dx^i(a)$ (cf. III.55) since, by the Chain Rule III.46,

$$\langle dx^j(a), \frac{d\xi_i}{dt}(0) \rangle = \frac{d}{dt}(a^j + \delta_{ij}t)_{|t=0} = \delta_{ij} .$$

The differential formalism can then be developed by using the Basic Principle V.22. In fact, if $\gamma : t \mapsto \gamma(t)$ is a curve whose expression in curvilinear coordinates is $(\gamma^i(t))$, then its velocity-vector at 0 can be written

$$\frac{d\gamma}{dt}(0) = \sum_{i=1}^{n} a^i \frac{d\xi_i}{dt}(0) \quad \text{where} \quad a^i = \frac{d\gamma^i}{dt}_{|t=0} .$$

Similarly, for a differentiable function φ in a neighbourhood of a point a, we can write

$$d\varphi(a) = \sum_{i=1}^{n} \phi_i \, dx^i(a)$$

where we denoted by ϕ_i the derivative at $t = 0$ of the partial function $t \mapsto \varphi(\xi_i(t))$. We thus recover the fundamental formulas of differential calculus

$$(V.47) \qquad d\varphi(a) = \frac{\partial \varphi}{\partial x^1}(a) \, dx^1(a) + \cdots + \frac{\partial \varphi}{\partial x^n}(a) \, dx^n(a) ,$$

where, unlike in III.56, the coordinates x^i are now *curvilinear* coordinates, which allows us to state an immediate corollary of Proposition V.34.

Corollary V.49. *A family of n functions $x^1, ..., x^n$ defines a system of curvilinear coordinates in a neighbourhood of a point a in $\mathbb{R}^n$ if and only if the n linear forms $dx^i(a)$ are linearly independent.*

Exercise V.50. Let φ be a function of class C^1 defined on an open subset of $\mathbb{R}^n$. Show that at every point where the differential of φ is non-zero, there exist $n - 1$ functions defined on a neighbourhood of this point which form with it a system of curvilinear coordinates.

V.51. The combined work of Donaldson[98] and Freedman[99], two of the recipients of the Fields Medal in 1986, has provided an astonishing solution of the problem of knowing whether vector spaces have a unique differentiable structure, in other words, to decide whether it is possible to define a coherent family of curvilinear coordinates which are incompatible with the linear family. In fact, they have shown that *"the space $\mathbb{R}^4$ has an uncountably infinite number of differentiable structures between which there exist no diffeomorphisms"*. It was already known that *for $n \neq 4$, the vector spaces $\mathbb{R}^n$ have only one differentiable structure.*

---

[98] The English mathematician Simon Kirwan Donaldson, born in 1957, has made major contributions to differential topology and to Kählerian geometry, using global analysis tools inspired by gauge theories to create new geometric invariants. Besides the Fields Medal in 1986, he received many other international honours, among which the Shaw Prize in 2009 and the Wolf Prize in 2020.

[99] The American mathematician Michael Hartley Freedman, born in 1951, is a topologist with outstanding results about 4-dimensional manifolds. He is now working at Microsoft Station Q on quantum computing.

The great originality of the work of Donaldson lies in the fact that it makes essential use of an auxiliary 5-dimensional space consisting of the solutions of a variational problem introduced by theoretical physicists in the study of gauge theories. Of course, theoretical physicists had their own motivation for formulating this problem, which had nothing to do with differential topology. In any case, this is a beautiful example of cross-fertilisation well-worth thinking about!

D. The Fundamental Examples

We give the fundamental examples of configuration spaces, namely the circle, tori, spheres and projective spaces.

V.52. We now come to the first true example of a configuration space. This is simply the circle[100] S^1 which serves, for example, to parametrise the positions of a plane pendulum moving around a fixed point.

The topology which we consider can be defined as that induced by the topology of the complex plane when the circle is viewed as the unit circle in $\mathbb{C}$,

$$S^1 = \{z \mid z \in \mathbb{C}, |z| = 1\}.$$

There is a natural map from $\mathbb{R}$ to S^1 given by the complex exponential map $\exp : \theta \mapsto e^{i\theta}$. It defines a bijection from $\mathbb{R}/2\pi\mathbb{Z}$ to S^1 (which one easily checks is a homeomorphism if $\mathbb{R}/2\pi\mathbb{Z}$ is endowed with the quotient topology.

V.53. To make a configuration space we must give a collection of compatible charts. Since, when θ varies in an open interval $]\alpha - \epsilon, \alpha + \epsilon[$ of length 2ϵ, assumed to be less than or equal to 2π, the map exp is a homeomorphism onto its image, we can consider its inverse. After adding $-\alpha$ to this inverse, we denote it by $\varphi_{\alpha,\epsilon}$ so that $\varphi_{\alpha,\epsilon}(e^{i\theta}) = \theta - \alpha$ and therefore $\varphi_{\alpha,\epsilon}(e^{i\alpha}) = 0$; this means that $\varphi_{\alpha,\epsilon}$ is a centred chart on the neighbourhood $\{e^{i\theta} \mid \theta \in]\alpha - \epsilon, \alpha + \epsilon[\}$ of $e^{i\alpha}$. (We have, of course, defined a kind of logarithm, but, to avoid any risk of confusion, we shall not make use of trigonometric functions in this elementary discussion.)

When the domains of two such charts $\varphi_{\alpha,\epsilon}$ and $\varphi_{\beta,\eta}$ have non-empty intersection, the change of coordinates map τ_β^α is simply translation by the vector $\beta - \alpha$ because the exponential map has the property of being a group homomorphism $e^{\gamma+\delta} = e^\gamma e^\delta$. (In fact, by definition $\varphi_{\alpha,\epsilon}(\alpha) = 0 = \varphi_{\beta,\eta}(\beta)$ and $\tau_\beta^\alpha = \varphi_{\beta,\eta} \circ (\varphi_{\alpha,\epsilon})^{-1}$; thus we have $\tau_\beta^\alpha(\theta) = \varphi_{\beta,\eta}(e^{i(\theta+\alpha)}) = \theta + \alpha - \beta$.)

Exercise V.54. Show that the polar coordinates define a global diffeomorphism between $\mathbb{R}^2 - \{0\}$ and $S^1 \times \mathbb{R}_*^+$ (for the notion of a product of configuration spaces, see V.21).

---

[100] The use of bold letters to denote a mathematical object, in this case the circle, is designed to emphasise its universal character.

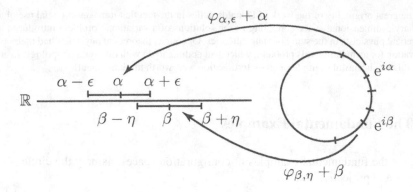

Fig. V.55 A change of chart on the circle

Project V.56. Prove that S^1 is the only compact connected configuration space of dimension 1.

Remark V.57. To describe the circle, we have chosen to speak in terms of angles, and this is quite natural. Note, however, that when we adjoin to the initial family of charts others whose coordinate changes are not linear, we can no longer consider the coordinates to be any kind of angles. But the logic of the notion of a configuration space is precisely to accept as coordinate systems all those which are admissible.

　This point may seem rather subtle, even byzantine, but it deserves serious reflection. The lesson to be learned from this is that the notion of a configuration space does not allow one to measure angles on the circle. To do this, one has to give an additional structure which we introduce now.[101]

V.58. Let us return to the exponential map from $\mathbb{R}$ to $\mathbf{S}^1$ which to θ associates $e^{i\theta}$. It is well-known that this map is a homomorphism from the additive group of $\mathbb{R}$ to the multiplicative group of $\mathbb{C}$ and that its kernel is the subgroup $2\pi\mathbb{Z}$ of $\mathbb{R}$. It follows that its image inherits the structure of an Abelian[102] group, which is precisely that of the *group of plane rotations*, concerning which we have the following result.

---

[101] This discussion, which some people might find rather trivial, goes to the heart of one of the functions of mathematics, namely to disentangle the various notions which play a role in a given situation. The first reaction of those who are practically minded is to find such considerations superfluous..... until they are forced to realise that in an analogous context this distinction can be concrete rather than abstract. The history of science repeatedly offers examples of such situations, most recently once again in high energy physics, particularly in the development of gauge theories and string theory.

[102] from the name of Niels Abel (1802–1829), Norwegian mathematician, who proved the impossibility of solving by radicals the general equation of the fourth degree (a result generalised by Évariste Galois) and who also made substantial contributions to analysis.

Proposition V.59. *The multiplication defined on the circle* S^1 *as the image under the complex exponential map of addition of real numbers is a differentiable map from* $S^1 \times S^1$ *to* S^1.

Similarly, taking the inverse for this multiplication is differentiable.

Proof. The proof is immediate since, in the linear charts, these maps can be written on the one hand as the sum of real numbers in certain intervals, and on the other hand as composites of translations and multiplication by -1. By Definition-Proposition V.24, this is enough as it suffices to check the result in some particular coordinate system.   □

Remark V.60. It is clear (not to mention tautological) that the group law we have defined on S^1 is the same as that which it inherits as a multiplicative subgroup of $\mathbb{C}$, but we wanted to define it using only *intrinsic* notions defined on S^1.

This would not have been the case if we had appealed to the differentiability of multiplication in $\mathbb{C}$ in order to justify that on S^1.

V.61. Proposition V.59 shows that on S^1 there is maximal compatibility between the group structure and the configuration space structure.

The standard terminology is to say that S^1 is a Lie[103] group, a notion which we shall now define formally.

Definition V.62. A configuration space M is a *Lie group* if there is a multiplication law defined on M which is differentiable as a map from $M \times M$ to M and for which taking the inverse is also a differentiable map from M to M.

V.63. The *Lie group* structure which we have introduced, combining that of a configuration space and that of a group, plays a central role in current developments in numerous branches of mathematics (differential geometry in particular) and theoretical physics, notably that which deals with elementary particles.

In Sect. E, we shall meet other examples of Lie groups in a context closer to classical mechanics.

V.64. The construction of products of configuration spaces described in V.21, when applied to the circle S^1, allows one to consider the product of n circles, traditionally called the *n-torus* T^n, as a configuration space of dimension n (compact by Proposition I.56).

It is clear that it is possible to define a generalisation of the complex exponential by associating to an n-tuple $(\theta^1, \cdots, \theta^n)$ of real numbers the n-tuple $(e^{i\theta^1}, \cdots, e^{i\theta^n})$ of points of S^1. This map induces a Lie group structure on the torus. To emphasise which structures are in play, the purists try to distinguish the torus as a Lie group (usually denoted by $\mathbf{T}^n$) from the torus viewed as a configuration space (then it is denoted by T^n, or S^1 if its dimension is 1)!

---

[103] The Norwegian mathematician Sophus Lie (1842–1899) was a professor at Oslo from 1872 until his death. He wrote fundamental works on continuous groups of transformations which were largely ignored during his lifetime.

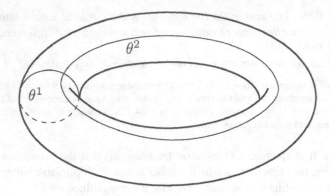

Fig. V.65 A "concrete" representation of the torus $\mathbf{T^n}$ with the angular coordinate system

In this context, it is often useful to think of the torus $\mathbf{T^n}$ as a parallelepiped whose faces are identified by means of translations parallel to the axes, the group structure being the addition in $\mathbb{R}^n$.

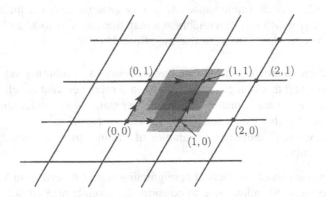

Fig. V.66 The torus $\mathbf{T^n}$ obtained by identifying the sides of a parallelogram in the plane, and the domains of three charts which cover it

V.67. The *double pendulum*, formed by two rods, one end of one rod being fixed, and the other rod being attached to the first at its other end, is an example of a mechanical system for which the torus $\mathbf{T}^2$ can be taken as the configuration space. In fact, one angle is needed to represent the position of the rod having one end fixed, and the other to represent the position of the second, either with respect to some fixed direction, or with respect to the first rod.

In fact, there are numerous mechanical or physical systems whose evolution can be described by the motion of an imaginary point on a torus of even dimension whose coordinates are called "*action-angle variables*". Unfortunately, we do not have time to go very far into their use in this course (see, however, the remarks in Chapter X).

V.68. We now come to another family of configuration spaces which generalise the circle and which are also of great importance in geometry and more broadly, namely the n-dimensional spheres $\mathbf{S}^n$.

The first picture of a sphere which comes to mind is of course that of the set of points of $\mathbb{R}^{n+1}$ whose distance to the origin is 1. It is possible to deduce the configuration space structure of the sphere from this realisation, as we shall see in Chapter VII, but, as we have done for the circle, we would like to define the configuration space structure on the sphere "*internally*", in some sense in an *intrinsic* way.

As for the circle, we begin with the sphere $\mathbf{S}^n$ as a space acquiring its topology from a Euclidean space by putting

$$\mathbf{S}^n = \{q \mid q \in \mathbb{R}^{n+1}, \ |q| = 1\} \, ,$$

which, as we have seen in Chapter I, is in fact a compact space.

In order to define a configuration space structure on $\mathbf{S}^n$ we must define a coherent family of local charts on it which will enable us to do differential calculus on it. This family is provided by a classical geometrical construction: that of *stereographic projections*, which we now describe analytically.

V.69. On the sphere $\mathbf{S}^n$ minus the north pole $p_+ = (0, \cdots, 0, 1)$, consider the map which to a point $q = (q^1, \cdots, q^{n+1})$ associates the point of intersection $\varphi_-(q)$ of the straight line p_+q with the plane

$$P_- = \{y \mid y \in \mathbb{R}^{n+1}, \ y^{n+1} = -1\}$$

tangent to the sphere at the south pole $p_- = (0, \cdots, 0, -1)$. Identifying the point $(y^1, \cdots, y^n, -1)$ of the plane P_- with the point $(y^1, \cdots, y^n)$ of $\mathbb{R}^n$ makes φ_- a chart on $\mathbf{S}^n - \{p_+\}$ since

$$\varphi_-(q) = \left(\frac{2q^1}{1 - q^{n+1}}, \cdots, \frac{2q^n}{1 - q^{n+1}} \right) .$$

In an analogous way, we construct a chart φ_+ on $\mathbf{S}^n - \{p_-\}$, where p_- is the south pole, by defining $\varphi_+(q)$ to be the point of intersection of the straight line p_-q with the plane $P_+ = \{x \mid x \in \mathbb{R}^n, \ x^{n+1} = 1\}$.

On the images of their common domain of definition $\mathbf{S}^n - \{p_+, p_-\}$, we can define the change of coordinates map τ_-^+ which to a point y in $\mathbb{R}^n - \{0\}$ associates $x = 4\,y/|y|^2$, i.e., it is an *inversion* with pole the origin and ratio 4. The map τ_-^+ is thus infinitely differentiable as is its inverse. We have thus made $\mathbf{S}^n$ into a configuration space.

It is easy to see that the sphere $\mathbf{S}^2$ is the configuration space of the *spherical pendulum* (if we accept the idealisation which amounts to forgetting that its point of attachment has been fixed in some way).

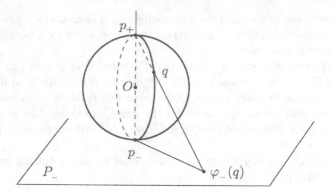

Fig. V.70 The stereographic projection of the sphere S^2 from its north pole

Exercise V.71. Show that on any open hemisphere, the coordinates of the projection onto the equatorial plane form an admissible system of local coordinates.

V.72. In 1966 Bessaga proved that the unit sphere of the Hilbert space l^2, which we know by II.90 is homeomorphic to the whole space l^2, is in fact diffeomorphic to it. (Recall that, by F. Riesz's Theorem I.77, the unit sphere of l^2 is not compact.) This emphasises once again the difference we have found between finite- and infinite-dimensional spaces.

Exercise V.73. Check that the preceding construction defines on S^1 the same configuration space structure as that defined by the complex exponential.

V.74. When $n = 2$ it is tempting to identify the plane $\mathbb{R}^2$ with the field $\mathbb{C}$ of complex numbers and to express the change of charts map τ_-^+ as an operation in this field. The only subtlety lies in the way the imaginary unit i in the planes P_- and P_+ are identified. If we want to be able to pass from one to the other by making them slide over the sphere while remaining tangent to it, we find that P_+ and P_- cannot be identified by a translation in $\mathbb{R}^3$, but by the symmetry about the origin. With this identification, we find that $\tau_-^+(z) = 4\,z^{-1}$ (using z, in the traditional way, to represent a complex number). The crucial point is that this function from $\mathbb{C}^*$ to $\mathbb{C}^*$ is *complex analytic* (one also says it is *holomorphic*).

   We have, in fact, defined the *Riemann sphere*, a concept which is the key to numerous problems relating to functions of one complex variable. On such an object, it is possible to do not only differential calculus but also holomorphic calculus. The configuration spaces of dimension 2 (and thus of complex dimension 1) where such a calculus can be developed are called *Riemann surfaces*. This theory, which played an important historical role towards the end of the XIXth century and the beginning of the XXth, has become of interest again at the present time for purely mathematical reasons, and because it is the natural context in which to develop *string theory*, which generated much excitement among theoretical physicists in the later part of the XXth century.

Project V.75. Show that the torus $\mathbf{T}^2$ can be given the structure of a Riemann surface. Is this structure unique?

V.76. If we restrict ourselves to the case $n = 2$, it is possible to use polar coordinates in the punctured plane (i.e., the plane minus the origin) where the chart takes its values. Since we know that these curvilinear coordinates give a diffeomorphism from $\mathbb{R}^2 - \{0\}$ to $\mathbb{S}^1 \times \mathbb{R}_*^+$, they allow one to define a new chart on $\mathbb{S}^2$ which to a point x in $\mathbb{S}^2$ associates (t, θ), where θ corresponds to the point on the equatorial circle which lies in the semi-circle with endpoints the poles p_+ and p_- which passes through x, and t is a function of the Euclidean distance from the poles which one can easily calculate.

This leads us naturally to define a chart on the sphere $\mathbb{S}^2$ *by suspension*, as follows. To any pair $(\theta, \psi) \in \mathbb{S}^1 \times [-\pi/2, \pi/2]$ one associates the point q in $\mathbb{S}^2$ with coordinates

(V.77)
$$\begin{cases} q^1 = \cos\psi \cos\theta \\ q^2 = \cos\psi \sin\theta \\ q^3 = \sin\psi \end{cases}$$

so that θ is the *longitude* and ψ the *latitude*. This map has an inverse on $]0, 2\pi[\times]\pi/2, \pi/2[$ which is the desired parametrisation.

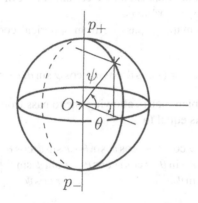

Fig. V.78 Geographical coordinates on the sphere $\mathbb{S}^2$

Exercise V.79. Study the map from T^2 to $\mathbb{S}^2$ (suggested by the construction V.76) which to the point of T^2 with coordinates (θ, ψ) (both varying in an interval of length 2π) associates the point of $\mathbb{S}^2$ with longitude θ and latitude ψ.

V.80. It is possible to extend the "angular" construction V.76 to the sphere $\mathbb{S}^n$ by using the fact that $\mathbb{S}^n$ can be obtained as the "suspension" of the equatorial sphere $\mathbb{S}^{n-1}$, namely the intersection of $\mathbb{S}^n \subset \mathbb{R}^{n+1}$ with the plane $q^{n+1} = 0$.

For example, we parametrise the sphere $\mathbb{S}^3$ by "suspending" the equatorial sphere of dimension 2, which gives the following formulas:

$$\text{(V.81)} \quad \begin{cases} q^1 = \cos \varphi \cos \psi \cos \theta \\ q^2 = \cos \varphi \cos \psi \sin \theta \\ q^3 = \cos \varphi \sin \psi \\ q^4 = \sin \varphi \end{cases}$$

where the angles ψ and φ vary in the interval $]-\pi/2, \pi/2[$ and θ in the interval $[0, 2\pi[$. As in V.76, to have a true chart, one must restrict to the open set of $\mathbb{R}^3$: $]0, 2\pi[\times] - \pi/2, \pi/2[\times] - \pi/2, \pi/2[$.

V.82. By making use of this analysis of the sphere, we can introduce another curvilinear coordinate system on $\mathbb{R}^n - \{0\}$ which is particularly suitable for the study of mechanical or physical systems which only depend on the distance from a point: the *spherical coordinates*.

A point $x \in \mathbb{R}^n$ is represented by its distance r from the origin and the point $x/|x|$ of the sphere $\mathbf{S}^{n-1}$.

If $n = 3$, this latter point can in turn be represented by its *latitude* and *longitude* (θ, ψ) as we have seen in V.76, giving the triple (r, θ, ψ) of spherical coordinates. As in the case of polar coordinates in the plane, for this coordinate system to be a chart in the strict sense, one must exclude a certain subset of $\mathbb{R}^3$ which in this case is a half-plane bounded by the x^3-axis.

The change of charts map τ_c^s, passing from spherical coordinates to Cartesian coordinates, is given by

$$\tau_c^s(r, \theta, \psi) = (r \cos \psi \cos \theta, r \cos \psi \sin \theta, r \sin \psi) .$$

In view of V.77, the Jacobian matrix allowing one to pass from ordinary coordinates to spherical coordinates is equal to

$$\begin{pmatrix} \cos \psi \cos \theta & -r \cos \psi \sin \theta & -r \sin \psi \cos \theta \\ \cos \psi \sin \theta & r \cos \psi \cos \theta & -r \sin \psi \sin \theta \\ \sin \psi & 0 & r \cos \psi \end{pmatrix} ,$$

from which one deduces easily that the Jacobian of the change of coordinates is equal to

$$\frac{D(x^1, x^2, x^3)}{D(r, \theta, \psi)} = r^2 \cos \psi$$

and which thus vanishes only for $\psi = \pm \pi/2$, in other words, along the axis passing through the two poles.

These explicit calculations are of course well known but of great importance in practical situations.

Exercise V.83. Show that spherical coordinates realise a global diffeomorphism from $\mathbf{S}^2 \times \mathbb{R}_*^+$ onto $\mathbb{R}^3 - \{0\}$.

V.84. As in V.51, we discuss briefly the distinct differentiable structures which can be defined on spheres. It is known that the spheres $\mathbf{S}^2$, $\mathbf{S}^5$ and $\mathbf{S}^6$ have only *one* differentiable structure. The question is open in dimensions 3 [104] and 4: this is the differentiable version of the famous *Poincaré conjecture* in these dimensions.

The construction of non-standard differentiable structures on all spheres of dimension greater than 6 is due to Milnor[105] in 1956; these are called *exotic spheres*. In collaboration with Kervaire, he showed that there are always a finite number of them; there are 28 in dimension 7, 2 in dimension 8, 992 in dimension 11, etc. This discovery is one of the high points of differential topology, but we should not think that these structures cannot be approached by explicit methods. In dimension 7, for example, the 28 exotic structures appear in a natural way on the topological spheres obtained by intersecting a standard sphere in $\mathbb{C}^5$ (and thus of dimension 9) with the complex hypersurfaces defined by the polynomial equations $(z^1)^{6k-1} + \sum_{i=2}^{5}(z^i)^2 = 0$, for $k = 1, 2, \cdots, 28$, among which appears the standard differentiable structure on the sphere $\mathbf{S}^7$.

V.85. We conclude our review of the fundamental examples with the (real) *projective spaces*, which appear naturally as objects of study, and which have a very rich internal structure.

V.86. By definition, the *real projective space* $\mathbb{RP}^n$ *is the space of straight lines passing through the origin in* $\mathbb{R}^{n+1}$. To make it into a configuration space, one must give a nice description of it, which is made possible by using the "homogeneous coordinates" which are *not* coordinates in the sense in which we have used the word in this course (but as this expression is traditional, we still retain it).

A straight line d in $\mathbb{R}^{n+1}$ passing through the origin is determined by any one of its points x_d other than the origin itself. The *homogeneous coordinates* of d are by definition the coordinates of the point x_d and are denoted by $[x_d^1; \cdots ; x_d^{n+1}]$ to emphasise that they are only defined up to a non-zero multiple.

We now define an open covering of $\mathbb{RP}^n$ by $n + 1$ open sets $(U_i)_{1 \leq i \leq n+1}$, often called *affine open subsets* of the real projective space, by putting

$$U_i = \{d \mid d \in \mathbb{RP}^n, x_d^i \neq 0\}.$$

This definition is compatible with the fact that the homogeneous coordinate x^i is only defined up to a non-zero multiple. On U_i we can define a chart φ_i by putting

$$\varphi_i(d) = \left(\frac{x_d^1}{x_d^i}, \cdots, \frac{x_d^{n+1}}{x_d^i} \right)$$

where the term which would have been in the ith position in the n-tuple $\varphi_i(d)$, and which would have had the value 1, has disappeared.

---

[104] Since the first edition of the notes, the question has been settled in 2003 by Grisha Perelman who solved the Poincaré conjecture in this dimension showing that any compact simply connected 3-manifold is diffeomorphic to $\mathbf{S}^3$. This led the Fields committee 2006 to propose him for the medal, a distinction that he refused.

[105] The American mathematician John Willard Milnor, born in 1931, worked in differential topology and more algebraic and dynamical aspects of topology and geometry. He is also well known for the clarity of his lectures and books. He received the Fields Medal in 1962, the Wolf Prize in 1989, and the Abel Prize in 2011.

On the image of $U_1 \cap U_2$, the change of charts map τ_2^1 sends the n-tuple $(y^1, \cdots, y^n)$ to the n-tuple $(1/y^1, y^2/y^1, \cdots, y^n/y^1)$. It is thus an infinitely differentiable map since, in $\varphi_1(U_1 \cap U_2)$, $y^1 \neq 0$. This collection of charts makes $\mathbb{R}P^n$ into a configuration space.

Since a line through the origin in $\mathbb{R}^{n+1}$ cuts the unit sphere in two diametrically opposite points, *the real projective space $\mathbb{R}P^n$ can be identified with the pairs of antipodal points on the sphere S^n*.

V.87. To add to this information, we can give a more explicit description of the projective spaces in low dimensions.

Exercise V.88. Show that $\mathbb{R}P^1$ is *diffeomorphic to the circle* S^1.

Exercise V.89. Show that the projective plane $\mathbb{R}P^2$, the space of lines through the origin in $\mathbb{R}^3$, has the following geometric description: starting with a *Möbius*[106] *band*, obtained physically by gluing the short edges of a paper band after giving it a single twist, $\mathbb{R}P^2$ is obtained by gluing a disc along the boundary of the Möbius band which, as is well-known, consists of a single closed curve.

Exercise V.90. Show that the map from S^n to $\mathbb{R}P^n$ induced by the identification described in V.86 is a local diffeomorphism. Is it a global diffeomorphism?

V.91. The construction of projective spaces can be carried out in a similar way over any field $\mathbb{K}$ (even finite ones!). We shall make a few remarks on the case of the field $\mathbb{C}$ of complex numbers.

We consider then the projective space $\mathbb{C}P^m$ which is by definition the space of complex lines through the origin in $\mathbb{C}^{m+1}$. The construction of its configuration space structure is formally identical to that which we presented in the real case, and it also makes use of *homogeneous coordinates* and *affine* open sets. The extra richness of complex projective spaces stems from the power of the theory of analytic functions on $\mathbb{C}$.

The geometry of these spaces lies at the heart of the fascinating theory of fibre bundles, which provides the correct language for the modern developments in algebraic geometry. A fundamental theorem of Serre[107] shows that on these spaces there is no difference between analytic and algebraic objects (often referred to as the GAGA theorem for "*Géométrie Algébrique-Géométrie Analytique*").

Project V.92. Show that the complex projective line $\mathbb{C}P^1$ can be identified as a complex analytic manifold with the Riemann sphere which we defined in V.74.

Project V.93. Show that the description of $\mathbb{R}P^n$ as the quotient of the sphere S^n given in V.86 generalises to the complex projective space $\mathbb{C}P^m$ in the following form: there exists a natural differentiable map from S^{2m+1} to $\mathbb{C}P^m$, called the *Hopf*[108] *fibration*, such that the inverse image of each point of $\mathbb{C}P^m$ is a circle.

---

[106] August Ferdinand Möbius (1790–1868) was a German mathematician and astronomer.

[107] Jean-Pierre Serre, born in 1926, made major contributions to several fields of mathematics: algebra, number theory, topology, group theory,... (the chair at the Collège de France he held for 38 years was named "Algèbre et géométrie"). He has been an active member of the Bourbaki group. The published account of the letters between him and Alexandre Grothendieck is a model of how carefully written exchanges can impact the progress of research in mathematics. He received the Fields Medal in 1954 (still the youngest ever to get it), the Wolf Prize in 2000, and he was the first recipient of the Abel Prize in 2003. The Gold Medal of CNRS was awarded to him in 1987.

[108] The German mathematician Heinz Hopf (1894–1971) made numerous contributions to algebraic and differential topology, and several geometric objects or invariants are named after him.

V.94. More generally, one can consider as configuration spaces the *Grassmann*[109] *manifolds* $\mathbb{K}G_{n,k}$ of order k, also called *Grassmannians*, which are the spaces of k-planes in $\mathbb{K}^n$ passing through the origin, for an arbitrary field $\mathbb{K}$. Projective spaces are thus Grassmannians of order 1.

These spaces play a central role in differential geometry, notably in the theory of fibre bundles, a theory which gave birth to K-theory, a theory in which considerable progress has been made in the second part of the XXth century and proved to be highly relevant in theoretical physics.

E. The Rotation Group

We devote a special section to the group of rotations for various reasons which we shall summarise first. In particular, we give it a configuration space structure for which it can be identified with the real projective space of dimension 3.

V.95. The group of rotations is first of all the proper context to study the mechanics of a rigid body moving about a fixed point (or for its motion relative to its center of gravity whose evolution is studied separately). From a more mathematical point of view, its internal structure is also interesting for other reasons: it is a compact space of dimension 3, which can be identified with the real projective space $\mathbb{R}P^3$, and which is at the same time one of the simplest non-Abelian Lie groups to study.

Moreover, there exists a natural map from $\mathbf{S}^3$ to this group which sends pairs of points of the sphere to points of the group of rotations. When the sphere $\mathbf{S}^3$ is itself viewed as a group, this ambiguity turns out to be related to the *spin* of a particle.

V.96. If we express in the form of an equation the fact that a linear map l from a vector space E to itself is an *isometry* (one also speaks of an *orthogonal transformation*) for a scalar product denoted by $(\ |\ )$, we obtain that, for two vectors v and w,

$$(v|w) = (l(v)|l(w)) = (l^*l(v)|w)$$

where l^* denotes the *adjoint* of l. (We recall that the adjoint l^* is defined as follows: $l^* = \# \circ {}^t l \circ \flat$ where $\flat$ and $\#$ are respectively the isomorphism from E to E^* defined by the scalar product and its inverse, and where ${}^t l$ is the map from E^* to E^* which is the transpose of l.)

A linear transformation is thus an isometry if and only if it satisfies the equation $l^* \circ l = id_E$. Note that the dependence on the scalar product is hidden in this equation in defining the adjoint.

Since the image of an orthonormal basis under $\flat$ coincides with the dual basis, when one describes the space in terms of this basis, the matrix representing l^* is exactly the transpose of the matrix L representing l. *The matrix L of an orthogonal transformation* (which, of course, is called an *orthogonal matrix*) *is thus characterised by the property* ${}^t L = L^{-1}$.

We deduce easily from this relation and the well-known multiplicative property of determinants that *the determinant of an orthogonal matrix is* ± 1.

---

[109] Hermann Günther Grassmann (1809–1877) had many interests besides mathematics, being also a linguist and an editor.

Definition-Proposition V.97. *The isometries of* $\mathbb{R}^3$ *equipped with its standard metric form a group, denoted by* O_3, *having two connected components which are distinguished by the value of the determinant.*

In the identity component, denoted by SO_3, *the transformations are* rotations *of determinant 1 which, except of course for the identity, have 1 as simple eigenvalue, the invariant direction being the* axis of the rotation.

Proof. The information contained in the statement has just been provided. ☐

V.98. When one wishes to parametrise the motion of a rigid body having one point fixed, one must first of all see how many coordinates are necessary to define its position. It will of course be much less than the $3n$ degrees of freedom of the n independent points of the body, since these points are constrained to remain at a constant distance from each other.

In fact, any point of the body is determined by its distance from three reference points distinct from the fixed point, provided these points form with the fixed point an affine frame. It is convenient to assume that this frame is *orthonormal*, so that the absolute values of the components in this frame of any vector having the fixed point as its origin coincides with the lengths of its projections on the axes.

The natural topology to put on this group is defined by saying that two rotations are close if the three vectors of their image frames are close.

V.99. The most classical system of coordinates for the rotation group is given by the *Euler angles* which we now define following the classical presentation, which from the strictly practical point of view is not in fact the best possible (see Exercise V.105).

The idea of their construction is very geometric since it exhibits a decomposition of any rotation as a product of three suitably chosen rotations, a property which arises from the structure of the group SO_3.

Taking an orthonormal frame $(\vec{\imath}, \vec{\jmath}, \vec{k})$ which defines the coordinates (x, y, z) as the frame of reference, its image under the rotation, say $(\vec{\imath}', \vec{\jmath}', \vec{k}')$, is obtained as follows.

The intersection of the plane $x'Oy'$ generated by $\vec{\imath}'$ and $\vec{\jmath}'$ with the horizontal plane $xOy = \mathbb{R}.\vec{\imath} \oplus \mathbb{R}.\vec{\jmath}$ is a line unless $\vec{k}'$ and $\vec{k}$ are collinear. If this is the case, we omit the first step; if not, we call the intersection of the two planes xOy and $x'Oy'$ the *nodal line*, and we make a rotation with axis $\vec{k}$ and angle $\varphi \in [0, \pi[$ which sends $\vec{\imath}$ to a vector $\vec{n}$ along the nodal line. The angle φ is the *first Euler angle*.

The second step is to make a rotation with axis $\vec{n}$ which transforms $\vec{k}$ to $\vec{k}'$, which is possible since $\vec{n} \in xOy \cap x'Oy'$. The *second Euler angle*, called θ, is the angle of this rotation and is thus determined up to a multiple of 2π. One usually assumes that $\theta \in [-\pi, \pi[$.

To transform the orthonormal frame $(\vec{\imath}, \vec{\jmath}, \vec{k})$ to the frame $(\vec{\imath}', \vec{\jmath}', \vec{k}')$, it remains to make a rotation with axis $\vec{k}'$ which sends $\vec{n}$ to $\vec{\imath}'$. The angle of this rotation, denoted by ψ, is the *third Euler angle* which we take in the range $\psi \in [-\pi, \pi[$.

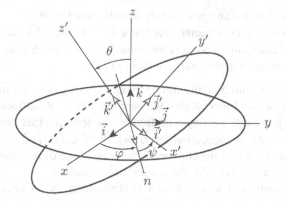

Fig. V.100 The Euler angles parametrising a point of SO_3, with the associated orthonormal frames

Remark V.101. Before modifying the definition of the Euler angles to obtain charts for the rotation group, several points deserve comment.

i) *There are various conventions in the literature*: the orientation of the angles may be taken in the clockwise (hence negative) sense; the nodal line may be taken with respect to the vector $\vec{\jmath}$; and the matrix which expresses the components of a vector in the old frame is sometimes called the "rotation matrix" (this outdated point of view is still used by some authors, cf. (Goldstein 1971)).

ii) The nodal line is indeterminate when $\vec{k}' = \pm\vec{k}$. The auxiliary vector $\vec{n}$ can then be chosen arbitrarily in the plane $\mathbb{R}.\vec{\imath} \oplus \mathbb{R}.\vec{\jmath}$. It suffices then to observe that only the sum $\varphi + \psi$ enters, the value of θ being 0 if $\vec{k}' = \vec{k}$ and π if $\vec{k}' = -\vec{k}$.

iii) With this description of a rotation, it is not immediately clear how to find its axis and angle. From an analytic point of view this is not a serious difficulty, but for geometric reasons this fact can be considered as a drawback.

V.102. To turn SO_3 into a configuration space, we must define charts. To define a chart in the sense in which we have used it throughout this chapter, the Euler angles must define an invertible map from an open subset of the rotation group onto a ball in the space of angular parameters. It is thus essential to restrict their domains, and this leads us to modify their definition as follows.

V.103. To remove the ambiguities, it is useful to define the vector $\vec{n}$ in the following way: if $\vec{\imath}' \in \mathbb{R}\vec{\imath} \oplus \mathbb{R}\vec{\jmath}$, one puts $\vec{n} = \vec{\imath}'$; if not, one takes for $\vec{n}$ the unit vector in $(\mathbb{R}\vec{\imath} \oplus \mathbb{R}\vec{\jmath}) \cap (\mathbb{R}\vec{\imath}' \oplus \mathbb{R}\vec{\jmath}')$ such that $\psi \in [0, \pi[$. In this construction, the angle ψ is determined only up to a multiple of π. (It is necessary to take an open interval with 0 as an endpoint since our construction is singular when $\vec{\imath}' \in \mathbb{R}\vec{\imath} \oplus \mathbb{R}\vec{\jmath}$ (i.e., when $\psi = 0$), hence the precautions we have had to take.)

In this way, $\vec{n}$ is in all cases a well-defined vector in the plane $\mathbb{R}\vec{\imath} \oplus \mathbb{R}\vec{\jmath}$, and one can define the angle φ up to a multiple of 2π; for example, one can take $\varphi \in [-\pi, \pi[$. It is then possible to invert the map we have defined on the image in SO_3 of the open domain in $\mathbb{R}^3$ defined by

$$(\varphi, \theta, \psi) \in\ ]-\pi, \pi[\ \times\ ]-\pi, \pi[\ \times\ ]0, \pi[.$$

We then obtain a chart on an open (in fact dense!) subset of SO_3.

In the chart we have constructed, the frame $(\vec{i}, \vec{j}, \vec{k})$ played a special role in the definition of the Euler angles, but it is the frame $(\vec{j}, -\vec{i}, \vec{k})$ which is its centre, i.e. the point whose coordinates in the chart are $(0, 0, \pi/2)$.

V.104. To make SO_3 into a configuration space by using only the modification of the construction of the Euler angles we have described, it suffices to repeat the construction we just made starting with frames of reference other than $(\vec{i}, \vec{j}, \vec{k})$. For example, one can take in succession the frames[110] $(\vec{j}, \vec{k}, \vec{i})$, $(\vec{k}, \vec{i}, \vec{j})$ and $(-\vec{j}, \vec{i}, \vec{k})$. Finally, one must check that the change of charts maps are differentiable. (We leave this verification as an exercise for the conscientious reader.)

It is in fact equivalent to check that multiplication by an orthogonal matrix parametrised in this way is a differentiable map, the matrix of a rotation in a new frame being POP^{-1} if O is its matrix in the frame of reference and P that of the new frame with respect to the old. (We also leave this verification as an exercise.)

Exercise V.105. Show that the rotation which sends the reference frame $(\vec{i}, \vec{j}, \vec{k})$ to the frame $(\vec{i'}, \vec{j'}, \vec{k'})$ can be obtained as the product of rotations whose axes are the vectors of the frame of reference: the rotation with axis $\vec{k}$ and angle φ, the rotation with axis $\vec{i}$ and angle θ and the rotation with axis $\vec{k}$ and angle ψ.

Deduce that, with respect to the frame $(\vec{i}, \vec{j}, \vec{k})$, the matrix of this rotation has the form

$$\begin{pmatrix} \cos\varphi\cos\psi - \sin\varphi\cos\theta\sin\psi & -\sin\varphi\cos\psi - \cos\varphi\cos\theta\sin\psi & \sin\theta\sin\psi \\ \cos\varphi\sin\psi + \sin\varphi\cos\theta\cos\psi & -\sin\varphi\sin\psi + \cos\varphi\cos\theta\cos\psi & -\sin\theta\cos\psi \\ \sin\varphi\sin\theta & \cos\varphi\sin\theta & \cos\theta \end{pmatrix}$$

(this formula might seem rather formidable, but it is not without its uses).

V.106. There is another way of describing rotations in $\mathbb{R}^3$ which is of great interest. To describe it, we must first make a detour into 4-dimensional space.

In fact, consider $\mathbb{R}^4$ and denote its canonical basis by $(1, i, j, k)$. We give it a multiplication law *denoted* by ".", assumed to be bilinear, by setting

$$(V.107) \qquad i.j = k, \quad j.k = i, \quad k.i = j, \quad i^2 = j^2 = k^2 = -1,$$

from which it follows that i, j and k anti-commute pairwise. We leave it to the reader to check that this gives $\mathbb{R}^4$ the structure of a field: this is the field $\mathbb{H}$ of *quaternions* (introduced by Hamilton) which extends the field $\mathbb{C}$ of complex numbers but, which, as one sees from Formulas (V.107), is *non-commutative*.

Like $\mathbb{C}$, the field $\mathbb{H}$ has a *conjugation* operation which associates to a quaternion $q = \alpha + \beta i + \gamma j + \delta k$ the quaternion $\overline{q} = \alpha - \beta i - \gamma j - \delta k$. We then have the fundamental relations $(q \mid q) = q.\overline{q}$ relating the multiplication law, the conjugation

---

[110] There are deep reasons why it is necessary to take at least *four* charts, reasons to which we shall return in Chapter VII.

and the standard scalar product on $\mathbb{H} = \mathbb{R}^4$, but one should note carefully that $\overline{q.q'} = \overline{q'}.\overline{q}$. Because of these relations, the unit sphere in $\mathbb{H}$ is a multiplicative subgroup (since $|q.q'|^2 = (q.q').\overline{(q.q')} = q.q'.\overline{q'}.\overline{q} = 1$ for $q, q' \in \mathbf{S}^3$) which we *denote* by $S\mathbb{H}$. We remark that the elements of $S\mathbb{H}$ are characterised by the identity $\overline{q} = q^{-1}$ and that, in view of the expression for the group law in terms of the components of a point of the sphere, $S\mathbb{H}$ is a Lie group.

V.108. We now introduce the *imaginary* elements of $\mathbb{H}$, i.e., the quaternions x such that $\overline{x} = -x$. They form the vector space $\Im m\mathbb{H} = \mathbb{R}.i \oplus \mathbb{R}.j \oplus \mathbb{R}.k$. If, for $q \in S\mathbb{H}$ and for $x \in \Im m\mathbb{H}$, we put $\rho_q(x) = q.x.q^{-1}$, we obtain an action of $S\mathbb{H}$ on $\Im m\mathbb{H}$ since $\overline{q.x.q^{-1}} = \overline{q^{-1}}.\overline{x}.\overline{q} = -q.x.q^{-1}$. Moreover, ρ_q is an isometry of $\Im m\mathbb{H}$ since

$$(\rho_q(x) \mid \rho_q(x)) = q.x.q^{-1}.\overline{q.x.q^{-1}} = (x|x) .$$

We have thus found a homomorphism ρ from the group $S\mathbb{H}$, which is topologically the sphere $\mathbf{S}^3$, and thus is a connected space, onto the group of rotations of the 3-dimensional space $\Im m\mathbb{H}$, and hence to the group SO_3.

It is easy to find the axis of the rotation ρ_q since this is, up to a scalar multiple, the unique vector v such that $\rho_q(v) = v$. Since $\rho_q(\Im mq) = \Im mq$ (for, if $q \in S\mathbb{H}$, we have $q.q.\overline{q} = q$ and $q.(\alpha 1).\overline{q} = \alpha 1$), $\Im mq$ is the required axis.

The *angle* ω of ρ_q is also easily obtained by evaluating the trace of the rotation, first in an orthonormal basis one of whose basis vectors is the axis of rotation, which gives $1 + 2\cos\omega$, and then in the orthonormal basis (i, j, k) which gives $4\alpha^2 - 1$ by a direct calculation, taking into account the fact that the quaternion $q = \alpha 1 + \beta i + \gamma j + \delta k$ is supposed to have norm 1, and hence $\alpha^2 + \beta^2 + \gamma^2 + \delta^2 = 1$.

Consequently, the angle ω of the rotation is determined up to a multiple of 2π by the relation $\cos\frac{1}{2}\omega = \pm\alpha$, the ambiguity in the sign arising from the choice of orientation of the axis.

Theorem V.109. *The group of rotations SO_3 is diffeomorphic to the real projective space $\mathbb{RP}^3$.*

Proof. We will show that the map from $\mathbf{S}^3$ to SO_3 which we have defined in V.108 is a continuous projection which identifies antipodal points. By V.86 it induces a bijection between SO_3 and $\mathbb{RP}^3$.

From V.108, it follows immediately that the map ρ is a continuous surjection from $S\mathbb{H}$ onto SO_3 since all axes of rotation can be obtained, as well as all angles of rotation, the continuity being a consequence of the formulas relating the unit quaternion q and the images of the basis vectors (i, j, k).

To complete the proof, it is enough to check that, given q_1 and q_2 in $S\mathbb{H}$ such that, for all $x \in \Im m\mathbb{H}$, $q_1.x.\overline{q}_1 = q_2.x.\overline{q}_2$, then $q_2 = \pm q_1$. Now, from the first relation we obtain $(\overline{q}_2.q_1).x.(\overline{q}_2.q_1) = x$, in other words the unit quaternion $q = \overline{q}_2.q_1$ commutes with all imaginary quaternions. But then it cannot have any component in $\Im m\mathbb{H}$ and hence must be equal to ± 1, which implies that $q_2 = \pm q_1$ as claimed.

We are not really in a position to show that the map from $\mathbb{RP}^3$ to SO_3 we have defined is a diffeomorphism. This will be much easier with the techniques introduced in Chapter VII. A direct proof can, however, be given by using Exercise V.114.   □

V.110. The map ρ clears the way for a second parametrisation of the rotation group SO_3, known as the *Cayley–Klein parametrisation*. In fact, if, for $q = \alpha + \beta i + \gamma j + \delta k$, we compute the matrix of ρ_q, we obtain by a careful calculation

$$\begin{pmatrix} \alpha^2 + \beta^2 - \gamma^2 - \delta^2 & 2(\beta\gamma - \alpha\delta) & 2(\alpha\gamma + \beta\delta) \\ 2(\alpha\delta + \beta\gamma) & \alpha^2 - \beta^2 + \gamma^2 - \delta^2 & 2(\gamma\delta - \alpha\beta) \\ 2(\beta\delta - \alpha\gamma) & 2(\alpha\beta + \gamma\delta) & \alpha^2 - \beta^2 - \gamma^2 + \delta^2 \end{pmatrix}$$

remembering that the real parameters α, β, γ and δ are connected to each other by the relation $\alpha^2 + \beta^2 + \gamma^2 + \delta^2 = 1$. (For an arbitrary element of $\mathbb{H}$, the matrix should be multiplied by $(\alpha^2 + \beta^2 + \gamma^2 + \delta^2)^{-1}$ since we have used the fact that, for a quaternion of norm 1, $\overline{q} = q^{-1}$.)

To have a chart in our sense, one must express one of the variables explicitly as a function of the others. For example, if we replace α by $\pm\sqrt{1 - (\beta^2 + \gamma^2 + \delta^2)}$, we are assuming that an open hemisphere is taken as the domain of the chart.

Finally, to show that this chart is admissible (i.e., can be adjoined to the others), one must check that the Jacobian of the change of charts map passing from (φ, θ, ψ) to (β, γ, δ) is non-zero which is, in the formulation we have given, a rather formidable calculation. We shall see a more efficient method of doing the calculation in Chapter VI, when we know how to calculate the tangent map to a differentiable map between configuration spaces.

V.111. It is traditional to view the construction given in V.110 through "complex" spectacles. In fact, let us return for a moment to the definition of $\mathbb{H}$. It would have been quite reasonable to present it as a 2-dimensional vector space over $\mathbb{C}$ since, by V.107, $q = \alpha + \beta i + \gamma j + \delta k$ can be written $q = (\alpha + \beta i) + (\gamma + \delta i).j$ (note the order of the factors!). If we now consider how to write right multiplication by the conjugate of the quaternion $q = \alpha + \beta i + \gamma j + \delta k$ on a quaternion $x = \xi + \eta.j$ viewed as a vector in $\mathbb{C}^2$, we find that it can be identified with the 2×2 complex matrix

(V.112) $$\begin{pmatrix} \alpha - i\beta & \gamma - i\delta \\ -\gamma - i\delta & \alpha + i\beta \end{pmatrix}$$

which is, as one easily checks, unitary and of determinant equal to the norm of q. We remark that i, j and k act respectively as the complex matrices

$$-i\sigma_3 = \begin{pmatrix} i & 0 \\ 0 & -i \end{pmatrix}, \quad i\sigma_2 = \begin{pmatrix} 0 & 1 \\ -1 & 0 \end{pmatrix}, \quad -i\sigma_1 = \begin{pmatrix} 0 & i \\ i & 0 \end{pmatrix},$$

where σ_1, σ_2 and σ_3 are traditionally called the *Pauli*[111] *matrices*, and in this line i is the fundamental imaginary number in $\mathbb{C}$.

---

[111] The Swiss physicist Wolfgang Pauli (1900–1958) was professor at the Eidgenössische Technische Hochschule in Zürich from 1928 until his death, with a break during World War II. He made it one of the great centres of research of his time. He received the Nobel Prize in physics in 1949 for the discovery of the exclusion principle which bears his name.

Consequently, the elements of $S\mathbb{H}$ can be identified with the 2×2 unitary matrices of determinant 1, which form the group denoted traditionally by SU_2. We leave it as an (easy) exercise for the reader to check that every element of SU_2 is actually of the form given by V.112, and hence that $S\mathbb{H}$ can be identified with the group SU_2.

Remark V.113. One can ask what is the nature of the 2-dimensional complex space "underlying" the ambient space, which appears so surreptitiously. It is the home of *spinors*, rather subtle objects the description of which we cannot go into here. They were introduced by É. Cartan in 1913 while he was doing a purely mathematical classification work. Note, however, that since the work of Dirac[112] the wave function of an electron is considered as a field which takes values in this space, whence the name *spinor field* which it is given. They are now fundamental objects in physics.

Exercise V.114. Show that the rotation with Euler angles (φ, θ, ψ) corresponds to the quaternion

$$q = \cos\frac{\theta}{2}\cos\frac{\varphi+\psi}{2} + (\sin\frac{\theta}{2}\cos\frac{\varphi-\psi}{2})i - (\sin\frac{\theta}{2}\sin\frac{\varphi-\psi}{2})j + (\cos\frac{\theta}{2}\sin\frac{\varphi+\psi}{2})k$$

(use the decomposition of the rotation into elementary rotations as in V.99).

From this complete the proof of Theorem V.109.

Exercise V.115. Let A be a skew-symmetric matrix of order n having eigenvalues of modulus less than 1. Show that the matrix $(I + A)(I - A)^{-1}$ is orthogonal, and that the set of orthogonal matrices which can be written in this form is a dense open subset of SO_n. Deduce a new parametrisation of SO_3.

V.116. This is not the place to discuss the nature of the *spin* of particles, a notion introduced into physics in the 1920s, but rather to see how the mathematical situation we have described gives a basis for the physical concept.

In V.108 we have seen the appearance of half the angle ω of a rotation in the expression of the quaternion which it defines in the Cayley–Klein parametrisation. At the time, we remarked that at the level of quaternions the angle ω was defined only up to a multiple of 4π. To be capable of distinguishing angles in space which differ by 4π seems to be impossible. In (Deheuvels 1981),[113] there is a description of an experiment showing that turning one object relative to another through 4π or 2π does not produce the same result.

---

[112] The English physicist Paul Adrien Maurice Dirac (1902–1985) received the Nobel Prize in physics in 1933 for the discovery of the positron. He played a leading role in theoretical physics by giving a particularly attractive form to quantum mechanics, which he studied from 1926. He wrote down the relativistically invariant equation (which today bears his name) satisfied by the wave function of a fermion such as an electron, introducing spinor fields in physics.

[113] but also in the remarkable book *"The Road to Reality"* published in 2004 by Roger Penrose. Born in 1931, Penrose is an English mathematician, physicist and philosopher of science with a huge breadth of interests. He contributed to a number of fields, from general relativity to cosmology. His work with Stephen Hawking about the development of space-time singularities (black holes) is a classic. He received the Wolf Prize in 1988 and the Nobel Prize in physics in 2020.

V.117. The intrinsic mathematical property of SO_3 which explains the phenomenon described in V.116 is the following. The curve γ formed by the rotations around an axis with angle varying from 0 to 2π, going from the identity to itself, cannot be deformed continuously to a point. When in a configuration space there exists such a closed curve, Mathematicians say that the space, here SO_3, *is not simply-connected*. On the other hand, if we traverse this curve twice in succession in the same sense, it is possible to shrink this new closed curve to a point. A curve in SU_2 which projects onto γ joins the matrix I to the matrix $-I$ and thus is not closed, but the lift of the curve γ traversed twice is closed. The sphere S^3 to which SU_2 is diffeomorphic being simply connected, this curve can be contracted to a point.

The projection from SU_2 to SO_3 which we have studied is a local diffeomorphism but not a global diffeomorphism. Such a map is called a *covering*, a notion which provides the modern setting to discuss *multi-valued functions*. Since the inverse image of any point of SO_3 consists of two points of SU_2, one speaks of a *covering with two sheets*. The transformations which interchange the points which project to the same point form a group, called the *covering group*. In our situation, the group which appears is the group $\mathbb{Z}_2$ of two elements.

V.118. For any configuration space M it is possible to define a simply-connected space which projects onto it. The group of the covering is then called the *fundamental group* of the space or its *Poincaré group*.

Further discussion of this would lead us into *homotopy theory*, a theory which has developed considerably since 1945. It has become of great interest to theoretical physicists today. It is a powerful tool for studying the *global form* of a space.

Exercise V.119. What is the configuration space of a rigid rod constrained to move in the plane? in n-dimensional space?

Project V.120. What is the configuration space formed by k jointed rods?

F. Historical Notes

V.121. The idea of describing geometric objects in terms of numbers goes back a long way in the history of science to the work of Descartes in the XVIIth century. To quote Weyl[114], *"the introduction of numbers as coordinates was an act of violence"*, but this

---

[114] Hermann Klaus Hugo Weyl (1885–1955) was a German mathematician, theoretical physicist and philosopher who had a huge impact on all these fields. He mostly shared his time between the ETH in Zürich and the Institute for Advanced Study (IAS) in Princeton. (He left the ETH for a position in Göttingen until he fled the Nazi regime to take a position at the newly established IAS.) His interest in mathematics and mathematical physics was very broad, from differential geometry to group representations, from number theory to analysis. He was one of the great propagators of general relativity from its onset. About his legacy, some speak of him as one of the last "universal" scientists in the tradition of Poincaré and Hilbert.

act has united algebra and geometry. The first expressions manipulated in this spirit were polynomials. The use of local charts and curvilinear coordinates to study the objects themselves is presented in the *"Disquisitiones Generales Circa Superficies Curvas"*[115] published in 1827 by Gauss. The passage to general coordinate systems was not made without difficulty.

V.122. The next step, introducing the notion of a configuration space (as we have already mentioned, *"manifold"* is the word English speaking mathematicians use, *"variété"* in French and *"Mannigfaltigkeit"* in German) is due mainly to Riemann. It is considered by the contemporary geometer Chern to be a true revolution which he compares with *"the introduction of clothing for mankind"*. We quote him: *"If geometry is the human body and coordinates are clothing, then the evolution of geometry has the following comparison.*

> Synthetic Geometry   Naked Man
> Coordinate Geometry   Primitive Man
> Manifolds   Modern Man ."

The development of the concept of a configuration space nevertheless lasted more than a century, from 1854 to 1940, and was accompanied by a change in point of view in which the emphasis shifted to the study of *global* questions. The concept itself is now well established as one of the cornerstones of geometry. This point was once again first made by Weyl in a text on the philosophy of mathematics in which he insisted on the fact that the concept of a configuration space enlarged the scope of geometry. In the extract we quote, he speaks of the space of colours which we described at the beginning of this chapter. He said: *"Es ist für die philosophische Diskussion der Geometrie wichtig, dass neben dem Raum noch ein ganz anderes Gebiet anschaulicher Gegebenheiten besteht: die Farben, welche ein geometrischer Behandlung fähiges Kontinuum bilden."*[116]

The concept of a configuration space has also turned out to be of central importance for theories in physics and mechanics. This is the case for general relativity, a remarkable construction which is the culmination of the idea of Einstein[117] that physical phenomena related to gravity should be interpreted as modifications of the

---

[115] for a bilingual edition, in Latin and English, of this historically important document, see (Dombrowski 1979) which contains in addition an analysis of the subsequent evolution of the concepts introduced by Gauss.

[116] *"For a philosophical discussion of geometry it is essential that, besides the physical space, there is another very different domain related to evident data, namely colours, which forms a continuum that can be dealt with geometrically."*

[117] The physicist Albert Einstein (1879–1955) taught in Zürich, Prague, Berlin and Princeton after having begun his professional life at the patent office in Zürich. He transformed a large part of physics at the beginning of the XXth century by proposing radically new theoretical models for the photoelectric effect leading to the wave-particle duality (1905), for the structure of spacetime by making the *relativity principle* one of the fundamental principles of physics (1905), and for the gravitational interaction by interpreting gravity as a modification of the geometry of spacetime (1913–1915). He made use of the newest mathematical concepts of his time in a very penetrating study of the fundamental concepts of physics.

geometry of space. Note though that this idea was already aired by Clifford[118] in 1869! It is also the case for gauge theories, to which we have often alluded in this chapter, theories whose basic concepts are precisely those which geometers had constructed throughout the first half of the XXth century, which led the physicist Yang[119] to say in 1975: *"That non-Abelian Gauge Theories are conceptually identical to ideas in a beautiful theory... developed by mathematicians without reference to the physical world was a great marvel to me... This is both thrilling and puzzling, since you, mathematicians, dreamed up these concepts out of nowhere."* Other scientists had already been astonished by these extraordinary coincidences: Wigner[120] spoke of *"the unreasonable effectiveness of Mathematics in the Natural Sciences"*, but we leave the last word to Chern, for whom the explanation of these connections lies in the fact that *"fundamental concepts are always rare"*.

V.123. The parametrisation of the group of rotations in three dimensions arose from the efforts of many workers in mathematics and mechanics, among which we may mention Euler in 1776, Gauss in 1819, Rodrigues[121] in 1840, Hamilton and Cayley in 1843, Klein in 1884, É. Cartan in 1913 and in 1938 to cite only the main contributors. The most striking feature of this list is the mixture of scholars whose motivations were extremely diverse and in which algebra and geometry were inextricably entwined.

---

[118] The English mathematician and philosopher William Kingdon Clifford (1845–1879) was in particular interested in geometric algebra. An important family of algebras bear his name. They play an important role in modern mathematics and theoretical physics, providing in particular the most efficient way of introducing spinors. At age 34, he died of tuberculosis in Madeira.

[119] The Chinese-American physicist Yang Chen Ning, born in 1922, is a theoretical physicist with a very broad coverage of the field. He and Lee Tsung-Dao stated that parity was violated in weak interactions for which they received the Nobel Prize in physics. He played a key role in the development of non-Abelian gauge theory, one of them being called Yang–Mills theory, after his collaboration in the 1950s with Robert Mills. In the second part of his life, he made great efforts to help the Chinese physics community develop at the highest level.

[120] Eugene Wigner (1902–1995) was a Hungarian-American mathematician and theoretical physicist who was interested in particular in introducing group theory in physics to deal with symmetries. In his youth, he went to the same high school as von Neumann. He was involved in the Manhattan Project in World War II. Later in his life, after a number of official advisory responsibilities, he developed an interest in more philosophical questions. He received the Nobel Prize in physics in 1963.

[121] Benjamin Olinde Rodrigues (1795–1851) was a French banker and mathematician who got very involved in the social reforms promoted by the 'Comte de Saint-Simon'. In mathematics, he is remembered for his formula concerning rotations, anticipating Hamilton's quaternions.

Chapter VI
Tangent Vectors and Vector Fields on Configuration Spaces

This chapter completes the background we need to be able to treat Variational Calculus by giving an intrinsic meaning to the notion of an *infinitesimal variation*.

In Sect. A, we introduce the notion of a *tangent vector* to a configuration space which is suggested by that of a velocity-vector to a curve. To extract all the necessary information, it is useful to make the tangent vectors operate as *derivations* on the algebra of local observables, a point of view that we have already introduced. in a more restricted setting. It is then easy to define operations which make the collection of tangent vectors into a *vector space*.

With these definitions, we are able (finally!) to define in Sect. B the *tangent space* at a point of a configuration space and in Sect. C the *tangent linear map* to a differentiable map between configuration spaces.

In Sect. D we generalise the concept of a *vector field* to configuration spaces. They provide a rigorous version of the notion of *virtual variation* in mechanics. We also give the formulas necessary to deal with vector fields in various local coordinate systems.

In Sect. E we again take up, in the setting of configuration spaces, the correspondence between vector fields and differential equations established in Chapter IV in the setting of open subsets of vector spaces. We examine, in particular, the effect of several global properties on this correspondence. We prove notably the completeness of vector fields defined on compact configuration spaces. We end the section by defining the *Lie derivative* which plays a central role in continuum mechanics under the name of *material derivative*. The basic formulas concerning the Lie derivative are stated as exercises.

A. Tangent Vectors to a Configuration Space

In this section, we introduce the *tangent vectors to a configuration space*. We define them initially as equivalence classes of curves having the same velocity-vector in at least one chart.

We then show that the tangent vectors at a point can also be viewed as *derivation operators* on the algebra of observables defined on a neighbourhood of the point.

VI.1. To define the notion of tangent vector to a configuration space, we shall recall the discussion we developed in Chapter III, leading to the notion introduced in III.15 of maps tangent to each other, which we extend to curves on a configuration space.

VI.2. To this end, let a be a point of a configuration space M. By a curve *originating at a*, we mean a curve γ on M defined in a neighbourhood of 0 such that $\gamma(0) = a$.

> *All the curves we shall consider are assumed to be*
> *differentiable in some chart*
> *and thus, by V.23, in any admissible chart.*

Definition-Proposition VI.3. *Let γ and η be two curves originating at a point a of a configuration space M of dimension n. These curves are said to be* tangent *at a if, in some local coordinate system x, the numerical curves (i.e. with values in $\mathbb{R}^n$) $(\gamma)_x (= x \circ \gamma)$ and $(\eta)_x (= x \circ \eta)$ are tangent in the sense of III.15, in other words if they have the same velocity-vector at 0.*

This property is independent of the system of local coordinates chosen to express it and defines an equivalence relation on the set of curves originating at a whose classes are called tangent vectors *at a. For a curve γ originating at a, its tangent vector at a is denoted by $\dot{\gamma}(0)$.*

Proof. For the proof, it is more convenient to call the curves γ_1 and γ_2 instead of γ and η.

If y is another chart containing a in its domain of definition, then after possibly restricting the neighbourhood of 0 where the curves are defined, we have $(\gamma_i)_y = \tau_y^x \circ (\gamma_i)_x$ (where $i = 1, 2$).

By the Chain Rule III.46, the equality of the velocity-vectors of the curves in the chart x implies the equality of the velocity-vectors in the chart y.

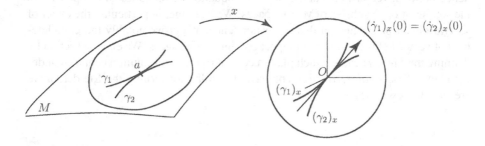

Fig. VI.5 Curves originating at a which are tangent at this point

The fact that the relation *"being tangent at a"* is an equivalence relation on the set of curves originating at a is immediate.

Reflexivity and symmetry follow from the fact that the relation which they express is an equality.

To prove transitivity, we consider three curves γ_1, γ_2 and γ_3 such that the first two and the last two are tangent to each other; it may be, however, that the equality of the velocity-vectors of γ_1 and γ_2 on the one hand, and of γ_2 and γ_3 on the other hand, are given in terms of different charts at a. But we have just shown that the equality in one chart implies the equality in any other chart.   □

Remark VI.4. A priori, there are two difficulties with Definition-Proposition VI.3:

i) to construct an object which we shall see very soon is simple enough, we use a rather monstrous space[122] consisting of the curves originating at a (with the complication, which we have not stressed very much, that the curves are not all defined on the same interval);

ii) it does not give any evidence why we are justified in calling the objects we have constructed *vectors*, i.e. the fact that the operations of addition and scalar multiplication can be defined on the set of tangent vectors at a point. We proceed to fill in these gaps in the sequel.

VI.6. To this end, it is useful to view a tangent vector at a point a of a configuration space M as an operator acting on the algebra $\mathcal{F}U$ of observables (i.e. the differentiable scalar functions) defined in a neighbourhood U of a.

Definition-Proposition VI.7. *Let $\varphi \in \mathcal{F}U$ and let X_a be a tangent vector at a. For all curves γ originating at a belonging to the equivalence class X_a, the value of the derivative at 0 of the function of a real variable $t \mapsto (\varphi \circ \gamma)(t)$ is the same. It is called the* directional derivative *of the observable φ in the direction of the tangent vector X_a and denoted by $\partial_{X_a} \varphi$.*

Proof. Let X_a be a tangent vector at a and φ an observable defined in a neighbourhood U of a. If γ is a curve originating at a belonging to the equivalence class X_a, note that there exists a neighbourhood of 0 on which the function of a real variable $t \mapsto (\varphi \circ \gamma)(t)$ is well-defined. We now remark that the derivative of this function at 0 is equal to $\langle d(\varphi^x)(0), (d\gamma_x/dt)(0) \rangle$ where φ^x denotes the local expression for φ in a chart x at a. This follows from V.47 since $\varphi \circ \gamma = \varphi \circ x^{-1} \circ x \circ \gamma = \varphi^x \circ \gamma_x$.

Since two curves originating at a which are tangent to each other have the same velocity-vector at 0 in every chart at a, we see that the value of the quantity $\langle d(\varphi^x)(0), (d\gamma_x/dt)(0) \rangle$ is independent of the chart used to evaluate it, and of the choice of representative γ in its equivalence class which is a tangent vector X_a at a. It is therefore legitimate to call $\partial_{X_a} \varphi$ the value of this derivative at 0.   □

Remark VI.8. It is important to verify that the notation we have just introduced is really compatible with the analogous notation we introduced in the setting of vector spaces. For this it suffices to take a linear chart and for the curve γ associated to a vector v in the vector space E, which is supposed to be attached to the point $a \in E$, the curve $t \mapsto a + tv$.

---

[122] Such a roundabout method is not unusual in mathematics, but it often gives the novice the impression that life is being made difficult for no good reason. It illustrates the power of the construction of objects as equivalence classes, but it goes to show that it is necessary to supplement it by devising a *concrete* picture of the objects.

Proposition VI.9. *Let a be a point of a configuration space M. For any tangent vector X_a at a, the map which, to any observable φ defined in a neighbourhood U of a, associates $\partial_{X_a} \varphi$ is linear and is a* derivation *at a, i.e. it satisfies the* Leibniz relation, *which means that, for any functions φ and ψ of $\mathcal{F}U$,*

$$\partial_{X_a}(\varphi\psi) = \psi(a)\, \partial_{X_a}\varphi + \varphi(a)\, \partial_{X_a}\psi \ .$$

Proof. To establish the proposition, it is enough to go back to the definition of the action of a tangent vector at a on a local observable at a. We have already seen above that the formula $\partial_{X_a}\varphi = \langle d(\varphi^x), (d\gamma_x/dt)(0)\rangle$ holds for any choice of chart x at a and any curve γ which defines X_a.

The linearity then follows directly from the obvious fact that the expression in a chart of a linear combination of observables is the same linear combination of their expressions in the chart.

The Leibniz formula for the action of the tangent vectors is then also a direct consequence of the defining formula which we recalled above and the classical formula for the derivative of a product. □

Remark VI.10. The notion of a derivation that appears in the statement of Definition-Proposition VI.9 can be introduced for any algebra, as one easily verifies.

One elementary consequence of the definition is the following: for any derivation X of the algebra of observables, $\partial_X 1 = 0$ where 1 denotes the unit element of the algebra (in our case, the function which is constant and equal to 1 is evidently an observable) and consequently $\partial_X(\alpha 1) = 0$ for any scalar α by linearity. (In fact, for any element φ in the algebra, we have $\partial_X\varphi = \partial_X(\varphi 1) = 1\,\partial_X\varphi + \varphi\,\partial_X 1$, and hence the result stated.)

VI.11. Given any tangent vector X_a, it will be useful to extend the notation $\partial_{X_a} f$ to maps f with values in a numerical space $\mathbb{R}^n$ (which we usually call "numerical functions"). Thus, if $x = (x^1, \cdots, x^n)$ is a chart at a, we shall use $\partial_{X_a} x$ to denote the n-tuple $(\partial_{X_a} x^1, \cdots, \partial_{X_a} x^n)$.

V.12. To deepen our understanding of tangent vectors (which are a rigorous version of *infinitesimal variations*), the following lemma of Hadamard[123] is very useful. It can be interpreted as a Taylor formula without remainder, something that the context of configuration spaces allow.

Hadamard's Lemma VI.13. *Let φ be an observable defined in a neighbourhood of a point a in a configuration space M of dimension n. For any chart $x = (x^1, \cdots, x^n)$ centred at a, there exist n local observables $\chi_1, ..., \chi_n$ (possibly defined on a smaller neighbourhood of a) such that*

$$\varphi = \varphi(a) + \sum_{i=1}^{n} x^i \chi_i \ .$$

---

[123] Jacques Hadamard (1865–1963) taught simultaneously at the Collège de France, at the École Polytechnique and at the École Centrale until his retirement in 1937. He gave the first proof of the prime number theorem and made a number of contributions to the theory of partial differential equations, notably to the propagation of singularities of the wave equation, and also to geometry, notably to the behaviour of geodesics in manifolds.

Proof. We fix a chart x at a whose image in $\mathbb{R}^n$ is assumed to be convex and we work with the expression φ^x for φ in this chart. By the fundamental theorem of integral calculus and the Chain Rule III.46, we can write

(VI.14)
$$\varphi^x(q^1, \cdots, q^n) - \varphi^x(0, \cdots, 0) = \int_0^1 \frac{d\varphi^x}{dt}(tq^1, \cdots, tq^n)\, dt$$
$$= \sum_{i=1}^n q^i \int_0^1 \frac{\partial \varphi^x}{\partial x^i}(tq^1, \cdots, tq^n)\, dt\, .$$

If we put $\chi_i^x(q^1, \cdots, q^n) = \int_0^1 (\partial \varphi^x/\partial x^i)(tq^1, \cdots, tq^n)\, dt$ (which is an observable by the theorems about differentiation under the integral sign), we have proved the lemma since the formula in the statement has Eq. (VI.14) as its expression in the chart x. □

Remark VI.15. In the proof of Hadamard's Lemma VI.13, we have used the fact that the observables are infinitely differentiable without saying so explicitly. In fact, it is because the functions φ are infinitely differentiable that the functions χ_i are also (and there lies the only subtlety in the lemma).

If φ had been a function of class C^k for k finite, the functions χ_i would only have been of class C^{k-1}, in other words they would not have been observables of the same class as φ. In this case, we would not have been able to carry out the operation of division *inside* the algebra of observables.

VI.16. Hadamard's Lemma VI.13 can be interpreted as a division theorem without remainder, with coefficients in the space of functions of class C^∞. In the case of the theory of analytic functions of one or several variables, there exists a much earlier division theorem which singles out one variable. It is due to Weierstrass and known as the *preparation theorem*. A generalisation to the differentiable case has been obtained much more recently. They have proved to be fundamental tools in the study of complex spaces, especially in the presence of singularities.

B. Tangent Spaces to a Configuration Space

We are now in a position to turn the collection of tangent vectors at a point of a configuration space into a vector space, the *tangent space* to the configuration space at the point.

Theorem VI.17. *The collection of tangent vectors at a point a of a configuration space M of dimension n can be identified with the real n-dimensional vector space $T_a M$ of derivations at a.*

The vector space structure thus defined on the equivalence classes of curves tangent at a coincides with the structure obtained by transporting that of $\mathbb{R}^n$ using a chart.

The collection of all tangent vectors to a configuration space M is denoted by TM and is naturally equipped with a natural projection $\pi_M : TM \longrightarrow M$ which to a tangent vector $v \in T_a M$ associates the point $a \in M$.

Proof. Before turning to the proof of the theorem proper, we examine some consequences of Hadamard's Lemma VI.13. Let X_a be a tangent vector at a. If, in a chart x centred at a (i.e. for which $x(a) = (0, \cdots, 0)$), $\varphi = \varphi(a) + \sum_{i=1}^{n} x^i \chi_i$ is the decomposition of a local observable at a, then by the Leibniz formula we have

$$(\text{VI.18}) \qquad\qquad \partial_{X_a} \varphi = \sum_{i=1}^{n} \chi_i(a) \, \partial_{X_a} x^i \, .$$

Let us show that the map which associates a derivation at a to a tangent vector at a is injective. In fact, suppose that two tangent vectors at a, say X_a and Y_a, have the same image under this map. This means that, for any observable φ, $\partial_{X_a} \varphi = \partial_{Y_a} \varphi$; in particular, for $i = 1, ..., n$, we have $\partial_{X_a}(x^i) = \partial_{Y_a}(x^i)$. Thus, in a chart x centred at a, the velocity-vectors of the expressions of two curves γ and η, which define the tangent vectors X_a and Y_a respectively, satisfy the n relations

$$\langle dx^i(0), (d\gamma_x/dt)(0) \rangle = \langle dx^i(0), (d\eta_x/dt)(0) \rangle \, ;$$

these equations imply that the curves γ and η are tangent, and hence that $X_a = Y_a$.

We now show that the map is surjective. It follows from Eq. (VI.18) that any derivation at a is determined by its values on the n functions x^i. If D is a derivation at a such that $D(x^i) = v^i$, then the tangent vector at a determined by the curve $t \mapsto x^{-1}(v^1 t, \cdots, v^n t)$ has the same action on the n functions x^i ($i = 1, \cdots, n$) as D, and hence coincides with it.

It remains to establish the coincidence of the vector space structures defined on $T_a M$ by the action on the observables and by transporting the structure on $\mathbb{R}^n$ using the velocity-vectors of curves in a chart. This is done using the correspondence established in the preceding paragraph. □

Remark VI.19. We could have defined the operations which make $T_a M$ a vector space in terms of the curves originating at a by showing that these definitions pass to the quotient by the equivalence relation which was used to introduce the notion of a tangent vector.

The approach we have taken has the advantage that it made us familiar with the manipulation of algebras of local observables, which are fundamental objects.

VI.20. Several notations are used simultaneously to denote tangent vectors in the literature. To better understand their significance, it might be useful to present some of them.

Recalling the notation of Lagrange, people working in mechanics often use the notation $(q^1, \cdots, q^n)$ to denote the coordinates in a chart. Since tangent vectors can be defined as equivalence classes of curves which are tangent to each other, and thus which have the same velocity-vector, an *infinitesimal variation* is often denoted by δq, where δ is considered as a "variation operator", but this point of view leads to a mathematical dead end. We comment on the notation $\dot{q}^i$ often used in mechanics at the end of VI.34.

Remark VI.21. From the two definitions of a tangent vector, it is clear that *"there does not exist any intrinsic means of comparing tangent vectors at two distinct points of a configuration space"* similar to the method available in vector spaces thanks to the existence of translations. We return to this point a little later, but *it is here that one of the main subtleties of the notion of a configuration space resides.*

C. Tangent Linear Maps to Differentiable Maps

We can now define (finally!) the *tangent linear map* to a differentiable map between configuration spaces, a task which we have had to put off since Chapter V due to the lack of a definition of the objects on which this map should act, namely tangent vectors.

Proposition VI.22. *A differentiable map f from a configuration space M to a configuration space N defines at each point a of M a linear map from $T_a M$ to $T_{f(a)} N$, called its* tangent linear map *and denoted by $T_a f$, such that, for any tangent vector X_a at a defined by a curve γ originating at a, $(T_a f)(X_a)$ is the tangent vector at $f(a)$ defined by the curve $f \circ \gamma$, i.e. $(f \overset{\cdot}{\circ} \gamma)(0) = (T_{\gamma(0)} f)(\dot{\gamma}(0))$.*

Proof. Let $f : M \longrightarrow N$ be differentiable. By V.23, in charts at a and at $f(a)$, the local expression of f is differentiable in the ordinary sense. Let X_a be a tangent vector at a and let γ_1 and γ_2 be two curves belonging to the equivalence class X_a.

We show that the curves $\eta_i = f \circ \gamma_i$ for $i = 1, 2$ are tangent to each other at $f(a)$. For this it suffices to take a chart x at a and a chart y at $f(a)$ and to note that $(\eta_i)_y = (f \circ \gamma_i)_y = f_y^x \circ (\gamma_i)_x$. Since by definition the curves γ_1 and γ_2 are tangent at a, we have $(\dot{\gamma}_1)_x(0) = (\dot{\gamma}_2)_x(0)$ and consequently

$$(\dot{\eta}_1)(0) = T_0 f_y^x((\dot{\gamma}_1)(0)) = T_0 f_y^x((\dot{\gamma}_2)(0)) = (\dot{\eta}_2)(0) .$$

The linearity of the map $T_a f$ results from the fact that the vector space structure of $T_a M$ is transported from that of $\mathbb{R}^n$ by the tangent linear map of the chart. $\qquad\square$

VI.23. It is necessary to do many exercises to familiarise oneself completely with tangent linear maps. We shall see in Chapter VII a powerful tool which enables one to calculate (often more simply!) the tangent linear map of a map between configuration spaces when it is obtained by restricting a map defined on a larger space. For this, we shall have to introduce another way of constructing configuration spaces, as sets satisfying *constraints*.

Exercise VI.24. State and prove a converse of Proposition VI.22.

VI.25. Proposition VI.22 provides the missing link which enables us to state the following fundamental results, which are illustrations of the Basic Principle V.21.

Extended Chain Rule VI.26. *Let f be a map from a configuration space M to a configuration space N which is differentiable at a point $a \in M$. If g is a map from a neighbourhood of $f(a)$ to another configuration space P which is differentiable at $f(a)$, then $g \circ f$ is differentiable at a and*

$$T_a(g \circ f) = (T_{f(a)} g) \circ (T_a f) .$$

Remark VI.27. Proposition VI.22 is a special case of the preceding rule.

Extended Local Inversion Theorem VI.28. *Let* $f : M \mapsto N$ *be a differentiable map between two configuration spaces. If there exists a point a in M where the tangent linear map* $T_a f$ *of* f *is a linear isomorphism from* $T_a M$ *onto* $T_{f(a)} N$, *then there exists open neighbourhoods U of a and V of* $f(a)$ *such that:*

i) $f|_U$ *is a diffeomorphism onto V,*
ii) *the inverse map g is differentiable and* $T_{f(a)} g = (T_a f)^{-1}$.

Remark VI.29. We have stated the results in Chapters III and IV in such a way that no modification is necessary for the case of configuration spaces. We could give other statements generalising the classical ones. To avoid overburdening the presentation, we refrain from doing so systematically.

VI.30. Among the fundamental examples of diffeomorphisms (defined only on open subsets of configuration spaces) figure, of course, the *admissible charts*. We are going to use them to introduce a notion which is extremely useful in applications. It allows one, in fact, to *extend calculus in local coordinates to tangent vectors*.

VI.31. For a chart $x = (x^1, \cdots, x^n)$ at a point a in M, it is traditional to denote by $((\partial/\partial x^1)(a), \cdots, (\partial/\partial x^n)(a))$ the basis of $T_a M$ which is the image of the basis (e_i) of $\mathbb{R}^n$ under $T_{x(a)} x^{-1}$. This basis is called the *natural basis* associated to the chart x.

Let us show that this notation is justified. In fact, since $\partial/\partial x^i = (T_{x(a)} x^{-1})(e_i)$, for any observable φ, we have by the Extended Chain Rule VI.26

(VI.32)
$$\left(\partial_{\frac{\partial}{\partial x^i}} \varphi \right)(a) = \frac{\partial \varphi^x}{\partial x^i}(x(a)) .$$

Just as we warned the reader in Chapter III against the dangers of the notation $\partial/\partial x^i$ in the setting of vector spaces, we once again insist on the fact that *each vector in this basis depends on all n of the given local coordinates* $x^1, ..., x^n$. This is clearly visible on the right-hand side of Eq. (VI.32) where it is the expression φ^x of the observable φ in the chart x which appears.

Corollary VI.33. *Let* X_a *be a tangent vector at a point a of a configuration space. If* $(X^1, \cdots, X^n)$ *are the components of* X_a *in the natural basis* $((\partial/\partial x^1)(a), \cdots, (\partial/\partial x^n)(a))$ *associated to a chart* $x = (x^1, \cdots, x^n)$ *at a, then*

$$X^i = \partial_X x^i .$$

Proof. This is a direct consequence of Eq. (VI.32) applied to the vectors in the natural basis appearing in the decomposition $X = \sum_{i=1}^{n} X^i (\partial/\partial x^i)$. □

VI.34. After having focused our attention on tangent vectors in terms of their action on the algebra of observables, it is useful to consider the induced map on the observables themselves.

Corollary VI.35. *Let φ be an observable defined in a neighbourhood of a point a of a configuration space M. The map which to a tangent vector $X_a \in T_a M$ associates $\partial_{X_a}\varphi$ is a linear form on $T_a M$, called the differential of φ at a and denoted by $d\varphi(a)$, which satisfies*

$$\langle d\varphi(a), X_a \rangle = \partial_{X_a}\varphi .$$

Proof. This is simply a reformulation of Proposition VI.22.   □

Remark VI.36. Here again, it is important to establish that the notation d we have introduced is compatible with that used in the setting of vector spaces.

The only point which might be troubling is the fact that one often uses the same name x^i to denote the observable defined on the configuration space and the linear coordinate in the numerical space where the chart takes its values. Thus, one should not be surprised by the *coexistence* of the two quantities dx^i, one being a linear form on $T_a M$, the other being an element of the dual of the numerical space. Fortunately, these two forms are interchanged by the transpose of the isomorphism $T_a x$.

This is a reason to *identify* objects defined on a configuration space and their representation in a chart *provided the expression involves only intrinsic quantities*.

Exercise VI.37. Show that the natural basis $(\partial/\partial x^i)$ associated to a given chart $x = (x^1, \cdots, x^n)$ is dual to the basis (dx^i) of the space of linear forms.

VI.38. We now consider briefly the special notations used when the source space or the target space are of dimension 1. They are clearly very close to those used in the setting of ordinary differential calculus.

VI.39. We are now in a position to speak of the *velocity-vector* of a curve γ in a configuration space. We can carry over almost word-for-word what we have written in the case of a curve in a vector space, except for the notation $\dot\gamma$ that we introduced in VI.3: the tangent vector $\dot\gamma = T_0\gamma(1)$ which coincides in any chart with the velocity-vector in the restricted sense above, but contains the point at which it is attached.

However, we believe it is useful to emphasise once again that, *by definition, a tangent vector is attached to a well-defined point of a configuration space* and that this point is often concealed in the way it is written. This is the case for the notation $\dot\gamma(t)$ where the point $\gamma(t)$ to which it is attached does not appear.

Of course, the formula

$$\langle d\varphi, \dot\gamma \rangle = \frac{d}{dt}(\varphi \circ \gamma)$$

(which we have established in the classical setting) relating the derivative of an observable φ along a curve γ, the differential of the observable and the velocity-vector of the curve, remains valid.

Once a chart $x = (x^1, \cdots, x^n)$ has been chosen, one often abuses notation by omitting the letter x from the notation used for a curve γ in this chart. One writes, for example, $\gamma(t) = (\gamma^1(t), \cdots, \gamma^n(t))$ so that $\dot\gamma(t) = \sum_{i=1}^{n}(d\gamma^i/dt)(\partial/\partial x^i)(\gamma(t))$. To avoid one more source of confusion, to denote the coordinates of a vector, we do not use the notation $\dot q^i$ so popular in mechanics. In fact, we reserve the "dot" to represent the derivative with respect to a well-defined real parameter, as in $\dot\gamma(t)$.

VI.40. It might be useful to comment briefly on the way a tangent vector can be *pictured mentally*.

In the case of curves, it is tempting to think of a tangent vector as an *arrow* attached to the curve at the point where the derivative is calculated. This representation can lead to error since it makes one think of a tangent vector as an *extended* object. It is better to think of them as infinitely small objects, a point of view which might seem archaic, but which has been revived by the recent developments in *non-standard analysis*.

From this point of view, the term *infinitesimal variation*, which was sometimes used, seems quite appropriate. Moreover, it underlies the action of tangent vectors on the algebra of local observables which we have defined.

Exercise VI.41. Show that, if f is a map from a configuration space M to a configuration space N which is differentiable at a point a of M, then for any observable φ on N

$$d(\varphi \circ f)(a) = {}^t(T_a f)(d\varphi(f(a))) .$$

Exercise VI.42. Show that the restriction to SO_3 of the "trace" map of matrices is a differentiable map, and calculate its differential.

VI.43. With this extension of our terminology, we can combine Proposition V.34 and Corollary V.50 to obtain the following statement.

Corollary VI.44. *Let x^1, ..., x^n be a family of n observables defined in a neighbourhood of a point a of a configuration space M. If the n linear forms $dx^i(a)$ are linearly independent, then $x = (x^1, \cdots, x^n)$ is a chart at a.*

D. Vector Fields on Configuration Spaces

In this section, we extend to configuration spaces the notion of a *vector field*. We examine how vector fields are represented in the natural bases associated to different charts.

So far we have only set the stage on which the play will unfold. It is now time to go into action. The fundamental notion is that of a *field*, i.e. of a quantity whose value depends on the point where it is considered. We have already encountered this idea in the form of *observables*, which are *scalar fields*. It served us well in the task of defining the concept of a configuration space and it also gave us a deep understanding of the notion of a tangent vector to a configuration space.

In this section, we shall restrict ourselves to fields which can be defined directly in terms of tangent vectors, namely *vector fields*, but in the following chapters we consider other fields the understanding of which is as fundamental as that of vector fields.

Definition VI.45. A vector field on a configuration space M is a map which to any point q in M associates a tangent vector at q.

VI.46. For an open subset U of an n-dimensional vector space E, this definition coincides with the usual notion. In fact, by using translations, a vector field can be considered as a map $X : U \longrightarrow E$ which, in a coordinate system (x^i) associated to a basis (e_i), can be written $X = \sum_{i=1}^{n} X^i(x^1, \cdots, x^n)\, e_i$, where the vectors e_i have also been denoted by $\partial/\partial x^i$. We shall generalise this to a curvilinear coordinate system.

Since there is no analogue of translations in a general configuration space, we shall have to carry out the construction in a chart, and then check that the object we have constructed is actually independent of the choices made.

To a chart $x = (x^1, \cdots, x^n)$ of M are associated n *natural vector fields*, the images under Tx^{-1} of the n constant vector fields e_1, ..., e_n on $\mathbb{R}^n$. At any point q in the domain of definition U of the chart, the natural vector fields $\partial/\partial x^1$, ..., $\partial/\partial x^n$ associated to it form the natural basis of $T_q U$ described in VI.30.

VI.47. We shall restrict our attention in what follows to vector fields which have certain regularity properties. We would like to be able to talk about differentiable vector fields. Since we have not yet made the collection of all tangent spaces into a configuration space into a configuration space itself (which we shall do in Chapter IX), we are constrained to use a local representation for the moment.

Definition-Proposition VI.48. *A vector field X on a configuration space M is said to be* differentiable *if, in a neighbourhood of any point q of M, there exists a chart x such that the components $(X^x)^i$ of X in the natural basis associated to x defined by*

$$X = \sum_{i=1}^{n} (X^x)^i \frac{\partial}{\partial x^i}$$

are differentiable functions.

The components $(X^y)^\alpha$ of X in the natural basis defined by any admissible chart $y = (y^1, \cdots, y^n)$ are then differentiable and satisfy the transformation formula

$$(VI.49) \qquad\qquad (X^y)^\alpha = \sum_{i=1}^{n} (X^x)^i \frac{\partial y^\alpha}{\partial x^i} \ .$$

Proof. Since the vectors $\partial/\partial x^i$ of the natural basis associated to the chart x form a basis at each point, every vector field can be expressed in a unique way in terms of these vector fields $\partial/\partial x^i$.

The fact that the components X^i with respect to the natural basis defined by any admissible chart will be differentiable functions follows from the formulas relating the natural bases of the two charts. In fact, if, for two charts x and y whose sets of indices are denoted respectively by Latin and Greek letters, we can write

$$X = \sum_{i=1}^{n} (X^x)^i \frac{\partial}{\partial x^i} = \sum_{\alpha=1}^{n} (X^y)^\alpha \frac{\partial}{\partial y^\alpha} \ ,$$

which implies Eq. (VI.49) since $\partial/\partial x^i = \sum_{\alpha=1}^{n} (\partial y^\alpha/\partial x^i)(\partial/\partial y^\alpha)$, and the remaining conclusion since the observables form an algebra. □

Remark VI.50. We shall be content to work with fields having the maximum regularity (i.e. C^∞). As for configuration spaces, the fundamental results are not changed if the regularity assumptions made are less draconian, but a minimal regularity, say C^1, is necessary (which means that the configuration space must be at least of class C^2). (Exercise: Explain why.)

From now on, all vector fields considered will be
assumed to be differentiable

Example VI.51. If one defines the configuration space structure on the circle $\mathbf{S}^1$ by two charts θ and θ' with values in $]-\pi, \pi[$ and related on the intersection of their domains of definition $\mathbf{S}^1 - \{-1, +1\}$ by $\theta = \theta' + \pi$, the natural vector fields $\partial/\partial\theta$ and $\partial/\partial\theta'$ associated to these charts coincide on this domain (this is an exceptional phenomenon!). The effect of this is to define a special vector field on the circle $\mathbf{S}^1$ when its configuration space structure is the one we have defined.[124] By abuse of notation, this vector field is often called $\partial/\partial\theta$ which suggests that θ *should be considered as a "generalised" chart, an angular variable defined modulo 2π.*

This construction can be extended to the torus of dimension n. It is possible, for example, to view the torus $\mathbf{T}^2$ of dimension 2 as the product of two circles $\mathbf{S}^1$ with angular coordinates θ^1 and θ^2 and to consider the "constant" vector fields $\alpha^1\,\partial/\partial\theta^1 + \alpha^2\partial/\partial\theta^2$ on it (where $\alpha^1, \alpha^2 \in \mathbb{R}$) with the same abuse of language as on $\mathbf{S}^1$.

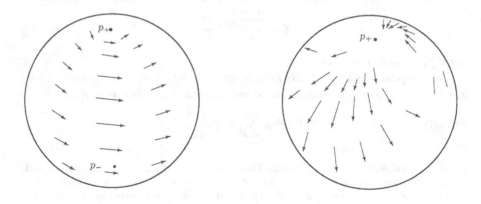

Fig. VI.52 Examples of vector fields on the sphere $\mathbf{S}^2$

Examples VI.53. The vector field featuring in the left half of Fig. VI.52 has the expression $\partial/\partial\theta$ in the geographical chart (θ, ψ) of $\mathbf{S}^2$. (Exercise: Propose a formula for the vector field on the sphere $\mathbf{S}^2$ featuring in the right half of the same figure.)

---

[124] In fact, the best way to think of this vector field is to relate it to the Lie group structure of S^1, but we leave this as an exercise.

On the sphere S^2 viewed as a subset of $\mathbb{R}^3$, one can define the vector field $X(q) = e_3 \times \vec{q}$ where $\vec{q}$ denotes the point $q \in S^2$ viewed as a vector in $\mathbb{R}^3$ and $\times$ is the vector product in $\mathbb{R}^3$, which is assumed to be oriented. We must, of course, verify that X is indeed a field of vectors tangent to the sphere (which, with the definitions we have at our disposal at this point of the course, is a little tricky) and that it is also differentiable.

VI.54. We can form the product of a vector field X by an observable φ by putting $(\varphi X)(a) = \varphi(a) X(a)$, the product on the right-hand side being the multiplication of the vector $X(a) \in T_a M$ by the scalar $\varphi(a)$. For this product, we have all the usual compatibility[125] properties between operations defined on the algebra of observables $\mathcal{F}M$ and on the vector space of vector fields on M, which we *denote* by $\mathcal{T}M$.

As we have seen in VI.9, there is another way to make a tangent vector act on an observable: for an observable φ and a vector field X, at a point a of M, we put $(\partial_X \varphi)(a) = (\partial_{X(a)} \varphi)(a)$. One easily checks that a vector field X is a *derivation* of the algebra $\mathcal{F}M$ of observables, i.e. it satisfies the Leibniz formula

$$\partial_X(\varphi\psi) = (\partial_X\varphi)\psi + \varphi(\partial_X\psi) \, ,$$

which extends the formula given as one of the definitions of a tangent vector at a point.

Project VI.55. Show that every derivation D of the algebra of observables $\mathcal{F}M$ of a configuration space M comes from a unique vector field on M

VI.56. If X_a is a tangent vector at a point a, Proposition VI.46 suggests that one can *prolong* it to a vector field X on the domain of definition of a chart $x = (x^1, \cdots, x^n)$ by putting, for example, $X = \sum_{i=1}^{n} X^i \, \partial/\partial x^i$ where the X^i denote the components of X_a in the basis $\partial/\partial x^i$ at the point a.

Of course, other extensions are possible since the only condition which has to be imposed on the components X^i of X is that they coincide at a with those of X_a. Note, in particular, that this extension procedure, consisting of "*fixing*" the components in a natural basis gives a *different* vector field when applied to another chart.

Remark VI.57. One should note that "*a vector field which is defined and differentiable in a chart is not necessarily the restriction to the domain of the chart of a differentiable vector field on the whole configuration space*". An example is given by the vector field $\partial/\partial\psi$ in the geographical chart (θ, ψ) on the sphere S^2.

VI.58. It is possible to define vector fields by means of the transformation rules relating their components in different coordinate systems, rules which are often called "*vectoriality criteria*": "*a vector field can be recognised by the fact that its components in two local coordinate systems are related by the transpose of the Jacobian matrix of the change of chart.*"

---

[125] The standard mathematical terminology is to say that the space $\mathcal{T}M$ is a *module* over the algebra $\mathcal{F}M$ of observables.

This property is the exact analogue of that which holds in a vector space for the components of a vector in one basis as a function of its components in another basis and the change of basis matrix. It is because of this that vector fields are often called *contravariant* objects.

Note also that a vector field is often referred to in the mechanics literature as a "*virtual variation*", which leads to the notation δq, or even δX.

VI.59. If we were to include the name of the chart in the notation for the components of a vector field in the natural basis associated to the chart (as we did in (VI.49)), the notation would become very cumbersome. Thus, we shall tend to refer to the chart only by the name given to the indices.

Formula (VI.49) is one of the *automatic formulas of vector calculus in curvilinear coordinates*, where the position of the indices often makes it unnecessary to think deeply about the objects themselves: "*to be summed, the indices must be in opposite positions, one above and one below*", a convention which is often called the *Einstein summation convention* because of its constant use in general relativity.

Solved Exercise VI.60. Among the formulas which one must know how to manipulate are those which allow one to express the components of a vector field in one curvilinear coordinate system in $\mathbb{R}^2$ or $\mathbb{R}^3$ in terms of those in another.

Thus, if X is a vector field defined on an open subset of $\mathbb{R}^2$, its components (X^x, X^y) in Cartesian coordinates (x, y) and (X^r, X^θ) in polar coordinates (r, θ) are related, in view of Formula (VI.49), by the relations

$$X^x = \cos\theta\, X^r - r\sin\theta\, X^\theta\,, \quad X^y = \sin\theta\, X^r + r\cos\theta\, X^\theta\,,$$

and conversely we have

$$X^r = \frac{x}{\sqrt{x^2 + y^2}} X^x + \frac{y}{\sqrt{x^2 + y^2}} X^y\,, \quad X^\theta = -\frac{y}{x^2 + y^2} X^x + \frac{x}{x^2 + y^2} X^y\,.$$

These formulas can also be deduced simply from the relations $r\,dr = x\,dx + y\,dy$ and $d\theta = (x\,dy - y\,dx)/r^2$ connecting the coordinate differentials and by using the duality which exists between the natural bases and the coordinate differentials.

Exercise VI.61. Give the formulas expressing the Cartesian coordinates of a vector field defined on an open subset of $\mathbb{R}^3 - \{0\}$ as a function of its components in spherical coordinates, and conversely.

VI.62. In the concrete example of the group of rotations SO_3, it is interesting to give precise formulas for the *left invariant* vector fields, i.e. those whose value at a rotation A is the image of their value at the identity under the tangent linear map of left multiplication by A.

Exercise VI.63. Give a matrix formula for the left invariant vector fields on the group of rotations SO_3 in the two presentations given in Chapter V.

Flow Box Theorem VI.64. *Let X be a vector field on a configuration space M. If at a point a of M we have $X(a) \neq 0$, then there exists a chart $x = (x^1, \cdots, x^n)$ at a in the domain of which $X = \partial/\partial x^1$.*

Proof. The theorem being local, we may suppose that we are working in $\mathbb{R}^n$ and that $a = 0$. We shall construct the required parametrisation by applying the Local Inversion Theorem III.70.

By making use of a linear change of coordinates, we may assume that at the point a we have $X(a) = e_1$ (if we had not made the identification with $\mathbb{R}^n$, we would have said that $X(a) = \partial/\partial x^1$). Consider the map $\Gamma : U \times I \longrightarrow \mathbb{R}^n$ provided by the Local Existence Theorem IV.36. Let $U' = U \cap (e_1)^\perp$ and let $\Gamma_0 : U' \times I \longrightarrow \mathbb{R}^n$ be the restriction of Γ. For $q \in U'$, we have $\Gamma_0(q, 0) = q$ and $(d\Gamma_0/dt)(0, t) = X(\Gamma_0(0, t))$ so that $(d\Gamma_0/dt)(0, 0) = e_1$. The tangent map of Γ_0 at $(0, 0)$ therefore maps the pair $(v, t) \in (e_1)^\perp \times \mathbb{R}$ to $v + te_1 \in \mathbb{R}^n$, and is therefore an isomorphism. By the Local Inversion Theorem III.70, Γ_0 defines a local diffeomorphism from a neighbourhood of $(0, 0)$ in $(e_1)^\perp \times \mathbb{R}$ to a neighbourhood of 0 in $\mathbb{R}^n$, giving the desired chart since by definition of Γ the curves $t \mapsto (x^2, \cdots, x^n, x^1 + t)$ in $(e_1)^\perp \times \mathbb{R}$ are sent to the integral curves of X. $\qquad\square$

Remark VI.65. The Flow Box Theorem VI.64 can be interpreted as a theorem giving the *normal form* of a vector field in a neighbourhood of a point where it does not vanish, namely that of a constant vector field.

The problem of finding a normal form for a vector field in a neighbourhood of a zero is infinitely more delicate. Since the field vanishes, it cannot of course be reduced to a constant vector field (the constant would have to be zero), but one can ask if, in a suitable parametrisation (as differentiable as the vector field), it is possible to reduce it to its first order linear part. The work of Henri Poincaré already showed the possible presence of *resonances* which prevents the reduction of even an analytic vector field to a *linear normal form*. Of course, progress has been made on this question, but it is always an active research topic.

VI.66. The global behaviour of a vector field is influenced (one could even say reflects) the global form of the configuration space on which it is defined. Thus, on certain spaces, such as the *even-dimensional spheres* S^{2m}, *every globally defined vector field necessarily has at least one zero.*

This remark is a gateway to an area of research which is very active today: *algebraic topology.*

Exercise VI.67. Show that *on the circle* S^1, *every nowhere vanishing vector field is the image of the natural vector field $\partial/\partial\theta$ under a diffeomorphism.*

VI.68. Remark VI.66 provides a good opportunity to make the distinction between properties which depend on the *differential structure* and those which depend only on the *topology* of the configuration space.

The notion of a vector field has meaning only when the differential structure is fixed. However, *the existence of zeros of a vector field is a property which is forced by the non-vanishing of a topological invariant* which one can attach to any topological space, namely its *Euler–Poincaré characteristic*. This number is equal to 0 for all topological manifolds of odd dimension and is, for example, equal to 2 for the even-dimensional spheres (note that this implies that every vector field on such a sphere, even if it has an exotic differential structure, has at least one zero). The interested reader should consult (Milnor 1965) for a discussion of the relation between the Euler–Poincaré characteristic and the "number" of zeros of a vector field.

E. Differential Equations on Configuration Spaces

In this section, we extend to configuration spaces the correspondence described in Chapter IV between vector fields and one-parameter groups of diffeomorphisms. As an example of the influence of the global properties of a space on the vector fields which are defined on it, we show that *every vector field on a compact configuration space is complete*.

This extension leads us naturally to examine a fundamental notion, the Lie derivative, which is used constantly in continuum mechanics, and for which we give several formulas.

VI.69. In Chapter IV, we gave a *"dynamical"* interpretation of a vector field defined on an open subset of a vector space by associating to it a one-parameter group of local diffeomorphisms which gave *differential equations* a geometric status.

Once again, we have taken care to state all the definitions and all the theorems in such a way that they extend to configuration spaces without any substantial modification of their wording. Above all, we shall naturally be interested here in the *global* properties of the solutions of these equations.[126]

VI.70. By Corollary IV.44, to any given vector field X on a configuration space M is associated a one-parameter group of local diffeomorphisms (ξ_t).

In fact, given a point a in M, it is always possible to work in a chart centred at a and to transport the local flow defined by the vector field X on the domain of definition of the chart. (Recall that we are only considering differentiable vector fields here.)

We state this result formally in the form of a theorem.

Theorem VI.71. *To any vector field X defined on a configuration space M is associated a one-parameter family of local diffeomorphisms (ξ_t) of M, the* flow *of X, defined on an open neighbourhood of $M \times \{0\}$ in $M \times \mathbb{R}$ and which satisfies at every point q in M the characteristic condition*

$$\frac{d}{dt}\xi_t(q) = X(\xi_t(q)) \, .$$

Proof. We must transport IV.43 and IV.44 to the case of configuration spaces.

At a given point a in M, it is possible to obtain the flow for sufficiently small times by working in a local chart. That the integral curves originating at a thus obtained coincide, at least for a certain length of time, is a consequence of the Uniqueness Theorem IV.28 since the images in M of the integral curves obtained in different charts must have the same velocity-vector at a, namely $X(a)$.

It then only remains, as in the case of open subsets of vector spaces, to define the maximal solution passing through each point by piecing together the local solutions. ☐

---

[126] Historically, it is actually the reverse of what took place: finding an appropriate setting in which to discuss the *global* properties of differential equations was a powerful stimulus to develop rigorously the notion of a *configuration space* and related notions.

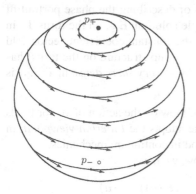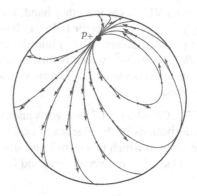

Fig. VI.72 Examples of flows of vector fields on the sphere S^2

Exercise VI.73. Determine the flow of the vector field shown on the left in Fig. VI.72 whose expression in the geographical chart (θ, ψ) is given by $\partial/\partial\theta$. In what way are the flows of the two vector fields shown in Fig. VI.72 special?

VI.74. We now give a global result of the greatest importance. It simplifies matters greatly for the family of spaces to which it applies, namely the *compact* configuration spaces.

Completeness Theorem VI.75. *Every (differentiable) vector field on a compact configuration space is complete, in other words, on a compact configuration space, there is a one-to-one correspondence between vector fields and one-parameter groups of globally-defined diffeomorphisms of the configuration space.*

Proof. This is in fact rather easy and follows from the classical uniformity properties of compact spaces.

Let X be a vector field defined on a compact configuration space M. We show first that, for some number $\epsilon > 0$, the life set Δ_X of X contains $M \times]0, \epsilon[$. By IV.36, we know that, for any point q in M, there exists $\epsilon_q > 0$ and an open set U_q such that every integral curve originating at a point in U_q is defined at least until time ϵ_q. The family of open subsets $(U_q)_{q \in M}$ form by definition an open covering of M.

Since M is assumed to be compact, there exists a finite subcovering of this covering, say $(U_{q_i})_{1 \le i \le k}$. This guarantees that the integral curve originating at every point of M is defined at least for time $\epsilon = \min_i \epsilon_{q_i}$. It also guarantees that the diffeomorphisms ξ_t for $|t| < \epsilon$ are globally-defined on M.

It remains to show that in fact $\Delta_X = M \times \mathbb{R}$. We take any $t_0 \in \mathbb{R}$ and verify that ξ_{t_0} is well-defined. Theorem IV.44 guarantees that $\xi(q, t + t')$ is defined whenever $\xi(\xi(q, t), t')$ is (and is equal to it).

Consequently, it suffices to divide t_0 into pieces of length ϵ, for which we are sure of the uniform existence in time of the solutions, and to use the semi-group relation repeatedly. In fact, if $t_0 = \ell \epsilon' + r$ with $0 < r < \epsilon' < \epsilon$, one can take the diffeomorphism $(\xi_\epsilon)^\ell \circ \xi_r$ as the definition of ξ_{t_0}.

The inverse correspondence between one-parameter families of diffeomorphisms of M and vector fields on M is of course given by taking the infinitesimal generator of the family. $\quad\square$

Remark VI.76. On the other hand, the problem of describing the phase portrait of a globally defined vector field can have no simple solution. E.g. on the torus $\mathbf{T}^2$ in the parametrisation (θ^1, θ^2) introduced in V.64, the constant coefficient vector field $\partial/\partial\theta^1 + \alpha\,\partial/\partial\theta^2$ (which is thus invariant for the group structure on the torus) has *all* its orbits *periodic* if α is rational and *all* its orbits *everywhere dense* in $\mathbf{T}^2$ if α is irrational.

VI.77. Corollary VI.71 suggests another point of view on the action of vector fields as derivations of the algebra of observables: one adopts the *Eulerian viewpoint*, in the sense in which this term is usually understood in continuum mechanics.

In fact, if X is a vector field and (ξ_t) is its flow, we have

$$(\partial_X\varphi)(a) = \frac{d(\varphi \circ \xi_t)}{dt}(a) = \lim_{t\to 0}\frac{\varphi(\xi_t(a)) - \varphi(a)}{t},$$

since by definition the tangent vector to the curve $t \mapsto \xi_t(a)$ at a is $X(a)$.

We have calculated the *derivative following the motion of the particle*, or the *material derivative*, if we think of X as the velocity-field of a fluid. The standard mathematical term is the *Lie derivative* and the usual notation $\mathcal{L}_X$ (to recall Lie's name).

We shall thus use $\partial_X\varphi$ or $\mathcal{L}_X\varphi$ interchangeably to denote the action of a vector field on an observable.

VI.78. The great advantage of this point of view is that *it extends to fields much more general than scalar fields*, in fact to all fields one can naturally construct on a configuration space (see Chapter VIII for the case of volume elements).

This is certainly true for vector fields, whose Lie derivative will now be defined.

VI.79. For two vector fields X and Y, we put[127]

$$\mathcal{L}_X Y(a) = \lim_{t\to 0}\frac{T\xi_{-t}(Y(\xi_t(a))) - Y(a)}{t}.$$

This Lie derivative, which is often called the *Lie bracket* in the case of vector fields (and denoted by [X,Y] instead of $\mathcal{L}_X Y$), has many interesting properties which we do not have the space to develop here. We give two of them in the form of exercises.

Exercise VI.80. Show that the Lie derivative can be identified with the commutator of derivations of the algebra of observables, i.e. that if X and Y are two vector fields and φ an observable, then $\partial_{\mathcal{L}_X Y}\varphi = \partial_X(\partial_Y\varphi) - \partial_Y(\partial_X\varphi)$, a relation which is equivalent to the compatibility of the Lie derivative with the action of vector fields as derivations, $\mathcal{L}_X(\partial_Y\varphi) = \partial_{\mathcal{L}_X Y}\varphi + \partial_Y(\mathcal{L}_X\varphi)$, Deduce that the Lie bracket of two vector fields is skew-symmetric, i.e. that $[X, Y] = \mathcal{L}_X Y = -\mathcal{L}_Y X = -[Y, X]$.

---

[127] The choice of t rather than $-t$ in this formula seems at first just a matter of convention. This is not quite true, for the choice we have made is the one which gives the beautiful compatibility relations with the action of vector fields on functions given in Exercise VI.80.

Exercise VI.81. Show that the Lie derivative acts as a derivation (and thus deserves its name), i.e. that for three vector fields X, Y and Z, we have the so-called *Jacobi identity* $[X,[Y,Z]] = [[X,Y],Z] + [Y,[X,Z]]$.

VI.82. Corollary IV.44 also suggests that we should try to understand the group of diffeomorphisms of a configuration space, a difficult but very interesting question. This group is, of course, not commutative, and one can show that the lack of commutativity of the flows of two vector fields is precisely measured by the Lie bracket of the vector fields.

This theme is very important in many situations arising in mechanics and physics, as well as in mathematics proper, but again, because of lack of space, we give only an introduction in the form of two exercises which have very varied applications.

Finally, in the case where the configuration space is a Lie group, we shall consider the relation between this non-commutativity of flows for certain special vector fields and that of the Lie group itself in Chapters IX and XI.

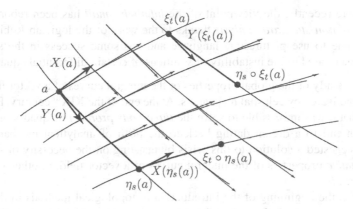

Fig. VI.83 A picture of the non-commutativity of two flows

Exercise VI.84. Show that a necessary and sufficient condition for the flows of two vector fields to commute is that their Lie bracket vanishes.

Deduce that two coordinate vector fields defined by the same chart have zero Lie bracket. Give an expression for the bracket of two vector fields in terms of their expressions in a natural basis associated to a chart.

Project VI.85. Determine the flow of the Lie bracket of two vector fields.

F. Historical Notes

VI.86. The history of the notion of an infinitesimal variation seems to be closely related to that of the Calculus of Variations. The birth of differential calculus was always accompanied by debates (often of a philosophical nature) about the *infinitely small*. This period of controversy and intense discussion is an excellent example of the development of *the scientific method* and of the possibility of *modelling* the real world.

This connection can be seen in the following text of Laplace[128], taken from his *Essai philosophique sur les probabilités*[129] published in 1814: *"Une intelligence qui pour un instant donné connaîtrait toutes les forces dont la Nature est animée et la situation respective des êtres qui la composent, si d'ailleurs elle était assez vaste pour soumettre ses données à l'Analyse, embrasserait dans la même formule les mouvements des plus grands corps de l'Univers et ceux du plus léger atome: rien ne serait incertain pour elle et l'Avenir comme le Passé serait présent à ses yeux."*[130]

VI.87. More recently, the viewpoint of the *infinitely small* has been reborn under the name of *non-standard analysis* thanks to the work of the logician Robinson[131]. It allows one to use picturesque language and has some success in the study of phenomena related to the instability of solutions of certain differential equations.

VI.88. The study of the global properties of the integral curves of a vector field was strongly motivated by celestial mechanics. At the end of the XIXth century, Poincaré showed that it was impossible to solve the *three-body problem* by quadratures, thus putting an end to a dream dating back to the birth of analytical mechanics. He himself suggested a solution to this crisis by insisting on the necessity of studying the *qualitative* properties of the integral curves of a vector field, in other words its *phase portrait*.

This was the beginning of the introduction of topological methods in the study of dynamics, a point of view which has been enriched considerably since that time, and has given birth to the modern theory of dynamical systems. Among those

---

[128] Pierre-Simon de Laplace (1749–1827) was at the same time a mathematician, an astronomer, a physicist and also a political figure of his time. His made major contributions to different fields of mathematics, in particular to probability theory, which gained him a lot of influence through his writings and actions. His first contribution submitted to d'Alembert was on the Calculus of Variations. Inspired by his deep knowledge of the evolution of the Solar system, he got interested in the general problem of the evolution of systems and approached it later in his life from a more philosophical point of view.

[129] *Philosophical Essay on Probabilites*

[130] Here is a translation of the statement by Laplace: *"An intelligence which, for a given moment, would know all the forces by which Nature is animated and the respective situation of the beings which compose it, if besides, it was large enough to submit its data to Analysis, would embrace in the same formulation the movements of the largest bodies in the Universe and those of the lightest atom: nothing would be uncertain for her and the Future as the Past would be present at her eyes."*

[131] The American mathematician Abraham Robinson (1918–1974) developed nonstandard analysis, based on a renewed use of infinitesimal and infinite numbers using his deep knowledge in logic. During his escape from Nazi Germany he widened his knowledge to include applied matters.

who have made important contributions to this development are: Lyapunov and Birkhoff[132] for the study of stability, Andronov[133] and Pontryagin[134], who introduced the study of *structural stability*, and more recently Thom[135] and Smale[136]. The field has now grown into a major field with new connections to probability theory and combinatorics for example.

We should also mention the connection made recently between the discrete approach to dynamical systems and statistical mechanics. A special branch, *ergodic theory*, arose out of this work. It has undergone considerable development thanks to the work of E. Hopf[137], and of an important school in the Soviet Union. We can mention in particular the contributions of Kolmogorov, Anosov[138], Sinai[139] and Arnol'd, and in France, Yoccoz[140].

---

[132] The American mathematician George David Birkhoff (1884–1944) contributed greatly to the study of dynamical systems through the development of ergodic theory. His influence in the American mathematical community has been considerable. He also made significant contributions to general relativity.

[133] Alexandr Alexandrovich Andronov (1901–1952) was a Soviet physicist who studied dynamical systems and in particular their stability.

[134] In spite of being blind from his adolescence, the Soviet mathematician Lev Semionovich Pontryagin (1905–1988) contributed to many fields of mathematics from topology (both algebraic and differential) to analysis, from dynamical systems to optimal control. Through his books he had a considerable influence on the next generations of mathematicians.

[135] René Thom (1923–2002), French mathematician, developed the tools of differential topology and the theory of singularities of maps. Some of his work, which considered the consequences of these theories in the modelling of complex systems in biology and the humanities, made a considerable impact in the media under the heading of *catastrophe theory*. He received the Fields medal in 1958.

[136] Stephen S. Smale, American mathematician born in 1930, worked in differential topology. He has also contributed to the study of dynamical systems. More recently, he has used the powerful ideas from this theory in numerical analysis and approximation theory. He received the Fields medal in 1966.

[137] Eberhard Hopf (1902–1983) was an Austrian-American mathematician who made significant contribution to ergodic theory and the theory of partial differential equations.

[138] A student of Pontryagin, the Russian mathematician Dmitri Victorovitch Anosov (1936–2014) contributed greatly to the differential geometric study of dynamical systems.

[139] The Russian-American mathematician Yakov Sinai, born in 1935, made epoch-making contributions to dynamical systems, at the boundary between deterministic and stochastic systems, and also to probability theory. They had a very significant impact in physics. A student of Kolmogorov, in a joint work, they introduced entropy as a measure of the unpredictability of the evolution of a system. Another contribution was his rigorous foundation of the renormalisation process that was so successful in physics. He received the Wolf Prize in 1997 and the Abel Prize in 2014.

[140] Jean-Christophe Yoccoz (1957–2016) was a French mathematician who made major contributions to the theory of dynamical systems, in particular to the theory of small denominators, often called KAM-theory. He got his PhD under Michael Herman at the Centre de mathématiques of École Polytechnique (which now bears the name of Laurent Schwartz). An accomplished technician, he was never blocked by a difficult calculation. He received the Fields Medal in 1994.

Chapter VII
Regular Points and Critical Points of Numerical Functions

This chapter is devoted to the information which can be extracted from the differential of a numerical function defined on a configuration space. Many of the developments which it contains can be considered as duals of those of Chapter VI.

In Sect. A, we take up systematically the discussion of *differentials of functions* which has been initiated in Chapter VI. We also define the analogues of vector fields, namely the *differential 1-forms*.

Section B is of considerable practical interest since it describes the conditions under which the imposition of *constraints* on a system leads to a new (constrained) configuration space free of singularities. This leads to the notion of a *submanifold* which is often used as a substitute for the abstract definition of a configuration space.

The notion of a critical point of a numerical function is introduced in Sect. C (in the case of an observable, such a point is also called an *extremum*). This is one of the keys to unlocking the Calculus of Variations. Thus, it is important to know how to write the relation which characterises critical points in various situations. In the case of a constrained system, this leads us to the notion of *Lagrange multipliers*. We begin studying the influence of the global form of a configuration space on the critical points of scalar functions defined on it.

In the final Sect. D, we introduce the notion of the *Hessian*[141] which gives the behaviour up to second order of an observable at a critical point. This notion enables us to give local models of scalar functions at generic critical points, these being deduced from the *Morse*[142] *Lemma*. We also give a rapid sketch of *Morse theory* (for a proper exposition, the interested reader is referred to the exemplary book (Milnor 1973) on this subject).

---

[141] Ludwig Otto Hesse (1811–1874) was a German mathematician who worked on the geometric theory of invariants.

[142] The American mathematician Harold Calvin Marston Morse (1892–1977) was interested in the interaction between the global form of a space and the existence of critical points of observables defined on it.

© The Author(s), under exclusive license to Springer Nature Switzerland AG 2022
J.-P. Bourguignon, *Variational Calculus*, Springer Monographs in Mathematics,
https://doi.org/10.1007/978-3-031-18307-2_7

A. Differentials of Functions

This section takes up systematically the notion of the *differential* of an observable. In particular, we examine the expressions for 1-forms in a local coordinate system.

VII.1. Let us recall briefly the definition of the differential of an observable which was the subject of Corollary VI.37: if φ is an observable on a configuration space M and if X_a is a tangent vector at a point a, the real number $\partial_{X_a}\varphi$ denotes the action of this vector as a derivation of the algebra of local observables at a. We can thus regard φ as defining a linear form on $T_a M$, and hence an element of $T_a^* M$ which we denote by $d\varphi(a)$ and which is sometimes called a *cotangent vector* at a (whence also the fact that $T_a^* M$ is called the *cotangent space* at the point a), or a *covector* at a.

When we are in an open subset of a vector space, it is clear that this linear form coincides with the differential which we introduced in Chapter III.

By the same definition, if x is a chart at a, we have

$$d\varphi(a) = {}^t(T_a x)(d\varphi^x)(x(a))$$

where $\varphi^x = \varphi \circ x^{-1}$ is the expression of the observable φ in the chart x.

Exercise VII.2. Show that every 1-form on the tangent space $T_a M$ at a point a of a configuration space M is the differential at a of a local observable at a.

Definition VII.3. A map λ which to any point q of a configuration space M associates an element of $T_q^* M$ is called a *field of 1-forms* or a *differential 1-form on M*.

Remark VII.4. This definition is strictly analogous to the notion of a vector field introduced in the preceding chapter. The use of the numeral "1" before "form" is there to suggest that it is possible to define differential forms of higher order. We do not have space to go systematically into these developments in multilinear algebra (which are nevertheless elementary), often known collectively as *exterior calculus*. However, we shall meet n-forms on configuration spaces of dimension n in Chapter VIII, and a differential 2-form of considerable interest in Chapter X.

So as not to overburden the exposition, and because the danger of confusion here is small, we shall usually speak simply of "differential forms" without specifying that we mean 1-forms.

VII.5. We have already met differential 1-forms (without knowing it of course): if φ is an observable, $d\varphi$ is by definition a differential 1-form. For a chart $x = (x^1, \cdots, x^n)$, the dx^i are n differential forms which, at each point a of the domain where the chart is defined, form a basis of $T_a^* M$, the dual basis to the natural basis $(\partial/\partial x^i)$. This leads us to the following definition which is once again completely analogous to Definition-Proposition VI.48.[143]

Definition-Proposition VII.6. *A differential 1-form λ on a configuration space M is said to be* differentiable *if, in a neighbourhood of each point q in M, there is a chart $x = (x^1, \cdots, x^n)$ such that the components λ_i of λ in the basis dx^i defined by*

$$\lambda = \sum_{i=1}^{n} \lambda_i \, dx^i$$

are differentiable functions.

Moreover, the components of λ in the natural basis defined by any admissible chart are then differentiable, and we have

$$\lambda_i = \langle \lambda, \frac{\partial}{\partial x^i} \rangle \, .$$

Proof. The proof is completely analogous to that given for vector fields. If we denote the sets of indices of the two charts x and y by Latin and Greek letters respectively, we have

$$\lambda = \sum_{i=1}^{n} \lambda_i \, dx^i = \sum_{\alpha=1}^{n} \lambda_\alpha \, dy^\alpha \, ,$$

so that

(VII.7)
$$\lambda_\alpha = \sum_{i=1}^{n} \lambda_i \, \frac{\partial x^i}{\partial y^\alpha}$$

since $dy^\alpha = \sum_{i=1}^{n} (\partial y^\alpha / \partial x^i) \, dx^i$.   □

VII.8. Just as for vector fields, there is a *"vectoriality criterion"* for 1-forms. One recognises them by the way their components transform from one chart to another: *a differential form is characterised by the fact that its components in a chart y are obtained from those in a chart x by multiplication by the Jacobian matrix of the change of chart.* Consequently, differential forms are called *covariant* objects.

To obtain Formula (VII.7), one needs only allow oneself to be guided by the position of the indices (which, we recall, must be in opposite positions to be summed). But for differential 1-forms, this rule is not new since it is precisely that of the calculus of differentials which we have recalled in III.56 and V.47.

From now on, all differential 1-forms
which we consider are assumed to be differentiable.

VII.9. The point of Exercise VII.2 was to show that every cotangent vector at a point a of a configuration space M is the value of the differential of a local observable at a. *However, this does not mean that, if λ is a differential 1-form, there is necessarily an observable φ such that $d\varphi = \lambda$ on an open set.*

For this property to be true, λ must satisfy a differential relation which expresses the vanishing of its *exterior derivative*, a fundamental operation to which we shall

---

[143] Again we work with a local representation because we have not yet made the collection of all cotangent spaces into a configuration space. We shall take up this task only in Chapter X.

return in Chapter X. Any form satisfying this condition is called a *closed* differential form. But, *this condition is sufficient only on sufficiently small open sets, or when the configuration space satisfies a global condition*, such as the vanishing of its fundamental group for example (one then says that the configuration space is *simply-connected*; this is the case for the sphere S^n for $n \geq 2$, for example).

B. Submanifolds and Constraints

In this section, we give a very powerful method of defining new configuration spaces in terms of old ones, namely their *submanifolds*. This is done by means of *constraints* which are required to satisfy some non-degeneracy conditions. The construction of *charts adapted* to a submanifold is described, as are its tangent spaces.

VII.10. As we have pointed out since the beginning of our study of configuration spaces, one drawback of the approach we have chosen is that it excludes systems subject to constraints. In this section, we work towards removing this embarrassing restriction.

Having done this, we shall have a powerful method for constructing new configuration spaces out of existing ones. We shall also have new tools to treat concretely the tangent and cotangent spaces of some configuration spaces we have met.

VII.11. We restrict ourselves here to *differentiable* constraints defined on a configuration space M of dimension n by a numerical function ϕ taking values in $\mathbb{R}^p$. We assume that the constraint is of the form $\phi(q) = 0$.

Even in the simplest case, that of a numerical space, elementary examples show that certain conditions must be imposed on the constraint function ϕ in order for it to satisfy the general dimensional principle "*every system of p equations in n unknowns has a general solution depending on $n - p$ parameters*", true for "*good*" linear equations.

In fact, let $(x^1, \cdots, x^n)$ be a coordinate system on $\mathbb{R}^n$ and let us define $\phi(x^1, \cdots, x^n) = \sum_{i=1}^n (x^i)^2$; then the point $(0, \cdots, 0)$ is the only point of the space which satisfies the constraint $\phi(x^1, \cdots, x^n) = 0$, so that the solution of a system having one equation and n unknowns is a single point, which it is reasonable to consider as a 0-dimensional space, hence the fact in this case we are confronted with a degenerate situation.

We give a definition to avoid these difficulties.

Definition VII.12. Let M be an n-dimensional configuration space. A numerical function $\phi : M \longrightarrow \mathbb{R}^p$ with components $(\phi^i)_{1 \leq i \leq p}$ is a *non-degenerate constraint* on a subset A of M if the p differentials $d\phi^i(a)$ are linearly independent at every point a of A (necessarily $p \leq n$), in other words if $T_a\phi : T_aM \to \mathbb{R}^p$ is surjective.

Remark VII.13. The condition stated in Definition VII.12 is clearly an open condition. If it is satisfied at a point a (in which case one says that a is a *regular point* of ϕ), then it is also satisfied in a whole neighbourhood of a. Consequently, the set A on which the constraint is assumed to be non-degenerate can always be assumed to be open, *which we shall do from now on.*

VII.14. The traditional terminology used to describe the property of the components of ϕ stated in Definition VII.12 is to say that the ϕ^i are *functionally independent.*[144]

Nevertheless, it might be considered too restrictive to require that the constraint takes values in a numerical space. Of course, it is possible to give a more general definition using differentiable maps with values in an arbitrary configuration space. We leave such an extension to the conscientious reader.

In what follows, we concentrate our attention on the set of points of the configuration space defined by the constraint.

Adapted Charts Theorem VII.15. *Let M be an n-dimensional configuration space. Given a point a of the subset C defined by a non-degenerate constraint ϕ with components ϕ^1, ..., ϕ^p with $p \le n$ (so that $C = \phi^{-1}(0)$), there exists at least one chart $x = (x^1, \cdots, x^n)$ of M at a such that, for $1 \le i \le p$, $x^{n-p+i} = \phi^i$.*

Such a chart is said to be adapted *to the constraint ϕ defining C in a neighbourhood of a.*

Proof. Take an arbitrary chart $y = (y^1, \cdots, y^n)$ of M at a. The map y being of rank n at a, the n forms $dy^j(a)$ form a basis of T_a^*M.

Since, by hypothesis, the p forms $d\phi^i(a)$ are linearly independent, by the theorem of incomplete bases, there are $n - p$ forms among the $dy^j(a)$ which form with these a basis of T_a^*M; to simplify the notation, suppose these are $dy^1(a)$, ..., $dy^{n-p}(a)$. Now define the map $x = (y^1, \cdots, y^{n-p}, \phi^1, \cdots, \phi^p)$ on a neighbourhood of a in M with values in the space $\mathbb{R}^{n-p} \times \mathbb{R}^p = \mathbb{R}^n$; its differentials then satisfy Criterion VI.42 and hence it is a chart at a. □

VII.16. Theorem VII.15 enables us to make the set C defined by a non-degenerate constraint ϕ into a configuration space in the following way. First, each point a in C has an open neighbourhood B in C (so that $B = C \cap U$ for an open subset U of M) homeomorphic to an open subset of $\mathbb{R}^{n-p}$: it suffices to take B to be the inverse image under x of the intersection of U with the subspace of $\mathbb{R}^n$ generated by the $n - p$ vectors e_1, ..., e_{n-p}.

As admissible charts of C, we *agree* to take the parametrisations defined by the first $n - p$ coordinates of a chart of M whose last p coordinates are identically zero on C. A typical example of such a chart is provided by Theorem VII.15. That such charts are compatible with each other follows from Proposition V.27, since the change of chart maps are obtained from those of the charts of M by restricting to a linear subspace of $\mathbb{R}^n$. In such a chart, C appears as an open subset of the linear subspace of the space $\mathbb{R}^n$ generated by the $n - p$ vectors e_i for $1 \le i \le n - p$.

The global version of the preceding discussion is provided by the following definition.

---

[144] Note also that the term *relation* is often used (especially in mechanics) instead of *constraints* because this word has another meaning there.

Definition VII.17. Let M be an n-dimensional configuration space. A subset N of M is called a *submanifold* of M if N can be defined in a neighbourhood of each of its points by a non-degenerate constraint.

Remark VII.18. It is useful to insist on the phrase "*be defined in a neighbourhood of each of its points*" used in Definition VII.17. It means that, if a is a point of N, there is a neighbourhood U of a in M and a constraint ϕ such that $\phi^{-1}(0) \cap U = N \cap U$.

It also means that *an open subset of a configuration space is a submanifold*. It suffices to take the empty constraint in a neighbourhood of each point of the open set. Of course it is not a very interesting one!

VII.19. From the discussion given in VII.16, it follows that *a submanifold is naturally a configuration space*, if one takes as its charts the restrictions of the adapted charts to the submanifold. Moreover, it also follows immediately that, for any submanifold N of M, the inclusion $\iota : N \longrightarrow M$ is a differentiable map.

If the constraint defining a submanifold N of a configuration space M of dimension n has values in $\mathbb{R}^p$, the codimension of N in M is p and its dimension is $n - p$.

Proposition VII.20. *Let N be a submanifold of a configuration space M. In order for a map f from a configuration space P with values in N to be differentiable, it is necessary and sufficient that $\iota \circ f : P \longrightarrow M$ be differentiable, where ι denotes the inclusion of N in M.*

Proof. In charts of M adapted to the submanifold N, the map f takes values in a linear subspace. The proposition thus reduces to the analogous proposition for linear subspaces of a vector space, in which case it is obvious.   □

VII.21. Among the examples of submanifolds of numerical spaces, one has of course the *spheres* $\mathbf{S}^{n-1}$ which can be defined, for example, by the numerical constraint $\phi(x^1, \cdots, x^n) \equiv \sum_{i=1}^{n} (x^i)^2 - 1 = 0$ in $\mathbb{R}^n$, which one checks is non-degenerate since at the point $(x^1, \cdots, x^n)$ we have $d\phi = 2 \sum_{i=1}^{n} x^i \, dx^i$ which vanishes only at the point $x = 0$, which does not belong to $\mathbf{S}^{n-1}$.

Remark VII.22. It is easy to see that the exact form of the constraint used to define the sphere as a configuration space is not very important: we could have taken for ϕ any function close to the one which we used, replacing the Euclidean quadratic form, for example, by another more general positive-definite quadratic form.

Exercise VII.23. Let ϕ be a C^∞ function defined on $\mathbb{R}^n$ and homogeneous of degree $k \neq 0$. Show that, for $\alpha \neq 0$, the set $\phi^{-1}(\alpha)$ is a submanifold. Give an example of a function ϕ for which $\phi^{-1}(0)$ is not a submanifold.

VII.24. Another family of submanifolds which is of great practical importance consists of graphs.

If f is a differentiable map from a configuration space M with values in a numerical space $\mathbb{R}^p$, its *graph* is the subset Γ of $M \times \mathbb{R}^p$ defined as follows:

$\varGamma = \{(q, f(q)) \mid q \in M\}$. The graph $\varGamma$ of f is automatically a submanifold of $M \times \mathbb{R}^p$ since $\varGamma$ can be defined by the constraint $\phi(q, x) \equiv x - f(q) = 0$, the differentials of whose components ϕ^i are equal to $dx^i - df^i(q)$ at the point $(q, f(q))$ and which thus form a system of rank p since the dx^i, $1 \leq i \leq p$, are by definition linearly independent.

VII.25. Let us return for a moment to the rotation group SO_3 which was the subject of Sect. E of Chapter V.

In V.95 we gave the constraint which defines SO_3 in the 9-dimensional vector space $\mathcal{M}_3$ of 3×3 matrices. The constraint satisfied by a matrix L in SO_3 can be written $L^t L - I = 0$ and is thus polynomial of degree 2 in the elements of the matrix L, which are the linear coordinates on $\mathcal{M}_3$. There are 9 equations in 9 variables, yet we know that SO_3 is a configuration space of dimension 3. Thus, the constraint cannot be non-degenerate.

To find a non-degenerate constraint to replace a degenerate one is not always a trivial exercise. However, Exercise V.114 gives an indication of the way to proceed in the case of SO_3. In fact, one should keep only the part of the equation $L^t L = I$ which is not satisfied tautologously. Now, for any matrix L, it is trivial to check that the matrix $L^t L$ is symmetric. Requiring that $L^t L$ be a symmetric matrix thus imposes no condition on L. This leads us to consider the constraint $L^t L - I = 0$ as taking values in the 6-dimensional space of 3×3 symmetric matrices, and thus to take as conditions on L only the 6 equations

$$(\text{VII.26}) \qquad \begin{cases} \sum_{j=1}^{3} L_j^i L_j^k = 0 & \text{for } 1 \leq i < k \leq 3, \\ \sum_{j=1}^{3} L_j^i L_j^i = 1 & \text{for } 1 \leq i \leq 3; \end{cases}$$

the verification of the non-degenerate character of these conditions is left to the reader. (There would be $n(n + 1)/2$ in dimension n, this number being, as is well-known, the dimension of the space of symmetric $n \times n$ matrices.)

The system (VII.26) defines the orthogonal group O_3 since so far we have not introduced any condition on the sign of the determinant. It follows from Eq. (VII.26) that the determinant is ± 1. To obtain SO_3, one must take the connected component of the submanifold containing the identity.

VII.27. In general, a constraint cannot be made non-degenerate simply by selecting the *same* components of the constraint at *every point* of the submanifold.

An example of a situation where one has to work harder is given in Exercise VII.28.

Exercise VII.28. In the space $\mathcal{M}_3$ of 3×3 matrices, study the set of orthogonal *projections* defined by the constraint $P^2 - P = 0$ and $^t PP - I = 0$. How many connected components does it have for the topology induced from $\mathcal{M}_3$? Compare with the space of orthogonal symmetries.

Project VII.29. Repeat Exercise VII.28 in the space of $n \times n$ matrices. Show that its connected components are all submanifolds diffeomorphic to Grassmannians. Show that, for $n = 4$, its components are diffeomorphic to $\mathbb{RP}^3$ and to a manifold covered by $S^2 \times S^2$.

Project VII.30. Show that the subset $\mathcal{L}_k$ of linear maps of rank k from $\mathbb{R}^m$ to $\mathbb{R}^n$ is a submanifold of $L(\mathbb{R}^m, \mathbb{R}^n)$ of codimension $(m - k)(n - k)$.

VII.31. One can give many examples of submanifolds of numerical spaces. But one can do better than this: an important theorem proved in 1944 by Whitney established that "*every n-dimensional configuration space can be identified with a submanifold of a 2n-dimensional numerical space*". In fact, it is rather easy to prove this theorem if one allows the numerical space in which we are embedding to have one dimension greater.

VII.32. The dimension $2n$ of the numerical space in which one can embed every configuration space of dimension n is actually the smallest possible. Of course, there are configuration spaces which are not so demanding, such as open subsets of numerical spaces, or spheres, or more generally hypersurfaces defined by a single equation. But others need all these dimensions because of their global properties. This is the case for the real projective space of dimension 2^n which does not embed in dimension $2^{n+1} - 1$, starting with $\mathbb{R}P^2$ which does not embed in $\mathbb{R}^3$.

The real projective plane admits an immersion in 3-dimensional space having a triple point whose most symmetric form is called the *Boy's*[145] *surface*. It admits embeddings in $\mathbb{R}^4$ of which the most famous, known as the *Veronese*[146] *embedding*, can be constructed as follows. To a point of the sphere $\mathbf{S}^2$ with coordinates (x^1, x^2, x^3) in $\mathbb{R}^3$, one associates the point in $\mathbb{R}^5$ having the following coordinates: $(2x^1x^2, 2x^2x^3, 2x^3x^1, 3^{-1/2}((x^1)^2 + (x^2)^2 - 2(x^3)^2), (x^1)^2 - (x^2)^2)$. It is easy to see that antipodal points have the same image, and that the map we have defined takes values in $\mathbf{S}^4$. Hence, we have defined a map from $\mathbb{R}P^2$ to $\mathbb{R}^4$ (for example, by stereographic projection from a point which is not in the image of the projective plane), and this is the required embedding.

For a given configuration space, the problem of finding the smallest embedding dimension is always an active research topic, and one which involves a wide range of techniques from algebraic and differential topology.

Project VII.33. Is it possible to define every submanifold of a numerical space by means of a single constraint (possibly vector-valued)? by a single non-degenerate constraint?

VII.34. It is useful to go rapidly through several elementary properties of submanifolds.

First of all, given two submanifolds N_1 and N_2 of configuration spaces M_1 and M_2, the *product* $N_1 \times N_2$ is naturally a submanifold of the product configuration space $M_1 \times M_2$.

On the other hand, if f is a *diffeomorphism* from a configuration space M to a configuration space M', the image under f of any submanifold n of M is a submanifold N' of M'. In fact, the constraints are obtained by composing the local constraints defining M with the inverse of the diffeomorphism f.

---

[145] A student of Hilbert, the German mathematician Werner Boy (1879–1914) found this non-singular immersion of $\mathbb{R}P^2$ in $\mathbb{R}^3$ but could not find a parametric representation. He died as a soldier during World War I.

[146] Veronese (1854–1917) was an Italian mathematician who made contributions to logic and algebraic geometry, some of which were controversial at his time.

Exercise VII.35. If N is a submanifold of a configuration space M and if P is a submanifold of N, show that P is a submanifold of M.

VII.36. We now come to a crucial point in the theory of submanifolds, namely the description of their tangent spaces in terms of the tangent spaces of the ambient configuration space.

Proposition VII.37 is of great practical significance, and simplifies substantially the determination of the tangent spaces of many examples of configuration spaces. It applies to any configuration space which can be described as a submanifold of a simpler configuration space, such as a numerical space, for example.

Proposition VII.37. *At a point a of a submanifold N of codimension p of an n-dimensional configuration space M, the image of its tangent space $T_a N$ under the inclusion of N into M is the $(n - p)$-dimensional vector subspace of $T_a M$ given by the intersection of the kernels of the differentials of a constraint ϕ defining N in a neighbourhood of a, i.e. $T_a N = \ker T_a \phi \subset T_a M$.*

Proof. Let $\phi = (\phi^1, \cdots, \phi^p)$ be a non-degenerate constraint at a. Let us introduce $K = \bigcap_{i=1}^{p} \ker d\phi_i(a)$.

If γ is a curve in N originating at a (and hence also a curve in M since the inclusion ι is differentiable), we necessarily have $\phi \circ \gamma \equiv 0$. Calculating the derivative of this identity at 0, we obtain $\langle d\phi(a), \dot{\gamma}(0)\rangle = 0$, whence $\iota(T_a N) \subset K$ since any tangent vector to N can be written $\dot{\gamma}(0)$ for a curve γ lying in N. It remains to check that every vector in K is tangent to a curve in N originating at a.

The Adapted Charts Theorem VII.15 gives a chart x of M at a which can be written in the form $x = (x^1, \cdots, x^{n-p}, \phi^1, \cdots, \phi^p)$ at any point of N. Every vector X_a in K can thus be written in the form $X_a = \sum_{i=1}^{n-p} X^i \partial/\partial x^i$. This is clearly the velocity-vector at a of a curve in N defined in a neighbourhood of a and lying in N, for example the curve expressed in the chart by $t \mapsto (tX^1, \cdots, tX^{n-p}, 0, \cdots, 0)$. $\square$

VII.38. If we apply Proposition VII.37 to the case of the sphere S^{n-1} in $\mathbb{R}^n$, we obtain that, at a point a of S^{n-1}, the subspace $T_a S^{n-1}$ can be identified with the kernel of the linear form $a^\flat$ (where $\flat$ refers to the standard euclidean structure of $\mathbb{R}^n$), in other words $T_a S^{n-1}$ can be identified with the *hyperplane orthogonal to the vector a.*

Recall that, even if it is tempting to think of this tangent hyperplane as the affine space perpendicular to a at this point, it is better not to imagine it being in the same space $\mathbb{R}^n$ as S^{n-1}.

Exercise VII.39. Show that *a vector field X defined on a configuration space M is a field of vectors tangent to a submanifold N of M defined by a non-degenerate constraint ϕ if and only if ϕ is a first integral of X.*

Solved Exercise VII.40. Proposition VII.37 gives a particularly simple description of the tangent space at a point L in SO_3.

We begin with the case where $L = I$, the identity matrix. The constraint defining O_3 in $\mathcal{M}_3$ can be written $L^t L = I$. If $t \mapsto L_t$ is a curve in $SO_3 \subset \mathcal{M}_3$ originating at I, its velocity-vector $\dot{L} = (dL_t/dt)(0)$ satisfies the relation $\dot{L} + {}^t\dot{L} = 0$ since transposition is a linear map. Hence, $T_I SO_3$ can be identified with the 3-dimensional vector space of *antisymmetric* matrices. (One should compare this with Exercise V.114.)

At a general point L of SO_3, where a tangent vector is again denoted by $\dot{L}$, the derivative of the constraint can be written $\dot{L}\,^t L + L\,^t \dot{L} = 0$. By using the fact that $^t L = L^{-1}$, and multiplying on the left by L^{-1} and on the right by L, this relation can be put in the form $L^{-1}\dot{L} +^t (L^{-1}\dot{L}) = 0$. This relation can be thought of in the following way: after transporting it to I by left multiplication by L^{-1}, a tangent vector to SO_3 at L becomes a tangent vector to SO_3 at I, and hence an antisymmetric matrix. This is no surprise since, SO_3 being a group, multiplication (on the left or right) by an element of SO_3 preserves this subgroup of $\mathbb{R}Gl_3$.

Everything we said in the previous paragraph applies almost without change to SO_n, the rotation group in n dimensions.

Project VII.41. Show that the tangent space at a map $a \in \mathcal{L}_k$ (the space of linear maps of rank k from $\mathbb{R}^m$ to $\mathbb{R}^n$ introduced in Exercise VII.30) consists of the linear maps $l \in L(\mathbb{R}^m, \mathbb{R}^n)$ such that $l(\ker a) \subset \operatorname{im} a$.

VII.42. We do not have space here to develop the theory of submanifolds further, but it is essential for us to have criteria which ensure that the *intersection* $N \cap N'$ of two submanifolds N and N' of a configuration space M is again a submanifold of M. Consideration of simple examples shows that this is not automatic.

The appropriate notion is that of *transversality*. Two submanifolds are said to be *transverse* if, at each point a in $N \cap N' \subset M$, the tangent subspaces $T_a N$ and $T_a N'$ generate the whole of $T_a M$. Thus, two intersecting curves in $\mathbb{R}^3$ are never transverse, while in $\mathbb{R}^2$ they are transverse provided they do not have the same tangent vector. Similarly, two surfaces (i.e. 2-dimensional submanifolds) in $\mathbb{R}^3$ are transverse at a point of intersection provided they do not have the same tangent plane.

Although elementary (and very natural), this condition took a long time to formulate. It arose in the work of Thom in differential topology in the 1950s. It provided the means with which to study the "*general position*" of mathematical objects and the "*genericity*" of solutions of equations depending on parameters. (For a less sketchy presentation, see (Demazure 1992).)

C. Critical Points and Critical Values of Functions

In this section, we define the fundamental notion of a *critical point* of an observable. We also give the criterion of *Lagrange multipliers* for determining the critical points of the restriction to a submanifold of an observable defined on an ambient configuration space.

We also examine the *global* properties of a configuration space which ensure that an observable necessarily has critical points.

VII.43. In Sect. B, we studied the properties of the level sets of a differentiable numerical function in a neighbourhood of a regular point. In this section we shall be concerned, on the contrary, with the points which are *not* regular.

Definition VII.44. Let M be a configuration space and $\varphi : M \longrightarrow \mathbb{R}^p$ a differentiable numerical function. A point a in M is said to be a *singular* or *critical* point of φ if the differential of φ at a, namely $d\varphi(a)$, is not surjective, and thus of rank strictly less than p.

Remark VII.45. By definition, the notion of a critical point is a *local* notion, in the sense that it only involves the behaviour of the function in a neighbourhood of the point where the property is considered.

However, the existence, on a given configuration space, of critical points depends on the global form of the space and of course on the global behaviour of the function considered. For example, we shall see a little later that any function on the torus $\mathbf{T}^n$ or on the projective space $\mathbb{R}\mathbf{P}^n$ has at least $n + 1$ critical points.

VII.46. If the components of φ are φ^1, ..., φ^p, one possible practical formulation of Definition VII.44 is the vanishing of a non-trivial linear combination of the differentials of the observables φ^i, i.e. the existence of p coefficients, α_1, ..., α_p, not all zero, such that $\sum_{i=1}^{P} \alpha_i d\varphi^i(a) = 0$.

In the case where $p = 1$, a is a critical point of an observable φ if and only if $d\varphi(a) = 0$. One also says that a is an *extremum* of φ. This notion generalises that of a maximum and a minimum, since as we have seen in III.4, the differential of an observable at a maximum and at a minimum is zero. But we shall see at the end of Sect. C and in Sect. D that *there can be global topological reasons which force every observable to have critical points other than maxima and minima.*

If M has dimension n, another expression of the fact that a point a is a critical point of a numerical function φ is the existence of a subspace E of $T_a M$ of dimension $d > n-p$ on which $d\varphi(a)$ is identically zero. This formulation is a direct consequence of the fundamental formula of linear algebra relating the dimensions of the kernel and the image of a linear map l defined on a space of dimension n, namely the relation $\dim(\operatorname{im} l) + \dim(\ker l) = n$.

VII.47. When one is only interested in determining the extrema of a function (as will be the case for us in Chapter IX), one usually proceeds by calculating the differential of the observable φ and determining its critical points simultaneously.

One then requires that, for any curve $t \mapsto \gamma(t)$, we have

$$\frac{d}{dt}\varphi(\gamma(t))_{|t=0} = 0 \,.$$

The explicit form which this formula takes for a given observable φ is often called the *First Variation Formula* of φ. In this formulation, the differential of φ at the point a, assumed critical, does not appear explicitly, and its calculation and its vanishing are merged into a single step.

This point of view is particularly fruitful when, for any tangent vector at a, it is possible to "guess" a simple curve originating at a of which it is the tangent vector (such as the curve $t \mapsto a + tX$ in an open subset of a vector space, for example, or more generally a curve corresponding to one of this type in a chart). (For applications of this remark, see Chapter IX.)

Exercise VII.48. On the rotation group SO_3, determine the critical points of the "trace" function.

VII.49. We can also consider what happens in the target space. A value α taken by a differentiable numerical function φ defined on a configuration space M is called a *critical value* if there exists at least one critical point a such that $\varphi(a) = \alpha$.

The set $C(\varphi)$ of critical values of φ cannot be too large, for a theorem of Sard[147] asserts that "$C(\varphi)$ *can be covered by a union of cubes whose measure can be made arbitrarily small*" (one says that $C(\varphi)$ is *negligible*).

VII.50. It is interesting to note that this theorem is an example of a result which is sensitive to the degree of differentiability of the function considered. In fact, if φ takes values in $\mathbb{R}^p$ and is only of class C^k on a configuration space of dimension n, then $C(\varphi)$ is negligible only if $k > n - p$ and this bound cannot be improved. Whitney has, for example, constructed a map of class C^1 from $\mathbb{R}^2$ to $\mathbb{R}$ for which all the points of an interval are critical values.

To conclude this paragraph, we note that this theorem says nothing about the critical level set (in the source space).

VII.51. Before we end this section, we still have to examine a situation which comes up in applications (and hence is very important): that of determining the critical points of the restriction to a submanifold N of a differentiable numerical function defined on a configuration space M. In a sense, the problem is to understand how a constraint affects the critical points of a numerical function.

Proposition VII.52. *An observable φ defined on a configuration space M has a critical point a when restricted to a submanifold N of M, defined in a neighbourhood of a by a non-degenerate constraint $(\phi^1, \cdots, \phi^p)$ if and only if there exist p real numbers λ_i, $1 \le i \le p$, called Lagrange multipliers of φ at a, such that*

$$d\varphi(a) = \sum_{i=1}^{p} \lambda_i \, d\phi^i(a) \,.$$

Proof. If a is a critical point of the restriction of φ to N, we have $\partial_{X_a} \varphi = 0$ for every vector $X_a \in T_a N$. This relation can be written $\langle d\varphi, X_a \rangle = 0$.

If the submanifold N can be defined in a neighbourhood of a by the non-degenerate constraint $(\phi^1, \cdots, \phi^p)$, a vector $X_a \in T_a M$ belongs to $T_a N$ precisely when the p conditions $\langle d\phi^i(a), X_a \rangle = 0$ hold. We deduce that the linear form $d\varphi(a)$ must be a linear combination of the p linear forms $d\phi^i(a)$, $1 \le i \le p$. $\qquad\qquad\square$

---

[147] Richard Sard, American mathematician born in 1909.

VII.53. In the form of a solved exercise we give a classical application of the notion of Lagrange multipliers. It concerns the search for eigenvectors and eigenvalues of a symmetric $n \times n$ matrix S.

Solved Exercise VII.54. We consider the symmetric bilinear form s whose matrix in the standard basis of $\mathbb{R}^n$ is the symmetric matrix S which we want to study, namely, for any vector $v = (v^1, \cdots, v^n)$ in $\mathbb{R}^n$, $s(v,v) = \sum_{i,j=1}^n S_{ij} v^i v^j$. It is immediate that s defines an observable φ_s on $\mathbb{R}^n$ if we put, for $v \in \mathbb{R}^n$, $\varphi_s(v) = s(v,v)$ (or, in matrix notation, $\varphi_s(v) = {}^t v S v$).

Since the sphere is defined everywhere by the non-degenerate constraint $\phi = \varphi_e - 1$ (where e denotes the standard Euclidean scalar product on $\mathbb{R}^n$), by VII.52 a point a in $\mathbf{S}^{n-1}$ is an extremum of the restriction of φ_s to $\mathbf{S}^{n-1}$ if and only if there exists a real number λ such that $d\varphi_s(a) = \lambda \, d\varphi_e(a)$.

Since $d\varphi_e(a) = 2a$ and $d\varphi_s(a) = 2s^{\#}(a)$ (where $s^{\#}$ denotes the linear map associated by the scalar product e to the bilinear form s, whose matrix in the orthonormal basis (e_i) is precisely (S_{ij})), we obtain the equation for the extrema of φ_s restricted to the sphere in the form

$$s^{\#}(a) = \lambda \, a \, .$$

We have thus shown that *a point a in $\mathbf{S}^{n-1}$ is a critical point of the observable φ_s, which at a point $v \in \mathbb{R}^n$ is equal to $\varphi_s(v) = s(v,v) = \sum_{i,j=1}^n S_{ij} v^i v^j$, if and only if a is an eigenvector of the symmetric matrix S associated to s, and the Lagrange multiplier λ is then the associated eigenvalue.*

It is interesting to note that, in this very special case, where many of the objects considered have several interpretations, λ is also the value which φ_s takes at a and can thus be identified with a *critical value* of the function φ_s.

Exercise VII.55. Re-do Exercise VII.48 using the Lagrange multipliers technique.

VII.56. The presentation which we have given provides an example of an *algebraic notion* which *turns out to be accessible to a variational approach*. There are many others.

What we have done is very modest, since we did not always have available the tools necessary to *prove* the existence of extrema.

In Sect. D, we sketch a method whose origins are in topology which allows one to answer these questions for many observables, namely Morse theory. Despite being of great interest, this method is not applicable in all situations since it requires that the critical points are "*non-degenerate*" in a certain sense. Other theories give (less precise) information on the number of critical points (possibly degenerate) which an observable can have. We now describe some aspects of one of them.

VII.57. This theory is based on the use of a topological invariant associated to a configuration space, namely its *Lyusternik[148]–Schnirelman[149] category*.

For a topological space M, its Lyusternik–Schnirelman category, *denoted* by cat M, is *the minimum number of elements in a covering of M by open subsets each of which can be contracted to a point*. This category is related to the dimension n of M since for any space one has cat $M \leq n + 1$. For us, the main result is the one which states that *"for any observable φ defined on a configuration space M, the number of critical points of φ is at least cat M"*.

It is thanks to this result that we can state, for example, that *"every observable on the torus $\mathbf{T}^n$ has at least $n + 1$ critical points"*, since one can prove that cat $\mathbf{T}^n = n + 1$.

Similarly, one can prove that cat $\mathbb{RP}^n = n + 1$. This is why one must have three charts to define an atlas on SO_3 which, as we know, can be identified with $\mathbb{RP}^3$.

VII.58. If we return to the discussion which we started in VII.54 on the eigenvectors of symmetric matrices, we thus obtain by a variational method a proof of the fact that *"every symmetric $n \times n$ matrix has at least n eigenvectors"*.

In fact, in VII.54, the eigenvectors appear as critical points of an observable on the real projective space $\mathbb{RP}^{n-1}$ which thus has at least n critical points.

The advantage of this approach is that it can be applied to perturbations of the observable φ_s which need not be the restriction of a quadratic form for which the result is valid. For example, if we take φ to be the sum of a quadratic function and a quartic function,

$$\varphi(x) = \sum_{i,j=1}^{n} b_{ij}\, x^i x^j + \sum_{i,j,k,l=1}^{n} a_{ijkl}\, x^i x^j x^k x^l$$

(where the coefficients a_{ijkl} and b_{ij} are assumed to be symmetric), we deduce the existence of at least n solutions $x = (x^1, \cdots, x^n)$ of the non-linear equation

$$\sum_{j=1}^{n} b_{ij}\, x^j + 2 \sum_{j,k,l=1}^{n} a_{ijkl}\, x^j x^k x^l = \lambda\, x^i$$

since this is the form which the equation for the critical points takes in this case.

Remark **VII.59.** In the preceding discussion, it is crucial that the observable φ_s which we defined on $\mathbf{S}^{n-1}$ is invariant under the antipodal map, and thus defines an observable on $\mathbb{RP}^{n-1}$. In fact, the topology of the sphere is *too simple* to guarantee the existence of sufficiently many extrema, which is not the case for the real projective space, the richness of whose topology we have emphasised on many occasions in this course.

---

[148] Lazar Aronovitch Lyusternik (1899–1981), Russian mathematician, was the author of numerous articles on the Calculus of Variations.

[149] The Russian mathematician Lev Genrichovich Schnirelman (1905–1938) made contributions to number theory and gave applications of topology to the Calculus of Variations.

D. Hessians and Normal Forms at Generic Critical Points

In this section we examine the normal forms which can be used for observables at a critical point. We show that, at a critical point, one can give an intrinsic meaning to the second order part of the observable, which one calls *its Hessian form*. The normal form of the observable in a neighbourhood of the critical point is completely determined by this Hessian form if it is non-degenerate.

VII.60. We already know from VI.42 that at a point where the differential of an observable is non-zero, it can be taken as a local coordinate, and thus made linear.

At a critical point of an observable, one must define the intrinsic version of the expansion of the observable up to second order. The bilinear form which we are going to introduce to this end will be the key to the theory of normal forms.

VII.61. We have already remarked that the notion of the expansion of a function up to second order has no intrinsic meaning on a configuration space. In fact, at any non-critical point of an observable φ, it is possible to take φ as a coordinate and thus to make the second order term in its expansion vanish.

This fact is also related to the impossibility of defining intrinsically the acceleration of a curve without giving additional information, a point already mentioned before in this course.

At a critical point of a numerical function, this difficulty disappears, as we show.

Exercise VII.62. Show that *at any non-critical point a of an observable φ defined on a configuration space M there exists a chart x centred at a in which the matrix of second derivatives of φ^x at 0 is any given symmetric matrix.*

Proposition VII.63. *Let a be a critical point of an observable φ defined on a configuration space M of dimension n. Let X_a, $Y_a \in T_aM$. If Y is a vector field such that $Y(a) = Y_a$, the quantity $\partial_{X_a}(\partial_Y\varphi)$ depends only on Y_a and defines a symmetric bilinear form on T_aM, called the* Hessian form *of φ at a, denoted by $\mathrm{Hess}_a\varphi$; by definition $\mathrm{Hess}_a\varphi(X_a, Y_a) = \partial_{X_a}(\partial_Y\varphi)$.*

Proof. The main point is to show that $\partial_{X_a}(\partial_Y\varphi)(a)$ is independent of the field Y used to extend Y_a to a neighbourhood of a. Hadamard's Lemma VI.13 allows us to do this easily.

In fact, take a chart $x = (x^1, \cdots, x^n)$ at a. By VI.13, we can write $\varphi = \sum_{i=1}^n x^i \chi_i$ for n observables χ_i. Since a is a critical point of φ, we have

$$0 = d\varphi(a) = \sum_{i=1}^n \chi_i(a)\, dx^i(a)\,,$$

and, since the n forms $dx^i(a)$ constitute a basis of T_a^*M, we deduce that $\chi_i(a) = 0$ for $1 \le i \le n$. Since Y is a derivation, we can explicitly write the function $\partial_Y\varphi$ as $\partial_Y\varphi = \sum_{i=1}^n\left((\partial_Y x^i)\chi_i + x^i(\partial_Y\chi_i)\right)$. When we evaluate X_a on this function using the fact that X_a is a derivation at a, we obtain

$$\partial_{X_a}(\partial_Y\varphi) = \sum_{i=1}^n(\partial_Y x^i)(a)\,(\partial_{X_a}\chi_i) + \sum_{i=1}^n(\partial_{X_a}x^i)(\partial_Y\chi_i)(a)\,,$$

because two series of terms vanish. From this formula, it is clear that only the value of Y at a matters, since by definition $\partial_Y \varphi(a) = \partial_{Y_a} \varphi$. Moreover, we see from the same formula that $\partial_{X_a}(\partial_Y \varphi)$ depends bilinearly on X_a and Y_a.   □

Remark VII.64. Proposition VII.63 contains a version of Hadamard's Lemma valid up to second order. In fact, in the course of the proof we have shown that the observables χ_i appearing in the decomposition of φ vanish at a. It is thus possible to apply Hadamard's Lemma to them, and thus to write in a neighbourhood of a

$$\varphi - \varphi(a) = \sum_{1 \leq i \leq j \leq n} x^i x^j h_{ij}$$

for $n(n+1)/2$ observables h_{ij} indexed by the pairs of symmetric indices (i, j). Consequently, we necessarily have

$$h_{ij}(a) = (\mathrm{Hess}_a \varphi)\left(\frac{\partial}{\partial x^i}, \frac{\partial}{\partial x^j}\right) = \frac{\partial^2 \varphi^x}{\partial x^i \partial x^j}(0) \,,$$

whose symmetry is guaranteed by Schwarz's Lemma III.97 since φ is of class C^2.

Corollary VII.65. *If γ is a curve originating at a point a of a configuration space M on which an observable φ is defined which has a critical point at a, then*

$$\frac{d^2}{dt^2}(\varphi \circ \gamma(t))_{|t=0} = \mathrm{Hess}_a \varphi\,(\dot{\gamma}(0), \dot{\gamma}(0)) \,.$$

Proof. At each point $\gamma(t)$ of the curve, we have $d(\varphi \circ \gamma)/dt = \partial_{\dot{\gamma}(t)} \varphi$. The stated formula now follows directly from that given in Proposition VII.63.   □

Exercise VII.66. On the rotation group SO_3, find the Hessian form of the "trace" at one of its critical points (cf. VII.48).

Definition VII.67. A critical point a of an observable φ is said to be *non-degenerate* or *of Morse type* if its Hessian form $\mathrm{Hess}_a \varphi$ is non-degenerate.

The observable φ is said to be a *Morse function* if all its critical points are non-degenerate.

Exercise VII.68. Show that a point a is a critical point of Morse type of an observable φ defined on a configuration space M if and only if $0 \in T_a^* M$ is a regular value of the differential 1-form $d\varphi$ viewed as a map from M to $T^* M$.

Morse Lemma VII.69. *Let a be a non-degenerate critical point of an observable φ. There exists a chart $x = (x^1, \cdots, x^n)$ at a and an integer k, called the* index *of the critical point, such that*

$$\varphi = \varphi(a) - (x^1)^2 - \cdots - (x^k)^2 + (x^{k+1})^2 + \cdots + (x^n)^2 \ .$$

Proof. The lemma is proved in two parts. The first consists in finding a chart at a in which φ coincides with the quadratic observable associated to its Hessian form in these coordinates. The second is the classical diagonalisation of a non-degenerate quadratic form.

To find the chart in which φ reduces to its expansion up to second order, we start from the proof of VII.63. We already know that there exists a chart y and a bilinear form $h = \sum_{i,j=1}^{n} y^i y^j h_{ij}$ (in these coordinates), whose coefficients h_{ij} are observables in a neighbourhood of 0 in $\mathbb{R}^n$, such that $\varphi^y = h$. Moreover, we have $h(a) = y^*(\mathrm{Hess}_a)$. We use this chart to identify a neighbourhood of a with a neighbourhood of 0 in $\mathbb{R}^n$. Accordingly, we write $h(a) = h_0$.

We show that there exists a differentiable map f defined in a neighbourhood of 0 and taking values in the automorphisms of $\mathbb{R}^n$ such that $f^*h = h_0$. (Recall that, by definition, for any vector v in $\mathbb{R}^n$, $f^*h(v, v) = h(f(v), f(v))$.) The map η which to $f \in L(\mathbb{R}^n, \mathbb{R}^n)$ associates the symmetric bilinear form $\eta(h) = f^*h$ is bilinear and thus differentiable. Moreover, it is a submersion in a neighbourhood of the identity map Id since its tangent linear map $T_{Id}\eta$ sends the linear map F to the bilinear form $(v, w) \mapsto h_0(F(v), w) + h_0(v, F(w))$. Now, being non-degenerate by hypothesis, h_0 defines a duality in the same way as does a scalar product. Consequently, every symmetric bilinear form h can be represented by a linear map l_h such that $h(v, v) = h_0(l_h(v), v)$, so that every bilinear form can indeed be found in the image of $T_{Id}\eta$. By the Submersion Theorem IV.9, there exists a differentiable map f defined in a neighbourhood of h_0 with values in $L(\mathbb{R}^n, \mathbb{R}^n)$ such that $f(h_0) = Id$ and $(f(h))^*(h) = h_0$. If we set $x = f(h(y)) \circ y$, we have a chart at a such that $\varphi^x = f(h(y)) \circ \varphi^y = (f(h(y)))^*h = h_0$, and this is the formula we wanted.

It is well-known that every symmetric bilinear form can be written as a sum or difference of squares. Morcover, by rescaling the coordinates, one can make the coefficients of all the squares equal to ± 1. It is thus enough to make a linear change of coordinates with constant coefficients to obtain a chart at a in which the observable takes the form stated in the lemma.   □

Corollary VII.70. *Every critical point of Morse type is isolated and the critical set of a Morse function is necessarily discrete, hence finite if the configuration space on which it is defined is compact.*

Proof. In a chart x as in the Morse Lemma VII.69, the differential of the Morse function can be written $d\varphi = -\sum_{i=1}^{k} x^i \, dx^i + \sum_{i=k+1}^{n} x^i \, dx^i$ and thus vanishes only at the critical point.

The second part of the corollary is also clear.   □

VII.71. In view of the importance of the Morse Lemma VII.69, it is worthwhile to pause and survey its significance in various situations.

A critical point of index 0 is a strict local minimum, a point of index equal to the dimension of the configuration space a strict local maximum. The critical points having an intermediate index are neither maxima nor minima. They are called *saddle-points* to indicate that at such points there are points arbitrarily close to the critical point where the observable takes values greater and smaller than its value at the critical point. (Of course, the picturesque name comes from the form of the graph of such a function defined on a surface as shown in Fig. VII.72.)

The types of local behaviour just described are all present in the example of the height observable on the torus considered as a hollow tube (see Fig. VII.80).

VII.72. One of the natural questions one can ask about Morse functions is: on a given configuration space, do they always exist? The answer is particularly simple since "*any observable taken at random is a Morse function*".

In fact, for a natural topology on the space of observables, the Morse functions form a dense open set so that, even if a given observable is not Morse, we can find a Morse function as close to it as we wish. One often refers to this property by saying that Morse functions are *generic*. Note also that, under a small perturbation in the space of observables (equipped with the above topology), a Morse function remains a Morse function of the same index, a property which is expressed by saying that Morse functions are *stable*.

Of course, there are observables which have degenerate critical points and, in a given situation, it is not always convenient to make them into Morse functions, especially if their degeneration is related to symmetry properties (see Chapter XI). There exists a *theory of singularities* of functions which provides a complete hierarchy of local models of which the Morse model is the simplest.

This theory, and the study of the deformations of the models it provides, constitutes the basis of *catastrophe theory*, created by Thom.

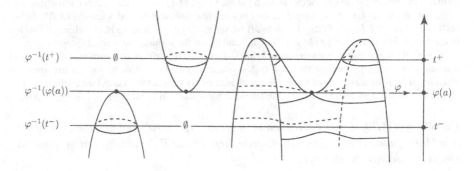

Fig. VII.73 Three local models of critical points of Morse type: a maximum, a minimum, a saddle-point

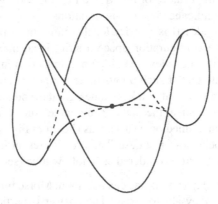

Fig. VII.74 The *"monkey saddle"*, an example of a degenerate critical point

VII.75. More generally, there is a theory of singularities of differentiable maps which aims to find models of maps in a neighbourhood of their singular points. It has been developed mainly by Whitney, Thom and Mather[150]. It aims to embed a map having a given singularity in a sufficiently large family to allow one to take account of every perturbation of the map. This family is called its *universal deformation*. This is possible only if the singularity is not too degenerate. The tools used often appeal to algebraic concepts which allow one to analyse the fine structure of the algebra of observables in a neighbourhood of the singular observable.

An example of such an algebraic invariant is the *Jacobian ideal* $J(\varphi)$ generated by the partial derivatives of the observable φ. If φ is non-singular at a point, then $J(\varphi)$ contains the constants, and thus coincides with the whole algebra of observables; if φ is a Morse function, it follows from the Morse Lemma VII.69 that $J(\varphi)$ is the maximal ideal consisting of the functions which vanish at the critical point. The Jacobian ideal becomes an interesting invariant for critical points which are more degenerate than Morse points. (For more details on this point, see (Demazure 1992).)

VII.76. To deal with problems in the Calculus of Variations, it is very important to have extensions of the Morse Lemma for observables defined on infinite-dimensional configuration spaces, such as Banach spaces or Hilbert spaces. Unfortunately, we cannot go into these extensions in this course.

Project VII.77. After giving a definition of a non-degenerate critical point for an observable defined on a Hilbert space, establish for such critical points an extension of the Morse Lemma (known as the Palais-Morse Lemma).

VII.78. The Morse Lemma is essentially all that can be said about a critical point of Morse type from a *local* point of view. *Morse theory* aims to exploit the information extracted from the local models at the different critical points of a Morse observable to deduce information about the *global* structure of the configuration space. We do not have space here to develop this theory in detail, but we would like to give its main features.

VII.79. We thus start with a configuration space M of dimension n, which we assume to be compact for simplicity, and an observable φ defined on M which we assume to be of Morse type and such that no critical value of φ is the image of more than one critical point. The basic idea is to *reconstruct* M by following its "*filling up*" by the pieces determined by φ. To this end, we associate to any real number λ the open subset $M^\lambda = \{q \mid q \in M, \ \varphi(q) < \lambda\}$. As long as λ is less than the minimum of φ, we have of course $M^\lambda = \emptyset$. When we have just passed the minimum, M^λ can be identified with a disc of dimension n.

The crucial property which underlies this theory is that if, in the interval $[\lambda_1, \lambda_2]$, φ has no critical value, then the subsets M^{λ_1} and M^{λ_2} are diffeomorphic, a diffeomorphism being constructed from the flow of a vector field such that the observable φ increases along its integral curves (a sort of "gradient" field). For each passage through a critical value, it is possible to extract from the Morse Lemma VII.69 a model of an attached disc whose dimension is equal to the index of the critical point

---

[150] The American mathematician John N. Mather (1942–2017) made decisive contributions to the theory of singularities of mappings in the first part of his career, and then moved on to study Hamiltonian dynamics.

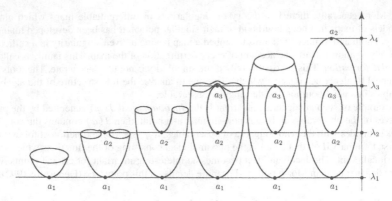

Fig. VII.80 The reconstruction of a torus using a Morse observable

which has been passed. The process ends with the adjoining of a disc of dimension n corresponding to passage through the maximum.

VII.81. From a careful examination of this construction, one can extract precise information about a certain number of invariants which algebraic topology attaches to a configuration space (such as inequalities on the dimensions of its *homology* groups, often called the *Morse inequalities*). In this way, Reeb[151] proved that a compact configuration space on which there exists a Morse function having only two critical points (which are necessarily a maximum and a minimum) is homeomorphic to a sphere (but, without additional information, one does not know how to prove that it has its standard differentiable structure).

Pursuing this approach in reverse, one can show that every Morse function defined on a known configuration space necessarily has a certain number of critical points and that these are necessarily of certain types. Thus, every Morse function on the torus $\mathbf{T}^n$ has at least 2^n critical points, e.g., on the 2-dimensional torus every Morse function has at least one maximum, one minimum and two saddle-points.

Project VII.82. Construct an observable on the torus $\mathbf{T}^2$ which has only 3 critical points (a maximum, a minimum and a monkey saddle).

VII.83. One can get a good understanding of complex projective spaces $\mathbb{CP}^m$ by studying their "filling up" by particular Morse functions. One interesting family can be defined as follows: if $[z^1; \cdots ; z^{m+1}]$ is a system of homogeneous coordinates on $\mathbb{CP}^m$, one simply takes

$$\varphi([z^1; \cdots ; z^{m+1}]) = \frac{\sum_{j=1}^{m+1} \lambda_j |z^j|^2}{\sum_{j=1}^{m+1} |z^j|^2}$$

where the λ_j are distinct real constants.

---

[151] Georges Reeb (1920–1993), French mathematician, taught at Grenoble and Strasbourg. He was one of the pioneers of the theory of foliations, which is intimately related to the study of dynamical systems. More recently, he has been an ardent advocate of non-standard analysis.

VII.84. The theory of critical points which we have developed, although rather satisfactory for compact configuration spaces, is less than satisfactory in two situations: non-compact configuration spaces and configuration spaces with "*boundaries*".

In the non-compact case, Morse theory can again be developed provided the observable studied satisfies a condition which ensures that "*the critical points do not escape to infinity*", often called the "*Palais*[152]*–Smale condition*". This situation typically arises when one considers variational problems on infinite-dimensional configuration spaces.

The case of "boundaries" turns out to be quite like the preceding case as far as the existence of interior critical points is concerned. Thus, if the space with "boundary" considered is compact, the observable might attain its maxima and minima in the interior (and then we are in the situation of an ordinary critical point), or on the "boundary". In this case, we can no longer speak of its "differential" at such a point as the readers will convince themselves. The search for critical points must take account of this dichotomy by using, if possible, the method of Lagrange multipliers to detect critical points coming from the "boundary" if it is a submanifold, or if one can reduce to this case by excluding successively the bad points corresponding to "corners", "hypercorners", and other singular points of the boundary.

E. Historical Notes

VII.85. The different ways of defining a configuration space already appear in the fundamental memoir of Gauss "*Disquisitiones Generales Circa Superficies Curvas*" (cf. (Dombrowski 1979)) published in 1827. We quote a passage from this work, where he lists the different ways of giving the tangent plane to a surface, as it is so illuminating: "*...Duae habentur methodi generales ad exhibendam indolem superficiei curbae. Methodus prima utitur aequatione inter coordinatas x, y, z, quam reductam esse supponemus ad formam W = 0, ubi W erit functio indeterminarum x, y, z... Methodus secunda sistit coordinatas in forma fonctionum duarum variabilium p, q... His duabus methodis generalibus accedit tertia, ubi una coordinatarum, e.g. z exhibetur in forma functionis reliquarum x, y: haec methodus manifesto nihil aliud est, nisi casus specialis vel methodi primae, vel secundae...*"[153]

VII.86. The global methods which assert the existence of critical points have been extensively developed thanks to the stimulus provided by the Calculus of Variations.

---

[152] Richard S. Palais, American mathematician born in 1931, was one of the founders of global analysis.

[153] Here is the translation of Gauss' text: "*... There are two general methods for representing a curved surface. The first method uses an equation between coordinates x, y, z, which we will suppose to be reduced to the form W = 0, where W will be a function of the indeterminate x, y, z... The second method represents the coordinates in the form of functions of two variables p, q... With these two general methods one can access a third, where one of the coordinates, e.g. z is presented in the form of a function of the remaining x, y: this method is manifestly nothing but a special case of the first or second method...*"

It was to prove the existence of three non-self-intersecting closed geodesics on the sphere S^2 that Fet[154] and Lyusternik developed the notions of category (cf. Sect. C).

Morse theory has been very successful and found a wide range of applications in very diverse areas of mathematics, such as in geometry (cf. (Milnor 1965) for a spectacular application to the study of the homotopy of orthogonal and unitary groups), and in analysis, where it is used in infinite-dimensional spaces in an extended form due to Palais and Smale, usually to solve non-linear partial differential equations.

---

[154] The Russian mathematician Abram Ilyich Fet (1924–2007) is also known as a Soviet dissident who did many translations for the Samizdat. In the first part of his career he worked on the Calculus of Variations, later moving to theoretical physics and chemistry.

Part III
THE CALCULUS OF VARIATIONS

"On a déjà plusieurs Traités de Méchanique, mais le plan de celui-ci est entiérement neuf. Je me suis proposé de réduire la théorie de cette Science, & l'art de résoudre les problêmes qui s'y rapportent, à des formules générales, dont le simple développement donne toutes les équations nécessaires pour la solution de chaque problème...

On ne trouvera point de Figures dans cet Ouvrage. Les méthodes que j'y expose ne demandent ni constructions, ni raisonnements géométriques ou méchaniques, mais seulement des opérations algébriques, assujetties à une marche régulière & uniforme. Ceux qui aiment l'Analyse verront avec plaisir la Méchanique en devenir une nouvelle branche, & me sauront gré d'en avoir étendu ainsi le domaine."

M. DE LA GRANGE,
in *Méchanique Analitique.*

Chapter VIII
Configuration Spaces of Geometric Objects

This first chapter of the third part, which is devoted to the *Calculus of Variations*, aims to present those spaces on which the functionals whose extrema we shall seek in Chapter IX are defined. These spaces are infinite-dimensional generalisations of the configuration spaces which have been the focus of our attention in the second part. In carrying out this extension, we shall not try to develop a general theory, which explains why this chapter is the shortest in the book (and why it does not contain substantial theorems).

In Sect. A, we study the *space of curves* on a configuration space, and we discuss in what sense it can be considered to be a configuration space in its own right. In particular, we examine how a *vector field along a curve* defines an infinitesimal variation of the curve, i.e., a tangent vector to the space of curves. We also discuss briefly the problem of *approximating* curves by broken curves.

We turn to the examination of the *space of surfaces* in Sect. B. To simplify the presentation, we restrict ourselves to surfaces in 3-dimensional numerical space. We present the notion of a *normal vector field*. We show how a vector field along a surface can be identified with an infinitesimal variation. We also say a few words on the problem of representing surfaces by *finite elements*.

As an example of a space of geometric objects, in Sect. C we give a concise presentation of the *group of diffeomorphisms* of a configuration space and the space of vector fields which can be thought of as its tangent space at the identity.

In Sect. D, we introduce another example of a field of geometric quantities, namely the *volume elements*. We are thus led to consider diffeomorphisms and vector fields which preserve a given volume element, i.e., those whose *divergence is zero*. These objects are the fundamental constituents of mathematical models of the mechanics of incompressible continuous media.

© The Author(s), under exclusive license to Springer Nature Switzerland AG 2022
J.-P. Bourguignon, *Variational Calculus*, Springer Monographs in Mathematics,
https://doi.org/10.1007/978-3-031-18307-2_8

A. Spaces of Curves

In this section, we study the space of *curves* on a configuration space. We define the tangent vectors to this space, which turn out to be *vector fields along a curve*.

VIII.1. We study the *space of curves* on a configuration space M of dimension n, which we denote by $C(M)$. We assume that these curves are maps defined on a fixed interval $[\alpha, \beta]$ which are differentiable and have no singular point (i.e., their velocity-vector is never zero).

When the curves lie in a numerical space $\mathbb{R}^n$, by using the vector space structure of the target numerical space, the space of curves can itself be given the structure of a vector space (clearly infinite-dimensional). We can then make it a metric or normed space in many ways as we have seen in Chapter I. Since we wish to consider curves on a configuration space, we shall not make use of this linear structure which plays the role of the model space for $C(M)$ in the same way as the numerical space $\mathbb{R}^n$ plays that role for M.[155]

Most of the developments we are going to present are in fact independent of the distance chosen (as long as it defines a sufficiently fine topology), but we shall try to avoid giving details of this.

VIII.2. We want to introduce the notion of a *tangent vector* at a point γ of $C(M)$. In Chapter II, we have seen that, in infinite-dimensional spaces, one must be careful in using duality. Consequently, the definition of tangent vectors in terms of their properties as derivations of local observables can no longer be used.

We proceed mainly by analogy with the other definition of a tangent vector as an equivalence class of curves tangent at a point, a definition we have already used in the finite-dimensional case. In VIII.3, we establish first of all what will be our "*curves of curves*".

Definition VIII.3. A *variation* of a curve $\gamma \in C(M)$ is a differentiable map Γ, taking values in M and defined on $[\alpha, \beta] \times] - \epsilon, \epsilon[$ for some $\epsilon > 0$, such that $\Gamma(., 0) = \gamma$.

VIII.4. This map of the rectangle $[\alpha, \beta] \times] - \epsilon, \epsilon[$ into M can be thought of as a family of maps in two different ways:

i) we can consider the family of curves $s \mapsto \gamma_s$ which at a point τ of the interval $[\alpha, \beta]$ takes the value $\gamma_s(\tau) = \Gamma(\tau, s)$, and which resembles the "curve of curves" we had in mind;

ii) we can also consider the maps $\eta_\tau(s) = \Gamma(\tau, s)$ which we call the *transverse curves* to distinguish them from the curves $t \mapsto \gamma_s(t)$ (in fact, they are not curves in the sense defined at the beginning of this chapter, for they are not defined on the interval $[\alpha, \beta]$).

---

[155] There are difficulties here arising from the non-uniqueness of the norm topology which one encounters in infinite-dimensions. But to extend the definitions of these distances to the case of curves on a configuration space, or to give a systematic theory of infinite-dimensional configuration spaces, would take us too far afield.

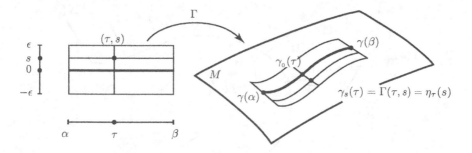

Fig. VIII.5 An ϵ-variation of a curve

In this second approach, we lose the picture of the points of γ together forming an object, and we are interested only in their individual behaviour.

Definition-Proposition VIII.6. *We say that two variations Γ_1 and Γ_2 of a curve $\gamma \in C$ are* tangent at γ *if, for all $\tau \in [\alpha, \beta]$, the transverse curves $(\eta_1)_\tau$ and $(\eta_2)_\tau$ are tangent at $\gamma(\tau) = (\eta_1)_\tau(0) = (\eta_2)_\tau(0)$.*

The equivalence classes defined by this relation are called infinitesimal variations *of γ and define the vector fields along the curve γ, i.e., the maps X which to any value of the parameter $\tau \in [\alpha, \beta]$ associate a tangent vector at the point $\gamma(\tau)$ in a differentiable way.*

For a variation Γ of a curve γ, the vector field $X = (T_{(.,0)}\Gamma)(\partial/\partial s)$ is called the transverse vector field of the variation.

Proof. The fact that the relation "being tangent at a curve" is an equivalence relation is proved in exactly the same way what was done in VI.3.

The fact that an equivalence class can be identified with a vector field results simply from the fact that at each point $\gamma(\tau)$ we have $d\eta_\tau(s)/ds_{|s=0} = (T_{(\tau,0)}\Gamma)(\partial/\partial s)$, which is thus a tangent vector at $\gamma(\tau)$ depending differentiably on τ.   □

Remark VIII.7. One point (which can seem subtle at first sight) deserves our attention. We have never demanded that the curves we are considering do not intersect themselves. Thus, it may well be that the curve γ has the form shown in Fig. VIII.8.

An infinitesimal variation along such a curve can thus take different values at the same point of M, corresponding to the curve passing through this point at different instants. This does not prevent the map $\tau \mapsto \partial\Gamma/\partial s = d\eta_\tau(s)/ds_{|s=0}$ from being differentiable. In Fig. VIII.8 it is even bijective since, at the only point where the two values can be compared, they are distinct.

Nevertheless, we shall often abuse notation (sometimes dangerously) by denoting the value of a vector field along a curve γ at a point $\gamma(\tau)$ by $X_{\gamma(t)}$ instead of $X(t)$.

VIII.9. From the definition of a vector field along a curve on a configuration space M as a map X from $[\alpha, \beta]$ to TM such that $\pi_M \circ X = \gamma$, it follows easily that these vector fields form a vector space. The operations are the usual vector operations in

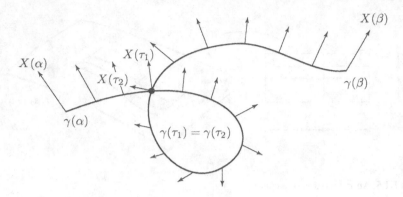

Fig. VIII.8 A non-simple curve and one of its infinitesimal variations

each tangent space, namely, for two vector fields X and Y, $(X+Y)(\tau) = X(\tau)+Y(\tau)$. Similarly, a vector field X along a curve can be multiplied by an observable φ defined along the curve by putting, as one must, $(\varphi X)(\tau) = \varphi(\tau)X(\tau)$.

> *From now on, we speak freely of a vector field X*
> *along a curve $\gamma \in C$ as*
> *a tangent vector to $C(= C^\infty([\alpha,\beta], M))$ at γ.*

Exercise VIII.10. Show that, conversely, for any vector field X along a curve γ, there exists a variation Γ of γ whose transverse vector field is X.

VIII.11. We can now ask what are the variations which change the *geometric* support of a curve. Since, according to our definition, a curve γ comes with its parametrisation, we can vary a curve by composing it with a diffeomorphism of the interval $[\alpha,\beta]$. For such variations, the image of the source rectangle $[\alpha,\beta] \times]-\epsilon, \epsilon[$ is completely "flattened" along $\gamma([\alpha,\beta])$. Such variations are not interesting since, for any point $\tau \in [\alpha,\beta]$, $\partial\Gamma/\partial s$ and $\dot{\gamma}(\tau)$ are collinear. They only reflect *reparametrisations* of the curve γ.

If we have a scalar product in each tangent space of M (giving such a field is called a *Riemannian metric*), we can define *normal* variations, i.e., those whose transverse vector is orthogonal to the velocity-vector of the curve at each point.

An abstract way of doing this without using a field of scalar products is to consider at each point $\gamma(\tau)$ the image of the transverse vector field in the quotient space $T_{\gamma(\tau)}M/\mathbb{R}.\dot{\gamma}(\tau)$, which amounts to forgetting the parametrisation.

VIII.12. For purely mathematical reasons, or for the purpose of graphical representation (for example, in connection with computer graphics), it is useful to develop methods to approximate curves.

VIII.13. The point of departure for approximation methods is to *discretise* the curve, i.e., to replace it by a *sequence*[156] *of points*. These points can be taken on the curve or to belong to a grid fixed once and for all (think of the screen of a graphics terminal). It is then necessary to have a procedure for joining two consecutive points of the sequence by a *conventional* curve. When the representation simulates a numerical space, the usual procedure is to use *line segments*, but the directions of these segments can themselves be chosen from among a finite number of fixed directions. This procedure can be refined and other interpolations can be used to ensure, for example, that the curve has great regularity.

There are other situations which really take place in a curved space; this is the case, for example, when one wishes to control the movements of a robotic arm, where the ambient space is the *group of displacements* in which we find our friend the *rotation group*.

Remark VIII.14. The discretisation we are going to describe for curves has its counterpart for the infinitesimal variations of curves. The vector fields along a conventional curve reduce, in the chosen model, to an *ordinary tangent vector at each point of the sequence* which replaces the curve, the values of which are perhaps restricted to a finite number of directions.

VIII.15. From a more mathematical point of view, the advantage of these methods is that they reduce the space of curves, which is a configuration space of infinite dimension, to an *approximate model* which is *finite-dimensional*.

To accomplish this, the set of points of the interval $[\alpha, \beta]$, which has the power of the continuum, is replaced by a finite number, say k, of points and the conventional curves joining these points are completely determined once their endpoints are given (possibly up to a finite ambiguity). The model of the space of curves is thus in this case the product of k copies of the configuration space in which the curves lie. Similarly, infinitesimal variations correspond to giving an ordinary tangent vector at each point of the sequence.

VIII.16. One might fear that too much mathematical information has been lost in the approximation. This depends on the problem under discussion, but it is not always the case (for an example of the use of such an approximation, see (Milnor 1973)).

B. Spaces of Surfaces in the Numerical Space $\mathbb{R}^3$

We consider the space of surfaces in $\mathbb{R}^3$ and determine its tangent vectors by a method analogous to that used for the space of curves.

VIII.17. The case of spaces of surfaces is in many ways analogous to that of curves. Nevertheless, there is a difference (and it turns out to be crucial in the Calculus of Variations) which derives from the fact that $\mathbb{R}$ possesses a natural *total order*, which is not the case for numerical spaces of higher dimension, in particular for $\mathbb{R}^2$.

---

[156] This word seems to be used more and more to mean a *finite ordered sequence*.

VIII.18. To simplify the presentation, we consider only spaces of surfaces in the numerical space $\mathbb{R}^3$. We *denote* by S the space of differentiable maps σ from a domain Ω in the plane $\mathbb{R}^2$ to $\mathbb{R}^3$ which have *no singular points*, i.e., those which we called *immersions* in Chapter III. At a point a of $\sigma(\Omega)$, by applying the Immersion Theorem IV.22, the image of the surface is *locally* a submanifold of $\mathbb{R}^3$. Here, "locally" means that it may be necessary to restrict the domain on which the surface is defined.

Remark VIII.19. For the domain Ω, one can take the unit disc, a rectangle or a domain with holes, and thus possibly a domain with complicated topology. In this approach, surfaces are often called *parametrised surfaces* to indicate that the geometric object which constitutes the surface in $\mathbb{R}^3$ is given together with a parametrisation. If the domain is closed (and thus contains its boundary), we have to face the fact that the surface possesses a boundary, which we have so far excluded from our discussions.

Because of the definition we have decided upon, the surfaces we consider cannot be both compact and without boundary. They are really *pieces of surfaces*. However, we can define a variation of a surface by translating directly that given for a curve.

Definition VIII.20. Let σ be a surface in $\mathbb{R}^3$. A *variation* of the surface σ is a map Σ defined on $\Omega \times\,] - \epsilon, \epsilon [$ for some $\epsilon > 0$, taking values in $\mathbb{R}^3$, such that $\Sigma(., 0) = \sigma$.

VIII.21. In a manner analogous to that which we used for curves, we distinguish the surfaces $\sigma_s = \Sigma(., s)$, which form the *curve of surfaces* which we are interested in, and the *transverse curves* $\eta_t = \Sigma(t, .)$, where $t = (t_1, t_2)$ is the point of the domain Ω under consideration.

Two variations Σ_1 and Σ_2 are said to be *tangent* at σ if at each point of Ω the transverse curves η_1 and η_2 are tangent at 0. They then define at each point the same vector in $\mathbb{R}^3$, say $X = (\partial \Sigma_1 / \partial s)_{|s=0} = (\partial \Sigma_2 / \partial s)_{|s=0}$, which is called the *transverse vector field* of the variation.

Definition VIII.22. A *vector field along a surface* σ is a differentiable map X defined on the domain Ω of the surface with values in $\mathbb{R}^3$, so that $X(t)$ is viewed as being attached to the point $\sigma(t)$.

VIII.23. As for curves, we consider separately the variations of a surface obtained by composing σ with a diffeomorphism of the domain Ω which we assume can be joined to the identity by a curve of diffeomorphisms (one says that it is *isotopic* to the identity). These variations do not change the geometric image of a surface, only its parametrisation. We call such variations *reparametrisations* of σ.

We can take advantage of the fact that the surfaces are in a numerical space which has a standard scalar product to normalise the variations in the following way.

VIII.24. We have seen that the image of a surface σ in $\mathbb{R}^3$ admits a tangent plane at each point (except possibly at multiple points); we define the *field of normals* on σ to be the map from Ω to $\mathbb{R}P^2$, the space of lines through the origin in $\mathbb{R}^3$, which to a point $t = (t_1, t_2)$ associates the translate of the affine line $(T_t \sigma)^{\perp}$ to the origin.

By using the ambient Euclidean metric, we can single out two vectors of length 1 on this line, the *unit normal vectors*. On any sufficiently small portion of the surface, we can make a coherent choice of one of these vectors in such a way that the resulting map is differentiable. (In fact, relative to a suitably chosen plane in $\mathbb{R}^3$, the surface can be defined as a graph $x^3 = f(x^1, x^2)$, and it suffices to choose, for example, the field along which the observable x^3 is increasing). As is well-known, this operation is called *orienting* the surface.

Note, however, that a *global* choice of oriented normals is not always possible, as the example of the Möbius band shows. When such a choice is possible, we say that the surface is *transversally orientable*, and we shall *assume that this is the case from now on*.

Exercise VIII.25. Show that the unit normal vector fields on an (oriented) surface are differentiable vector fields on the surface.

Definition VIII.26. A *normal variation* of a surface σ is a variation whose transverse vector field is normal at every point of σ.

Proposition VIII.27. *Every variation Σ of a surface σ is tangent to a variation $\tilde{\Sigma}$, defined on a domain which is arbitrarily close to that of σ, which is the composite of a reparametrisation and a normal variation.*

Proof. Let X be the vector field defined by the given variation Σ. By means of orthogonal projection onto the tangent plane to the surface σ, we can decompose X into its tangential part X^T and its normal part $X^\perp$.

The tangential part X^T defines a differentiable vector field on the domain of definition Ω of the surface (since the orthogonal projection depends differentiably on the point). For a compact domain strictly contained in Ω, the flow ξ_s of X^T is defined on a uniform time interval $]-\epsilon, \epsilon[$. If we consider the variation $\tilde{\Sigma}$ of σ defined on $\Omega \times]-\epsilon, \epsilon[$ by $\tilde{\Sigma}(t, s) = (\sigma + sX^\perp)(\xi_s(t))$, it follows from the Chain Rule III.46 that the transverse vector field to this variation is simply $X^T + X^\perp = X$ along σ (i.e., for $s = 0$). □

VIII.28. Let σ be an oriented surface in $\mathbb{R}^3$. The map ν from Ω to the sphere $\mathbf{S}^2$ which to a point of σ associates the positively oriented unit normal vector is called the *Gauss map* of the surface σ.

Many of the geometric properties of the surface can be expressed in terms of its Gauss map. For example, Gauss proved that the curvature of the surface can be obtained almost without calculation from the (oriented) area of the image of the surface under its Gauss map. This is also the case for variational properties, such as the behaviour of a *soap film*, which can be characterised mathematically as a surface of *extremal area* (also called a *minimal surface*), a topic which we take up in Sect. D of Chapter IX. The "minimality" is related to properties of the Gauss map to the sphere $\mathbf{S}^2$ regarded as the Riemann sphere.

VIII.29. The Euclidean scalar product on the ambient space $\mathbb{R}^3$ enabled us to define the field of normals. It also induces a scalar product on the *tangent space* to the surface σ, called the *first fundamental form* of the surface, and *denoted* by g (i.e., for two vectors v and w tangent to σ at $\sigma(t)$, we put $g(v, w) = (v|w)$, since we can view v and w as vectors in $\mathbb{R}^3$).

Exercise VIII.30. Show that, if $g = g_{11} (dt^1)^2 + g_{12} dt^1 dt^2 + g_{22} (dt^2)^2$ is the expression for the first fundamental form of a surface in terms of a chart (t^1, t^2) of a surface, the coefficients g_{11}, g_{12} and g_{22} which appear are differentiable functions on the surface.

VIII.31. The metric properties of σ which can be derived from the first fundamental form, such as, for example, the *curvature* (which can be interpreted as a deviation of the lengths of equidistant curves from a point from the lengths of Euclidean circles), enables one to obtain global information about the surface if it is compact and without boundary. Thus, a theorem due to Gauss and Bonnet[157] asserts that *"the integral of the curvature, divided by 4π, is an integer which gives the number of holes of the surface"*. The integer in question is in fact the number of times the Gauss map covers the sphere, provided of course this number is counted algebraically, taking into account the orientation.

VIII.32. The construction of approximations of surfaces is more subtle than that for curves. It is often referred to as the decomposition into *finite elements*.

In fact, one must first give a polygonal decomposition of the domain of definition of the surface, and then take, instead of the exact values of the map defining the surface, the values belonging to a predetermined grid. The final step consists of defining a *conventional* approximation to the surface by *facets*.

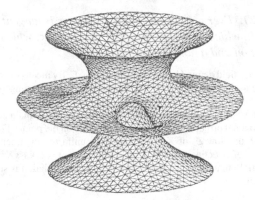

Fig. VIII.33 An approximation to a surface drawn using the computer software VPL

There are many ways of doing this: it is possible to have *plane* facets, either triangular if the domain of definition of the surface is decomposed into triangles; or polygonal if the approximation procedure consists of transporting to a point of the surface the normal vector at that point and replacing the surface near the point by its tangent plane. Nowadays, it is also feasible to use conventional surfaces which are less simple.

Problems related to these approximations are the subject of intense research at the moment, closely connected to the development of *computer graphics*.

---

[157] Ossian Bonnet (1819–1892) was a French mathematician. He made several important contributions to the differential geometry of surfaces. A former student at École Polytechnique he moved away from a career as engineer to teach there and in other places.

VIII.34. The graphics tools which are in the process of being developed (such as the software VPL mentioned above)[158] have already had a non-negligible impact on research in differential geometry. In particular, it has enabled Hoffman and Meeks[159] to discover new properties of certain surfaces, such as the Costa surface (of which an approximation is given in Fig. VIII.33) which, it has been proved, can be globally embedded in $\mathbb{R}^3$.

In time, many other geometric properties may be detected using this exploratory tool, and finally established by conventional methods.

C. On the Group of Diffeomorphisms

This section is devoted to understanding how the group of diffeomorphisms of a configuration space can itself be viewed as a configuration space whose tangent space at the identity is the space of vector fields on the original configuration space.

VIII.35. There are many other configuration spaces of geometric objects which deserve consideration, either for purely mathematical purposes, or because they turn out to be necessary for the construction of models. In this section and the next, we mention a few of them.

VIII.36. In the same way as the *rotation group* is forced upon us as the natural setting in which to discuss the mechanics of a rigid body moving about a fixed point, if we discuss deformable media it is natural to say that a *configuration* of such a medium, which initially occupies a domain Ω in $\mathbb{R}^3$, is the *group of diffeomorphisms* of Ω (which we *denote* by $\mathcal{D}\Omega$). Such a parametrisation of the medium expresses the *Lagrangian viewpoint* (in the terminology of mechanics) since the initial position serves to label each point of the medium which one then follows throughout its motion. A *trajectory of the medium* is then simply a *curve in $\mathcal{D}\Omega$*, the fact that we have *diffeomorphisms* corresponding to the hypothesis that the medium does not fracture, and is not subject to shock waves.

The group $\mathcal{D}\Omega$ is of course infinite-dimensional, and its structure contains much information about the space Ω. One other difficulty: it does not seem to be easy to construct approximations of it.

VIII.37. It is possible to make the group of diffeomorphisms an infinite-dimensional configuration space in a precise sense. But, because of the ω-Lemma III.64, one must be careful.

The topological structure of the group of diffeomorphisms of surfaces has been known for twenty years. The identity component of the group $\mathcal{D}S^2$ can be contracted onto the rotation group SO_3. Similarly, the identity component of the group of diffeomorphisms of the torus $\mathbf{T}^2$ can be contracted onto the largest connected group of isometries of the torus, namely the group of its translations, which is isomorphic simply to $\mathbf{T}^2$.

---

[158] This software was written by David Hoffman and Jim Hoffman, who both worked at the time at the University of Massachusetts at Amherst, and are not related. David Hoffman is an American mathematician, specialising in minimal surfaces, while Jim Hoffman is a computer programmer specialising in computer graphics. In fact, it was a geometric problem which has been the main motivation for Hoffman to commission Hoffman to create the VPL software.

[159] William Hamilton Meeks III, born in 1947, is an American mathematician working in differential geometry with major contributions to the study of minimal surfaces.

In higher dimensions, the situation is much more complicated and only a few results are known on the global structure of the group of diffeomorphisms. Note, however, that the exotic differentiable structures on the sphere S^n (cf V.83) can be put into one-to-one correspondence with the connected components of the group of diffeomorphisms of the equatorial sphere S^{n-1}, which describe the different ways of gluing together two standard n-dimensional discs along their common boundary to form a sphere.

Remark VIII.38. Let us show that *an infinitesimal deformation at the identity of the group of diffeomorphisms $\mathcal{D}M$ of a configuration space M can be identified with a vector field on M.*

In fact, let $s \mapsto \eta_s$ be a curve of diffeomorphisms which we interpret as a variation of the identity. This means we assume that, for any point q of M, the curve $s \mapsto \eta_s(q)$ is differentiable and $\eta_0(q) = q$. By taking the velocity-vector of each of these curves at each point of M, we obtain a vector field on M. On any compact subset of M, and for any vector field X, we can find a curve of diffeomorphisms of M whose velocity-vector at the identity is X: it suffices to take, for example, the flow of X and to apply Theorem IV.36 and Exercise VI.75.

VIII.39. Remark VIII.38 suggests the introduction of another configuration space of geometric objects, which turns out to be a vector space, namely *the space $\mathcal{T}M$ of vector fields on a configuration space M*. It leads us quite naturally to the *Eulerian point of view*, which is predominant in fluid mechanics. In this approach, one is not interested in what happens to an individual particle, but only in the motion of that particle which coincides with the observer at the instant the observation is made.

Analysing *the relation between the Lagrangian and the Eulerian viewpoints* brings into play a construction analogous to that which we used to show that finite-dimensional Lie groups are parallelisable: every tangent vector at a point ξ of $\mathcal{D}M$ can be brought to the identity by taking its image under the tangent linear map of multiplication by ξ^{-1}. By means of this construction, we can reduce the trajectory, in the Lagrangian sense, of a continuous medium to a curve in the tangent space at the identity of the group $\mathcal{D}M$, namely the space of vector fields on M. This is precisely the *Eulerian* description of the evolution of the system.[160]

Project VIII.40. Does every diffeomorphism of a configuration space M belong to the flow of a vector field on M?

D. Volume Elements

The *volume elements* which are studied in this section are first defined in a purely algebraic setting, and then extended to fields on a configuration space. By considering their Lie derivative, one arrives at the definition of the *divergence* of a vector field.

---

[160] This viewpoint turns out to be very powerful for proving the existence of solutions of the partial differential equation governing the motion of an incompressible fluid, which is traditionally called the *Euler equation*.

VIII.41. Linear algebra gives an adequate description of the notion of *volume* in a vector space E of dimension n in terms of the *alternating multilinear n-forms*, of which the prototype is the *determinant n-form* on the numerical space $\mathbb{R}^n$ which we denote by ω_n: it is defined on n elements of $\mathbb{R}^n$ as the determinant of the $n \times n$ matrix having these n-tuples as its column vectors. Recall that a multilinear map ω from $\times_{i=1}^n E$ to $\mathbb{R}$ is said to be *alternating* if, for any permutation π of $\{1, 2, \cdots, n\}$ and any system of n vectors v_1, ..., v_n, we have

$$\omega(v_{\pi(1)}, \cdots, v_{\pi(n)}) = (-1)^{\operatorname{sgn}(\pi)} \omega(v_1, \cdots, v_n)$$

where $\operatorname{sgn}(\pi)$ denotes the *sign* of the permutation π.

Proposition VIII.42. *In an n-dimensional vector space E, the alternating multilinear n-forms form a 1-dimensional vector space, which we denote by VE.*

Proof. In fact, if a system of n vectors $v_1, ..., v_n$ is dependent, we necessarily have $\omega(v_1, \cdots, v_n) = 0$ for any alternating multilinear n-form ω. The form ω is thus determined by its value on a basis of E. $\qquad\square$

Remark VIII.43. To choose one of the two half-lines in the space VE is equivalent to *orienting* E, the "*positive*" bases being those on which an alternating multilinear n-form of the chosen half-line takes a positive value.

VIII.44. Given a linear map l from a vector space E to a vector space F, assumed to be of the same dimension n, it is possible to associate to any element ω of VF an element $l^*\omega$ of VE by putting, for n vectors v_1, ..., v_n in E,

$$l^*\omega(v_1, \cdots, v_n) = \omega(l(v_1), \cdots, l(v_n)) \,.$$

This alternating multilinear n-form is non-zero precisely when the map l is bijective. A classical application of this remark is the following: if the space E is provided with a basis b (which, we recall, can be considered to be an invertible map from $\mathbb{R}^n$ to E), then the determinant of an endomorphism l of E is simply the value of the alternating multilinear n-form $l^*(b^{-1})^*\omega_n$ on the n vectors forming the basis b.

Definition VIII.45. On an n-dimensional configuration space M, a *volume element* is a choice of a non-zero alternating multilinear n-form in each tangent space whose value on the natural basis of a chart in a neighbourhood of each point is a differentiable function.

Remark VIII.46. As we have already remarked several times before for other fields, it follows directly from the definition of a volume element that its value on the natural basis of any admissible chart is a differentiable function.

VIII.47. In $\mathbb{R}^n$ considered as a configuration space and referred to linear coordinates $(x^1, \cdots, x^n)$, the simplest example of a volume element is of course given by the determinant n-form ω_n, which is a constant volume element. It is usually *denoted* by $\omega_n = dx^1 \wedge \cdots \wedge dx^n$, the $\wedge$ symbol indicating that the n-form is alternating.

When we take curvilinear coordinates (y^α), the determinant n-form may no longer be equal to 1 on the natural basis. One easily sees that the relation

$$\omega_n = \frac{D(x^1, \cdots, x^n)}{D(y^1, \cdots, y^n)} \, dy^1 \wedge \cdots \wedge dy^n$$

holds.

If we consider polar coordinates (r, θ) on $\mathbb{R}^2 - \{0\}$, we have $\omega_2 = r \, dr \wedge d\theta$; in cylindrical coordinates (r, θ, z) on $\mathbb{R}^3$ minus the z-axis, we have $\omega_3 = r \, dr \wedge d\theta \wedge dz$; and finally, in spherical coordinates (r, θ, ψ) on $\mathbb{R}^3 - \{0\}$, we have another expression of ω_3, namely $\omega_3 = r^2 \cos\psi \, dr \wedge d\theta \wedge d\psi$.

VIII.48. A volume element ω is a field which serves as a model for the *density* of a medium. Thus, it is natural, when studying the mechanics of incompressible media, to consider those diffeomorphisms which preserve a volume element. This property can be expressed very simply in terms of the *inverse image* operation on a volume element, which we describe now.

VIII.49. Let ω be a volume element on an n-dimensional configuration space N. If f is a differentiable map from a configuration space M to N, we define $f^*\omega$ on n vectors $v_1, ..., v_n$ in the tangent space at a point a of M by setting

$$(f^*\omega)(v_1, \cdots, v_n) = \omega(T_a f(v_1), \cdots, T_a f(v_n)) .$$

Thus, a differentiable map f from a configuration space M to itself such that $f^*\omega = \omega$ *preserves* the volume element ω. Since, for two differentiable maps f and g from a configuration space M to itself, we have $T_a(g \circ f) = T_{f(a)}g \circ T_a f$ by the Chain Rule III.46, it follows that $g^*(f^*\omega) = (g \circ f)^*\omega$. Consequently, the diffeomorphisms which preserve a volume element ω form a subgroup, which we *denote* by $\mathcal{D}_\omega M$, of the group $\mathcal{D}M$ of diffeomorphisms of M.

Exercise VIII.50. Show that a map f from a connected configuration space M to itself is a local diffeomorphism if one can find a volume element ω such that $f^*\omega$ is also a volume element.

Exercise VIII.51. Show that one can define the *quotient* of two volume elements, and that this is an observable.

VIII.52. There is one (tricky!) point which we have passed over in silence. *It is not guaranteed that, on a given n-dimensional configuration space, there exists a volume element.* Of course, one always exists locally since it suffices to transport by means of a chart the determinant n-form ω_n on $\mathbb{R}^n$. The problem is *global*. There is no obstruction to defining, on any n-dimensional configuration space, a field of alternating multilinear n-forms (even one that is not identically zero !), but *a volume element must never vanish.*

One can show that "*the necessary and sufficient condition for one to be able to define a volume element globally is that one can choose a coherent system of charts the Jacobians of whose transition diffeomorphisms are everywhere positive*". (Exercise: Prove it.) Such spaces are called *orientable*. One easily sees that "*parallelisable spaces are orientable*", as are *spheres* and *odd-dimensional projective spaces*.

If a configuration space is not orientable, there are two ways to proceed: one can pass to a weaker notion, that of *densities*, which are essentially positive quantities whose expressions in two charts are related by means of the *absolute value* of the Jacobian of the transition diffeomorphism; or one can change the space, using one of the fundamental results of the theory of *covering spaces*, according to which "*every configuration space has a covering space of two sheets which is orientable*". An example of a situation related to this theorem is provided by the coverings of even-dimensional projective spaces by spheres.

Project VIII.53. Show that the real projective plane $\mathbb{R}P^2$ is not orientable.

Proposition VIII.54. *Let M be a configuration space. The vector fields X on M whose flow preserves a volume element ω are characterised by the relation $\mathcal{L}_X \omega = 0$.*

In the numerical space $\mathbb{R}^n$ referred to linear coordinates (x^i) and for the determinant n-form ω_n, this condition is expressed by the vanishing of the divergence *of X, denoted by* div X, *i.e., by the relation*

$$\operatorname{div} X \equiv \sum_{i=1}^{n} \partial X^i / \partial x^i = 0.$$

Proof. By definition of the Lie derivative applied to the volume element ω,

$$\mathcal{L}_X \omega = \frac{d}{dt} \xi_s^* \omega_{|s=0}$$

where (ξ_s) denotes the flow of the vector field X. If $\xi_s^* \omega = \omega$ for all s, we thus obtain $\mathcal{L}_X \omega = 0$.

Conversely, suppose that $\mathcal{L}_X \omega = 0$. By using the homomorphism property of the group $s \mapsto \xi_s$, we see that

$$0 = \xi_{s_0}^* (\mathcal{L}_X \omega) = \xi_{s_0} (d\xi_s/ds)_{|s=0} = (d\xi_{s'}/ds')_{|s'=s_0} .$$

We thus find that $s \mapsto \xi_s^* \omega$ is a constant function on its domain of definition. Since ξ_0 is the identity, we thus have $\xi_s^* \omega = \omega$.

Consider now the case of the numerical space $\mathbb{R}^n$ with its standard volume element ω_n. The formula for the divergence follows from the calculation of the derivative of the Jacobian of the family of diffeomorphisms (ξ_s). The calculation of $\mathcal{L}_X \omega$ reduces to that of the Jacobian of ξ_s in the linear chart x. We know that the flow (ξ_s) admits an expansion $\xi_s(q) = q + sX(q) + o(q)$ at a point q of $\mathbb{R}^n$. Applying the Chain Rule III.46 and the fact that the differential of the determinant at the identity is the trace, we obtain that

$$\mathcal{L}_X \omega_n = \frac{d}{ds} \left(\frac{D(\xi_s(x^1, \cdots, x^n))}{D(x^1, \cdots, x^n)} \right)_{|s=0} = \operatorname{trace} \left(\frac{\partial X^i}{\partial x^j} \right) = \sum_{i=1}^{n} \frac{\partial X^i}{\partial x^i} .$$

$\square$

Remark VIII.55. In the proof of Proposition VIII.54, we have not been precise about the domain of definition in time of the flow of the vector field X ; if X is complete, the identity holds for any $s \in \mathbb{R}$, otherwise one must localise the result in space.

VIII.56. It is useful to know how to calculate the *divergence* of a vector field X defined on an open subset of a numerical space $\mathbb{R}^n$ in curvilinear coordinate systems.

These calculations, which are in fact very simple, depend on the *Leibniz formulas* $\mathcal{L}_X(\varphi\omega) = \varphi \mathcal{L}_X\omega + (\partial_X\varphi)\,\omega$ and $\mathcal{L}_{\varphi X}\omega = \varphi \mathcal{L}_X\omega + (\partial_X\varphi)\,\omega$, valid for any observable φ. (Exercise: Prove them.)

Exercise VIII.57. Express the divergence of a vector field X defined on an open subset of $\mathbb{R}^2 - \{0\}$ as a function of its components X^r and X^θ in polar coordinates.

Answer the same question for a vector field defined on an open subset of $\mathbb{R}^3 - \{0\}$ as a function of its components X^r, X^θ and X^ψ in spherical coordinates.

Project VIII.58. Let ω and ω' be two volume elements on a compact configuration space M. Show there exists a diffeomorphism ξ of M and $\lambda > 0$ such that $\omega' = \lambda\,\xi^*\omega$ (Moser's Theorem).

VIII.59. There are other configuration spaces of geometric quantities which merit study, usually because they are the natural setting for variational problems. To name just a few, we mention:

i) spaces of *exterior differential k-forms* (for $1 \le k \le n$) forming the setting of *de Rham theory*, with many applications in topology, differential geometry and complex analytic geometry;

ii) spaces of *differentiable maps* from one configuration space to another, which generalise directly the examples treated in this chapter; they form the setting for non-linear σ-models used by physicists, especially in gauge theories; string theory itself involves the maps from a Riemann surface to a numerical space of large dimension (the dimensions for which the theory simplifies the most are 10 and 26, because of the appearance of a huge group of unbroken symmetries);

iii) spaces of *Riemannian metrics*, objects which consist of the assignment of a scalar product in each tangent space depending differentiably on the point; the dynamical formulation of general relativity reduces the solution of the Einstein equations (which involve a *Lorentzian metric*) to an evolution equation on the space of Riemannian metrics on a space-like hypersurface in space-time;

iv) spaces of *connections*, objects originally introduced as tools for making calculations of a geometric nature, some of which were motivated by the Calculus of Variations (as an expression of the equation for geodesics); in the case of general vector bundles, they have become fundamental objects in gauge theories, since it is on the space of connections that is defined the *energy* functional of *Yang-Mills theory*, one of the gauge theories which have been studied most intensively.

E. Historical Notes (Contemporary)

VIII.60. Objects presented in this chapter feature among the basic spaces of global analysis. They have become essential for treating numerous geometric problems connected to the global form of spaces. The development of tools enabling one to obtain non-trivial mathematical results in this theory goes back to the 1930s for linear problems (one should mention here de Rham), and extensions necessary to treat non-linear variational problems was obtained at the beginning of the 1960s (as well as Morse, Atiyah[161] and Singer[162], Palais and Smale should be mentioned).

The geometric ideas we have developed very briefly must be combined with the analytic refinements of Chapters I and II to construct *chains of spaces* related to each other by *compact* inclusion properties which enable one to find sufficiently many convergent sequences. We cannot say more about it, for to be more specific we would need to have available the terminology of the theory of partial differential equations.

After developing the tools themselves came a period during which the fruits of this hard labour were harvested. Numerous geometric problems (some of which vainly searched for decades) were solved. This transformation of internal results into a systematic method for solving many problems was in particular achieved by Yau.[163]

VIII.61. As well as subtle results from analysis, the solution of geometric problems often requires one takes account of topological properties of the spaces studied, such as the construction of invariants, usually algebraic, ensuring a space is sufficiently simple, or complicated enough, according to the problem considered.

By a happy (but proper!) coincidence, we are witnessing at this moment a spectacular reaping of results of a topological nature derived by methods of global analysis. Among those most actively involved in this are Uhlenbeck[164], and Donaldson, whose methods combine geometry and analysis and lead to the creation of new objects.

---

[161] The British-Lebanese mathematician Michael Francis Atiyah (1929–2019) contributed essential results to many branches of mathematics, from algebraic geometry to differential geometry. His major achievement was to prove what is known as the Atiyah–Singer theorem that computes the analytic index of an elliptic differential operator in geometric terms. It has applications in many areas and has become an essential tool also in theoretical physics. Atiyah played a key role in linking mathematicians and theoretical physicists. He also had a number of public responsibilities, e.g. being the President of the Pugwash movement at the end of the XXth century. He received the Fields Medal in 1966 and the Abel Prize in 2004.

[162] Although initially interested in physics, Isadore Manuel Singer (1924–2021) became a mathematician and lived in the United States all his life. He produced major contributions in several fields of mathematics from operator algebras to differential analysis and global analysis. His most famous result is the index theorem he obtained with Atiyah (see Atiyah's note above). He had extensive discussions with physicists and was also a great promoter of joint work between mathematicians and theoretical physicists. He received the Abel Prize in 2004 (jointly with Atiyah).

[163] The Chinese-American mathematician Yau Shing Tung was born in 1949 in Mainland China but, shortly after his birth, his family moved to Hong Kong. He got his university education from the University of California Berkeley under Chern's supervision. In a short period of time in the 1980s he solved many major problems in differential geometry using global analysis techniques, leading him to establish the field at the world level by making the resolution of non-linear problems in partial differential equations a main tool. He also kept close contacts with theoretical physicists. He founded several mathematical institutes in China, having major success in finding financial resources to run them. Recently, he left his position at Harvard and is now at Tsinghua University. He received the Fields Medal in 1982 and the Wolf Prize in 2010.

[164] Karen Keskulla Uhlenbeck, born in 1942, is an American mathematician with many contributions to differential geometry and theoretical physics, solving problems in minimal surfaces, gauge theories and integrable systems. She is considered one of the founders of modern geometric analysis. The title of her doctoral dissertation *The Calculus of Variations and Global Analysis* under Palais was premonitory. She has been very active in promoting actions to make the mathematical community more welcoming to women. She received the Abel Prize in 2019, the first woman ever.

Chapter IX
The Euler–Lagrange Equations

This chapter and its companion, Chapter X, contain the *fundamental equations of the Calculus of Variations*. We begin with the Lagrangian point of view, which consists of determining the extremals of a function defined on the space of curves and obtaining by integration a function depending on the positions and velocities.

In Sect. A, we give the collection of all tangent spaces to a configuration space a natural configuration space structure.

We then show, in Sect. B, how an observable on this 'extension by velocities', the *Lagrangian*, defines the *action* functional on the space of curves on the configuration space, and we give the *First Variation Formula* for it. The fundamental equations which must be satisfied by the extremals, due to Euler and Lagrange, are established in Sect. C. This enables us to see how, in a configuration space, a Lagrangian allows one to define a notion of *momentum* of a curve. We stress the intrinsic character of the Euler–Lagrange equations, a property often not mentioned though it is used implicitly. They give rise to *differential equations of second order*.

These equations are applied in Sect. D to finding the curves of extremal length (and energy), namely the *geodesics*. For the shortest paths (which are particular examples of geodesics), we state an existence theorem when the configuration space is compact.

The study of the *motion of a rigid body* is the subject of Sect. E. The appropriate configuration space is the group of rotations. By making use of the group structure, the Euler–Lagrange equations can be rewritten as *non-linear differential equations* of the first order often called *Euler's equations*.

In Sect. F we give an outline of a variational problem defined on the space of surfaces, a space of objects of dimension greater than 1, namely the study of *surfaces of stationary area*. The equation which characterises them is the vanishing of the *mean curvature*.

J.-P. Bourguignon, *Variational Calculus*, Springer Monographs in Mathematics,
https://doi.org/10.1007/978-3-031-18307-2_9

A. The Extension by Velocities of a Configuration Space

The object of this section is to present a family of functionals defined on the space of curves on a configuration space. For this it is necessary to view the tangent vectors to a configuration space as the points of another configuration space, the *extension by velocities* of the configuration space we started with.

IX.1. The functionals on the space of curves on a configuration space which we are going to consider are special in that they depend only on the velocity-vector of the curve at each point.

The calculations we must carry out make it necessary to consider the tangent vectors to a configuration space themselves as points of a configuration space. We begin by giving the relevant definitions.

IX.2. Let M be a configuration space. In VI.17 we introduced $TM = \bigcup_{a \in M} T_a M$, which is the *disjoint* union of all the tangent spaces. (This means that a point of TM is an element of the vector space $T_a M$ for some point a in M).

We are going to define simultaneously a topology and a coherent family of charts on TM. It will be convenient to have available the map π_M, called the *natural projection*, which to an element $v \in T_a M$ associates its point of attachment $a \in M$,

To make TM itself into a configuration space, we are going to make use of the charts of M in the following way. Let $x = (x^1, \cdots, x^n)$ be a chart at a point a with domain an open subset U of M. We define its *natural extension Tx* at $u \in T_a M$ by putting, for $v \in TU(= \pi_M^{-1}(U))$,

$$(Tx)(v) = (x^1(\pi_M(v)), \cdots, x^n(\pi_M(v)), \partial_v x^i, \cdots, \partial_v x^n),$$

in other words, the last n coordinates are the components of the tangent vector v in the natural basis defined by the chart x. We *denote* the n additional coordinates on TM by the Greek letter corresponding to the Latin letter used for the chart: thus, the natural extension of the chart x will be denoted by (x, ξ)[165] and the $2n$ coordinates of a natural chart of the extension by velocities by $(x^1, \cdots, x^n, \xi^1, \cdots, \xi^n)$.

IX.3. We define a topology on TM by taking as a base of open sets the subsets A which are the images of open subsets under the natural extensions of charts. Proposition IX.4 says precisely that such a definition is viable.

It remains, however, to verify that the topology thus defined is Hausdorff. This is a consequence of the fact that, for two distinct points of TM (which thus have coordinates in the extension of a chart which are not all equal), it is possible to find disjoint open sets in the domain of this chart separating their images under the extension of the chart.

---

[165] Recalling the proof of VI.17, we can also say that the n functions ξ^i can be interpreted at a point a as the evaluation of the dual basis of the natural basis of the vector space $T_a M$ associated to the chart.

Proposition IX.4. *The natural extensions to TM of the charts of an n-dimensional configuration space M give TM the structure of a 2n-dimensional configuration space, its* extension by velocities.

Proof. To check this, let x and y be two charts of M and consider the change of coordinates map τ_{Ty}^{Tx} between the natural extensions of these charts. It can be expressed as a function of τ_y^x as follows: the point (q, χ) with coordinates $(q^1, \cdots, q^n, \chi^1, \cdots, \chi^n)$ is transformed by τ_{Ty}^{Tx} into $(\tau_y^x(q), (T_q \tau_y^x)(\chi))$. By the Chain Rule III.46, τ_{Ty}^{Tx} is an infinitely differentiable map at every point of the image of the common domain of definition of x and y. Moreover, τ_{Ty}^{Tx} has inverse τ_{Tx}^{Ty} by construction, which ensures both that the topology defined by the two natural extensions coincide on their common domain of definition and that the charts form a coherent set. □

Remark IX.5. There are many systems of notation in use in the literature to denote the natural extensions of charts.

In mechanics the notation of Lagrange is often used. This consists of denoting the coordinates of a point by $(q^1, \cdots, q^n)$ and the n additional coordinates[166] necessary to parametrise TM by $\dot{q}^1, ..., \dot{q}^n$.

If, as we have seen, tangent vectors are defined as equivalence classes of curves tangent at a point with the same velocity-vector there, this system of notation has the disadvantage of suggesting the existence of a distinguished real variable with respect to which derivatives are taken. To get around this, one often speaks of a *virtual time* or *virtual velocities*, but this way of thinking is also open to misinterpretation.

Exercise IX.6. Given two charts x and y, find the Jacobian matrix of the change of chart function τ_{Ty}^{Tx} associated to their natural extensions. (cf. Formula (VI.49).)

IX.7. One of the consequences of Proposition VI.22 is that, for a differentiable map f from an m-dimensional configuration space M to an n-dimensional configuration space N, one can write the following commutative diagram:

$$
\begin{array}{ccc}
TM & \xrightarrow{Tf} & TN \\
\pi_M \downarrow & & \downarrow \pi_N \\
M & \xrightarrow{f} & N
\end{array}
$$

where all the arrows are differentiable. Restricted to each "fibre" of the projection π_M, which is a vector space, the tangent linear map of a differentiable map is linear, which justifies another name for the extension by velocities, namely the *tangent bundle* of the configuration space which is the term generally used by mathematicians.

If x is a chart of M and y one of N, the local expression of f at the point $(q, \chi) = (q^1, \cdots, q^m, \chi^1, \cdots, \chi^m)$ can be written

$$
Tf_{Ty}^{Tx}(q^i, \chi^i) = \left((f_y^x)^j(q), \sum_{i=1}^{m} \frac{\partial (f_y^x)^j}{\partial x^i}(q)\, \chi^i \right).
$$

---

[166] We remark that this notation is consistent with the term *"extension by velocities"* which it suggests for TM, and which we have used.

Exercise IX.8. Show that *a vector field X on a configuration space M is differentiable if and only if X* : *M* ↦ *TM is a differentiable map.*

Exercise IX.9. Let *M* be a configuration space. Show that the natural projection $\pi_M : TM \longrightarrow M$ is a differentiable submersion.

IX.10. The extension by velocities of a configuration space has now been in use for two centuries, in more or less elaborate forms.

   More recently, it has been found necessary to study spaces which, like the extension by velocities, have a natural projection onto another configuration space, which is called the *base*, and such that the inverse image of each point has the structure of a vector space (not necessarily of the same dimension as the base, as in the case of the extension by velocities). Such spaces are called *vector bundles*. One needs not stop here, for it is not necessary to restrict one's attention to spaces having a projection whose fibres are vector spaces. One family of such spaces occupies a privileged position: the bundles whose fibres are orbits of a Lie group acting freely. These particular spaces are called *principal fibrations* and can be thought of as *generalised frame bundles*. By using *representations of Lie groups*, it is possible to construct new vector bundles. These notions go back to É. Cartan, but have been substantially clarified by Whitney and Ehresmann[167]. They are the basic tools of modern differential geometry (recall the statement of Chern quoted in Chapter V).

   For many, it was a surprise to see the modern physics of fundamental interactions making use of these concepts and formulating the classical laws of gauge theories in these terms (but one still has to "quantise", a step which is crucial for physical applications of these models; despite recent progress, this step raises difficult mathematical questions which are still poorly understood to this day). To emphasise the fact that additional dimensions are added to the configuration space (identified with space-time), physicists often speak of *external degrees of freedom* as opposed to the extension by velocities of the configuration space whose parameters are considered to be *internal*.

IX.11. We now examine the construction of the extension by velocities in the examples we have discussed in Chapter V.

   We begin with the (elementary!) case of vector spaces.

IX.12. At a point *a* of a vector space *E*, *the tangent space can be identified naturally with the vector space E itself viewed as the tangent space at 0.* The particular property of *E* which allows this identification is *the existence of a unique translation sending the origin to the point a.* We thus see that *TE* can be identified with $E \times E$ by associating to $V_a \in TE$ the element $(a, \partial_{V_a} x)$ of $E \times E$ (here, *x* denotes the tautological chart of *E*).

   If we were working in an open subset of *E*, we would in the same way be able to identify the tangent spaces $T_a U$ and $T_b U$ by using the translation by the vector $\overrightarrow{ab}$ and we would obtain that *TU* can be identified with $U \times E$. One says that *U* is *parallelisable.*

IX.13. The argument we have given for vector spaces (or open subsets of vector spaces) generalises to all Lie groups: thus, Lie groups are *parallelisable* configuration spaces.

---

[167] The French mathematician Charles Ehresmann (1905–1979) contributed to a number of concepts of bundle theory essential for modern differential geometry and topology. Later, he turned his interest to category theory. He was a member of the Bourbaki group when it was created.

IX.14. In fact, if G is a Lie group, multiplication by an element g is a diffeomorphism from G to itself which sends, for example, the identity element of the group, say e, to g. Consequently, the tangent linear map $T_e(L_g)$ of the map L_g given by left multiplication by g is an isomorphism from $T_e G$ to $T_g G$.

Among the fundamental examples of configuration spaces which we have presented, several were Lie groups, for example the circle $\mathbf{S}^1$, and more generally the torus $\mathbf{T}^n$, the rotation group SO_3 and its universal covering SU_2. For all these configuration spaces, the extension by velocities is the product of the space with its tangent space at a point, e.g., at the identity.

Exercise IX.15. Give an explicit matrix description of the parallelisation of the extension by velocities of SO_3 in the two classical representations.

IX.16. By a classical theorem proved in the 1930s, all 3-dimensional configuration spaces for which it is possible to define a coherent orientation of their tangent spaces (one says that such spaces are *orientable*) *are parallelisable*. This dimension is the only one in which such a property is true.

IX.17. One should not think that this situation is typical. On the contrary, the extension by velocities TM of a configuration space M has in general a global structure which cannot be written as the product of M with the tangent space at a point (we then say that the extension by velocities has a *non-trivial* structure.[168]) This is already the case for the sphere $\mathbf{S}^2$. (Exercise (rather difficult): Prove it). The fact that the extension by velocities is not globally a product (even though it is a product in the domain of any chart) is precisely the reason why it is *essential* to have global tools to prove the results of this chapter rigorously !

IX.18. The only spheres which are parallelisable are the spheres $\mathbf{S}^1$, $\mathbf{S}^3$ and $\mathbf{S}^7$ by a theorem due to Adams[169], Kervaire and Milnor in the 1960s.

For $\mathbf{S}^1$ and $\mathbf{S}^3$, we have already seen that they can be given a multiplication law which makes them into Lie groups. These laws are in fact inherited from the structure of a field on the vector spaces $\mathbb{R}^2 (= \mathbb{C})$ and $\mathbb{R}^4 (= \mathbb{H})$ in which the spheres are viewed as the spheres of radius 1 for the standard metric.

The case of $\mathbf{S}^7$ is more subtle, as there is no structure of a field on $\mathbb{R}^8$, but only a weaker structure corresponding to a product which satisfies the associativity axiom only for triples of the form (a, b, a). The vector space $\mathbb{R}^8$ provided with this multiplication law is called the space of *Cayley octonions*. There are beautiful geometric properties hidden in this object, such as the *principle of triality* (the interested reader is referred to (Besse 1978), for example).

Even though spheres are not parallelisable in general, the structure of their extension by velocities is rather simple. In fact, using the description of $\mathbf{S}^n$ as a subset of $\mathbb{R}^{n+1}$, we see that we have only to "add" a line to the extension by velocities for the new fibre space with $n+1$-dimensional fibres thus formed to be trivial, in other words, globally a product of $\mathbb{R}^{n+1}$ with $\mathbf{S}^n$. This is usually expressed by saying that the spheres $\mathbf{S}^n$ are *stably parallelisable*.

---

[168] The subtlety of the global structure of a configuration space derives from the fact that the parallelism defined by two distinct charts defined on sufficiently small open sets do not coincide.

[169] The British mathematician John Frank Adams (1930–1989) was a topologist with major contributions to the theory of homotopy.

IX.19. The structure of the extension by velocities of real projective spaces is actually rather subtle, except for the one which interests us the most, namely $\mathbb{RP}^3$, which can be identified with SO_3 and is thus parallelisable.

A description of the extension by velocities of the projective plane $\mathbb{RP}^2$ can be obtained from that of S^2 since the n-dimensional projective space can be identified with the set of pairs of antipodal points on the sphere of the same dimension. One deduces (easily!) that the projective plane is not parallelisable.

One of the essential properties of the extension by velocities of projective space is preserved under a limiting procedure which we now present. In fact, the natural injections of $\mathbb{R}^n$ into $\mathbb{R}^{n+1}$, which associates to an n-tuple $(x^1, \cdots, x^n)$ the $(n+1)$-tuple $(x^n, \cdots, x^n, 0)$, define natural injections of $\mathbb{RP}^{n-1}$ into $\mathbb{RP}^n$, and taking the inductive limit gives a "limit space" $\mathbb{RP}^\infty$ which is of course "infinite-dimensional", but whose geometric structure is actually much simpler that that of (finite-dimensional) projective spaces. This *stabilisation* procedure, which is in a sense analogous to that which we introduced in connection with the sphere in IX.18, still retains the essence of the non-triviality of the extension by velocities of real projective spaces in the form of a 1-dimensional cohomology class, called the *first Stiefel[170]-Whitney class*. This essentially reflects the fact that the image of the equator of S^n in the projective space of the same dimension cannot be contracted to a point in the projective space.

IX.20. A completely parallel construction works for complex projective spaces, where the field $\mathbb{R}$ of real numbers is replaced by the field $\mathbb{C}$ of complex numbers. One builds in the same way an infinite complex projective space $\mathbb{CP}^\infty$ whose structure is again rather simple. This object also retains the essential part of the non-triviality of the extensions by velocities of complex projective spaces in the form of a cohomology class (2-dimensional this time) called the *first Chern class*. This class corresponds to the fact that the image of the sphere S^2 (which we have seen in V.73 can be identified with the complex projective line $\mathbb{CP}^1$) in $\mathbb{CP}^\infty$ cannot be contracted to a point.

B. The Action and its First Variation

We now introduce the notion of a *Lagrangian* on a configuration space and we examine how it is possible to obtain from it a function on the space of curves on the configuration space.

We establish the *first variation* of this observable on the space of curves which involves the *momentum* of the curve at which it is calculated.

IX.21. In a problem of mechanics or physics, there are procedures for determining what the Lagrangian must be in terms of quantities classically attached to the system, such as the *kinetic energy* and the *potential energy*, but we shall return to this point a little later.

From the mathematical point of view, Lagrangians give rise to a very special geometry since they are observables defined on the extension by velocities of a configuration space, a space which has a special internal geometry.

Unfortunately, we do not have time to go into any detail about this, but it will be a (hidden!) guide to the developments in this chapter. We give an example of it in Exercise IX.36.

---

[170] A student of H. Hopf, Eduard Stiefel was a Swiss mathematician who contributed to several important questions in topology and geometry. Later he moved on to more applied areas.

> *A configuration space M being fixed, we suppose given*
> *a priori an observable A on the extension by velocities*
> *TM of M (or at least on a dense open subset of TM).*
> *This function is the Lagrangian of the problem.*

IX.22. On the space C ($= C^\infty([\alpha, \beta], M)$) of curves γ on a configuration space M, we define the *action* functional, *denoted* by $\mathbf{A}$, by

$$\mathbf{A}(\gamma) = \int_\alpha^\beta A(\dot\gamma(\tau))\, d\tau\,.$$

In the most elementary case where the curve is in a numerical space $\mathbb{R}^n$, we can use the isomorphism between $T\mathbb{R}^n$ and $\mathbb{R}^n \times \mathbb{R}^n$ to associate to the velocity-vector $\dot\gamma(\tau)$ of the curve γ at time τ the pair $(\gamma(\tau), \gamma'(\tau))$, representing the point on the curve and its derivative at the point. It is by thinking of this case that the action integral is often written $\int_\alpha^\beta A(\gamma(\tau), \gamma'(\tau))\, d\tau$, but this way of writing it is legitimate only when the extension by velocities decomposes *naturally* as a product which, we recall, is the case when a chart has been chosen, but which is not the case in general globally as we have seen in IX.17.

Examples IX.23. The simplest Lagrangians are those which come from an observable φ defined on M, i.e., those which can be written as $\varphi \circ \pi_M$. They thus depend on the velocity-vector of the curve only through the point of the space M to which the vector is attached.

One of the most classical Lagrangians is that whose action integral is the *length* of the curve. To define this notion in a general configuration space, we need a scalar product g in each tangent space, a so-called *Riemannian metric*. We assume that g is *differentiable*, i.e., that its coefficients g_{ij} in an admissible chart $x = (x^1, \cdots, x^n)$, defined by the relation $g = \sum_{i,j=1}^n g_{ij}\, dx^i dx^j$, are differentiable functions. The length of a curve γ is then

(IX.24)
$$\mathbf{L}(\gamma) = \int_\alpha^\beta \sqrt{g(\dot\gamma(\tau), \dot\gamma(\tau))}\, d\tau\,.$$

Note that, in view of the differentiability hypotheses we have made on the metric and on the curve, there is no problem defining the integral.

Another Lagrangian which is natural in this context is the *energy* of the curve in this metric, which is defined by

(IX.25)
$$\mathbf{E}(\gamma) = \frac{1}{2} \int_\alpha^\beta g(\dot\gamma(\tau), \dot\gamma(\tau))\, d\tau\,.$$

This terminology derives from the model in which the curve represents an elastic band and its "*potential energy*" is then proportional to the integral of the square of the length of the tangent vector to the elastic band.

It is very instructive to compare the variational problems associated to these two Lagrangians, as we shall explain in detail in Sect. D.

Solved Exercise IX.26. Show that *the length functional is invariant under reparametrisation.*

Let γ be an element of C. If we compose γ with a diffeomorphism ψ of the interval $[\alpha, \beta]$ preserving the endpoints,[171] we obtain a curve $\tilde{\gamma}$ having the same geometric image, but a priori traversing it at a different velocity. If, for greater clarity, we denote by $\tilde{\tau}$ the parameter of the curve $\tilde{\gamma}$, we have $d\tau = \psi'(\tilde{\tau})\, d\tilde{\tau}$ so that we obtain

$$
\begin{aligned}
\mathbf{L}(\tilde{\gamma}) &= \int_\alpha^\beta \sqrt{g(\dot{\tilde{\gamma}}(\tilde{\tau}), \dot{\tilde{\gamma}}(\tilde{\tau}))}\, d\tilde{\tau} \\
&= \int_\alpha^\beta |\psi'(\tau)|\, \sqrt{g(\dot{\gamma}(\tau), \dot{\gamma}(\tau))}\, d\tilde{\tau} \\
&= \int_\alpha^\beta \sqrt{g(\dot{\gamma}(\tau), \dot{\gamma}(\tau))}\, d\tau
\end{aligned}
$$

which is precisely $\mathbf{L}(\gamma)$.

IX.27. When a concrete situation is being modelled, it may be that time appears explicitly as one of the variables on which the Lagrangian depends. Since, in the solution of the problem, it will also play a role as a dynamical variable, it must be given special treatment. We leave this to the reader.

IX.28. Having fixed a Lagrangian A on a configuration space M, we propose to solve the following problem:

"describe the differential of the action functional (considered as an observable on the space C of curves on M) at a point γ of C."

Thus, given a *variation* $(\tau, s) \mapsto \Gamma(\tau, s) = \gamma_s(\tau)$ of a curve γ, we try to find the infinitesimal variation of the action $\mathbf{A}$,

$$
(IX.29) \qquad \frac{d}{ds}\mathbf{A}(\gamma_s)_{|s=0} = \int_\alpha^\beta \frac{d}{ds}(A(\dot{\gamma}_s(\tau)))_{|s=0}\, d\tau \ .
$$

In the preceding integral we must evaluate the derivative of a function along the curve $s \mapsto \dot{\gamma}(\tau)$ in TM, so it is necessary to consider the curve formed by its velocity-vectors, and this is what we shall do now.

The curve $(\tau, s) \mapsto (T_{(\tau,s)}\Gamma)(\partial/\partial\tau)$ is a variation of the curve $(\tau, s) \mapsto \dot{\gamma}_0(\tau)$ in TM.

> *In this chapter, we only use the local charts (x^i, ξ^i) of TM,*
> *namely the natural extensions of the charts (x^i) of M.*

---

[171] Exercise: Show that ψ is necessarily *increasing*.

IX.30. We are now in a position to continue our study of the variation of the action and to develop Formula (IX.29).

IX.31. To carry out the calculation completely without introducing supplementary notions, we shall assume *temporarily* that the curve γ along which we are taking the variation lies entirely in the domain of a single chart[172] (x^i). We obtain

$$\frac{d}{ds}\mathbf{A}(\gamma_s)_{|s=0} = \int_\alpha^\beta \frac{d}{ds}\left(A(\frac{\partial\Gamma}{\partial\tau}(\tau,s))\right)_{|s=0} d\tau$$

$$= \int_\alpha^\beta \left(\sum_{i=1}^n \frac{\partial\Gamma^i}{\partial s}(\tau,0)\frac{\partial A}{\partial x^i}(\gamma^k(\tau),\frac{d\gamma^k}{d\tau}) + \frac{\partial^2\Gamma^i}{\partial s\partial\tau}(\tau,0)\frac{\partial A}{\partial\xi^i}(\gamma^k(\tau),\frac{d\gamma^k}{d\tau})\right) d\tau .$$

By the Schwarz Lemma (cf. III.97), one can replace the second derivative term $\partial^2\Gamma/\partial s\partial\tau$ by $\partial^2\Gamma/\partial\tau\partial s$. The last term in the expression we have obtained can then be transformed using integration by parts, which gives

$$\frac{d}{ds}\mathbf{A}(\gamma_s)_{|s=0} = \int_\alpha^\beta \left(\sum_{i=1}^n \frac{\partial\Gamma^i}{\partial s}(\tau,0)\frac{\partial A}{\partial x^i}(\gamma^k(\tau),\frac{d\gamma^k}{d\tau}) - \frac{\partial\Gamma^i}{\partial s}(\tau,0)\frac{d}{d\tau}\frac{\partial A}{\partial\xi^i}\right) d\tau$$

$$+ \left[\sum_{i=1}^n \frac{\partial\Gamma^i}{\partial s}\frac{\partial A}{\partial\xi^i}\right]_\alpha^\beta .$$

Remark IX.32. In the preceding formula, the variation of the action finally involves only the transverse vector field $\partial\Gamma/\partial s$ of the variation Γ, which we denote by X.

Note that we evaluate the first derivative of the "action" functional $\mathbf{A}$ in the direction of a tangent vector at a point γ of the space C and that we have seen in VIII.6 that $T_\gamma C$ can be identified with the space of vector fields along γ.

First Variation Formula for the Action IX.33. *Let X be an infinitesimal variation of a curve γ on a configuration space M. The First Variation of the action functional $\mathbf{A}$ defined by the Lagrangian A is given by the formula*

$$\frac{d\mathbf{A}}{ds}(\gamma_s)_{|s=0} = \int_\alpha^\beta -\langle(\vartheta^A\gamma)(\tau), X(\tau)\rangle \, d\tau + [\partial_{X(\tau)}(A_{|T_{\gamma(\tau)}M})]_\alpha^\beta ,$$

where $(\vartheta^A\gamma)$ denotes the acceleration momentum[173] *of the curve γ defined by the Lagrangian A, whose expression in a local chart (x^i,ξ^i) is given by the differential 1-form on M defined along γ given by*

---

[172] We can always reduce to this case by cutting up the interval on which the curve is defined into intervals of sufficiently small length.

[173] We have adopted this terminology by analogy with that of the *momentum,* used to denote the product of the mass and the velocity.

$$(\vartheta^A \gamma)(\tau) = \sum_{i=1}^{n} -\left(\frac{\partial A}{\partial x^i} (\gamma^k(\tau), \frac{d\gamma^k}{d\tau}) - \frac{d}{d\tau} \frac{\partial A}{\partial \xi^i} (\gamma^k(\tau), \frac{d\gamma^k}{d\tau}) \right) dx^i ,$$

and where $A_{|T_q M}$ denotes the restriction of A to the tangent space at the point q.

Proof. This is just another way of writing the formula at the end of IX.31, after making the infinitesimal variation X appear explicitly and reinterpreting the integral term.

The term $[\sum_{i=1}^{n} (\partial \Gamma^i / \partial s)(\partial A / \partial \xi^i)]_{\alpha}^{\beta}$ can in fact be written as in the formula in the statement of the proposition, by bringing in the restriction of the Lagrangian to the tangent space and viewing the value $X(\tau)$ of the transverse vector field as a tangent vector at the point $\dot{\gamma}(\tau)$ of the vector space $T_{\gamma(\tau)} M$.

It remains finally to dispose of the hypothesis made for convenience in obtaining the expression given at the end of IX.31, namely that the curve lies entirely in the domain of definition of a local chart. In the general case, after cutting up the interval $[\alpha, \beta]$ into sufficiently small subintervals, we see that the contributions at the ends of the intermediate intervals evaluated in possibly different charts are intrinsically defined. Since they cancel out in pairs except for the actual ends of the curve, we obtain the stated formula. $\square$

Exercise IX.34. Deduce from the formula for the acceleration momentum the formulas giving the ordinary acceleration in polar coordinates on $\mathbb{R}^2$. (One should express the Euclidean metric e in polar coordinates (ρ, θ).)

The same question for spherical coordinates on $\mathbb{R}^3$.

Remark IX.35. i) The First Variation Formula contains the calculation of the *differential* of the function **A** at the point γ of the space C of curves on M, but we shall be deliberately vague about the fact that we should interpret this differential as a vector in the dual of $T_\gamma C$, which we have identified with the space of vector fields along γ.

It also shows that the acceleration momentum is an intrinsic quantity, a fact which is not clear from its expression in a local chart (see, however, IX.46).

ii) The infinitesimal variation of the action functional is often denoted by δ**A** and the momentum is called the *variational derivative* of the Lagrangian. We shall not use this terminology.

Exercise IX.36. Let A be a Lagrangian defined on the extension by velocities of an n-dimensional configuration space. Show that *the differential form on TM defined in a chart (x^i, ξ^i) by $\sum_{i=1}^{n} \partial A / \partial \xi^i \, dx^i$ is in fact an intrinsic quantity, often called the* impulse *associated to the Lagrangian A.*

C. The Euler–Lagrange Equations

In this section we discuss the fundamental problem of the Calculus of Variations, namely to give the equations satisfied by the extremals of a variational problem.

IX.37. Even though the problem we discuss in this section can be considered as a sub-problem of that taken up in Sect. B, its importance, both in theory and practice, is such that it deserves to be treated separately.

As in Sect. B, we suppose that a Lagrangian A is fixed on a configuration space M. We want to solve the following problem: "*given two points a and b in M, find the curves γ on M defined on the interval $[\alpha, \beta]$ such that $\gamma(\alpha) = a$ and $\gamma(\beta) = b$ for which the action integral defined by A is extremal.*"

Remark IX.38. In many practical situations, the actual problem is more restricted: it is to determine the *minima* or *maxima*[174] of the action (note that, to make sense of these notions of maxima or minima, one requires only the *topology* of the space on which the functional is defined). The same is not true for the other extrema, the *saddle-points* (also called *passes*), which make sense only for configuration spaces.

We have seen in Chapter VII that "*the equation for the critical points is the same for all the different types of critical points, whether maxima, minima or saddle-points*". The information contained in the equation for the critical points extends only up to the first order. To make finer distinctions, it is necessary to look at the terms of second order, and thus to consider the value of the Hessian form. Unfortunately, we do not have time to give this calculation for the action functional, classically called "*the Second Variation Formula*".

Note, however, that when one looks for generalised solutions (often called "*weak*" solutions when a certain number of derivatives must be taken, making use of function space duality), the "minima" (or "maxima") often have stronger regularity properties than the general solutions. Using these properties, one can often show that the generalised minimising solutions are in fact solutions in the classical sense, namely sufficiently differentiable solutions for which the differentiation operations involved have their usual meaning.

IX.39. Before moving on to the description of the equations giving the critical points of the action functional, we need a technical lemma.

Du Bois-Reymond's Lemma IX.40.[175] *Let λ be a differential form defined along a curve γ. If, for every vector field X defined along γ, we have $\int_\alpha^\beta \langle \lambda(\tau), X(\tau) \rangle \, d\tau = 0$, then $\lambda(\tau) = 0$ for all $\tau \in [\alpha, \beta]$.*

Proof. Suppose that there exists a value τ_0 of the parameter at which the differential form is not zero. Consider a local chart (x^i) at the point $\gamma(\tau_0)$. By VII.6, the differential form λ can be written $\lambda = \sum_{i=1}^n \lambda_i \, dx^i$ with at least one of the $\lambda_i(\gamma(\tau_0)) \neq 0$.

We construct a vector field X_0 along γ in the domain of definition U of the chart by putting $X_0(\tau) = \sum_{i=1}^n \lambda_i \, \partial/\partial x^i$. To get a differentiable vector field along γ, we multiply X_0, for example, by a non-negative differentiable function φ which vanishes outside an interval $]\tau_0 - \epsilon, \tau_0 + \epsilon[$ which

---

[174] Passing from one to the other is simply a matter of changing the sign of the Lagrangian, which is sometimes only a matter of convention.

[175] The German mathematician Paul David Gustav Du Bois-Reymond (1831–1889) made various contributions to the theory of functions. Motivated by mathematical physics, his work was concerned principally with the study of differential equations.

is sufficiently small to be contained in U (take, for example, $\varphi(\tau) = \exp(-((\tau - \tau_0)^2 - \epsilon^2)^{-1})$ for $\tau \in]\tau_0 - \epsilon, \tau_0 + \epsilon[$ and $\varphi \equiv 0$ for other values of $\tau \in [\alpha, \beta]$).

For the vector field $X = \varphi X_0$, the integrand is non-negative since, for each value of τ, it is equal to $\sum_{i=1}^n (\lambda_i)^2$. Moreover, we know that there exists an interval of non-zero length 2δ centred at τ_0 on which one of the components of λ, say λ_1, is at least $\lambda_1(\tau_0)/2$. Consequently

$$\int_\alpha^\beta \langle \lambda, X \rangle \, d\tau \geq \int_{\tau_0 - \delta}^{\tau_0 + \delta} \langle \lambda, X \rangle \, d\tau \geq \delta \lambda(\tau_0) > 0$$

which contradicts the fact that the integral of the differential form λ against every vector field along γ is assumed to be zero.   □

The Euler–Lagrange Theorem IX.41. *Let γ be a curve on a configuration space M defined on $[\alpha, \beta]$. To make the action functional $\mathbf{A}$ defined by a Lagrangian A stationary for all infinitesimal variations X of γ leaving its endpoints $a = \gamma(\alpha)$ and $b = \gamma(\beta)$ fixed (i.e., such that $X(\alpha) = 0$ and $X(\beta) = 0$), it is necessary that the acceleration momentum $\vartheta^A \gamma$ is zero, in other words that γ satisfies the Euler–Lagrange equations which, in a natural local chart (x^i, ξ^i), can be written*

$$(IX.42) \qquad \left(\frac{d}{d\tau} \frac{\partial A}{\partial \xi^i} - \frac{\partial A}{\partial x^i} \right)(\dot{\gamma}(\tau)) = 0, \quad 1 \leq i \leq n .$$

Proof. The proof is a combination of the First Variation Formula IX.33 and Du Bois-Reymond's Lemma IX.40.

In fact, if the curve γ makes the functional $\mathbf{A}$ stationary for all infinitesimal variations X such that $X(\alpha) = 0$ and $X(\beta) = 0$, the First Variation Formula says that the duality pairing between the acceleration momentum of γ and any vector field along γ defined on any subinterval I of $[\alpha, \beta]$ not containing α or β is zero. By Du Bois-Reymond's lemma, this implies that $(\vartheta^A \gamma)(\tau) = 0$ for all $\tau \in I$.

Since all the quantities under consideration are differentiable, one obtains by continuity that the acceleration momentum vanishes for all $\tau \in [\alpha, \beta]$.   □

Remark IX.43. We have only been able to say that the Euler–Lagrange equations give *necessary* conditions for the action functional to be extremal for the reason given in Remark IX.35.

In fact, by defining the variations as we have done, we have not considered the *most general* element of the dual of the tangent space to C at the point γ. However, in many of the situations we shall consider in the sequel, the Euler–Lagrange equations allow one to determine effectively the curves which make the action functional extremal.

Nevertheless, one should note that in our arguments we have always assumed that the curve γ is regular. There are, however, problems in the Calculus of Variations in which the data are regular but the solutions are *irregular* curves. An explicit (and appealing) example can be found on page 110 of (Avez 1983).

IX.44. The way we have obtained the Euler–Lagrange equations tended to keep in the background the intrinsic geometry of the extension by velocities TM as a configuration space, but we preferred to proceed in this way since otherwise we would have been forced to introduce several additional concepts. The interested reader will find a completely intrinsic derivation of these equations on pages 23 to 25 of (Besse 1978).

One of these additional notions which is essential is that of the *vertical differential*, often denoted by d_V, whose existence implies that the differential form $d_V A = \sum_{i=1}^{n} \partial A/\partial \xi^i \, dx^i$ is intrinsically defined; Exercise IX.36 asked for a verification of this.

IX.45. The form in which we have given the Euler–Lagrange equations in Theorem IX.41 is not the most practical one for solving effectively the problem of finding an extremal curve. Moreover, it tends to conceal the nature of the equation which the curve γ must satisfy.

Before expanding the "time" derivative contained in the Euler–Lagrange equations, we must comment on the notion of acceleration in configuration spaces.

IX.46. If we work in a chart (x^i) in which a curve γ can be written $\tau \mapsto (\gamma^i(\tau))$, the point at "time" τ of the curve $\dot{\gamma}$ is represented by the pair $(\gamma^i(\tau), d\gamma^i/d\tau)$ and its velocity-vector by the quadruple $(\gamma^i(\tau), d\gamma^i/d\tau, d\gamma^i/d\tau, d^2\gamma^i/d\tau^2)$ obtained by juxtaposing the derivatives of the preceding pair. We thus see the appearance of the ordinary acceleration $d^2\gamma/d\tau^2$ of the curve γ in this chart.

This is why it is natural to call the *acceleration-vector*, and *denote* by $\ddot{\gamma}$, the curve of velocity-vectors of the curve $\dot{\gamma}$ of velocity-vectors of γ which is a curve in TM. But one should note that the *acceleration-vector is not an element of the extension by velocities TM of the configuration space M, but belongs to the larger* (and at first sight rather monstrous) space TTM. Exercise IX.47 shows how the local expressions of the acceleration-vector correspond in two local charts, which shows clearly *the non-intrinsic character of the acceleration-vector*.

We remark that, in a chart (x^i, ξ^i) of TM, an arbitrary curve on TM, that we denote by $\tau \mapsto (x^i(\tau), \xi^i(\tau))$, does not have a velocity-vector represented by a quadruple whose two middle terms are equal as is the case for the curve formed by the velocity-vectors of a curve γ on M. This last property which lies at the heart of the special geometry of TM as a configuration space, distinguishes those elements of TTM which are called *jets*. Its intrinsic character is established in Exercise IX.47.

Exercise IX.47. Using the expression of the elements of the extension by velocities of a configuration space M as pairs, compare the expressions in two local charts of M of the acceleration-vector of a curve γ on M written in the form of quadruples.

Deduce that *the notion of a jet is intrinsically defined*.

IX.48. We can now expand the first term in Formula (IX.42). We have

$$(IX.49) \qquad \left(\frac{d}{d\tau}\frac{\partial A}{\partial \xi^i}\right)(\gamma^i(\tau), \frac{d\gamma^i}{d\tau}) = \sum_{j=1}^{n} \left(\frac{\partial^2 A}{\partial \xi^j \partial \xi^i}\frac{d^2\gamma^j}{d\tau^2} + \frac{\partial^2 A}{\partial x^j \partial \xi^i}\frac{d\gamma^j}{d\tau} \right),$$

which features the matrix $\partial^2 A/\partial \xi^i \partial \xi^j$ of second derivatives of the Lagrangian restricted to a tangent space in the coordinates ξ^i.

The equation takes a particularly simple form when this matrix is invertible, for if we multiply Eq. (IX.49) by the inverse of this matrix, we can isolate the second derivatives of the curve γ.

Definition IX.50. A Lagrangian $A : TM \to \mathbb{R}$ is said to be *regular* if the map from TM to T^*M which, to a tangent vector v at $q = \pi_M(v)$, associates the differential at v of the restriction of A to each tangent space $T_q M$ (i.e., the element $d(A_{|T_q M})(v)$ of T_q^*M) is of maximum rank.

IX.51. The formulation given in Definition IX.50 might seem rather complicated. It has the immediate advantage of being intrinsic, and in the future it will lead us to the Hamiltonian formulation which is described in Chapter X.

Of course, it is very important to know how to express the regularity of a Lagrangian in terms of a local description, and we deal with this now.

IX.52. Fix a point q in M. In a chart (x^i) defined in a neighbourhood of q, the map from $T_q M$ to T_q^*M which to v with local coordinates (q^i, v^i) associates $d(A_{|T_q M})(v)$ has the local expression $\sum_{i=1}^{n} (\partial A / \partial \xi^i)(q^i, v^i) \, dx^i$.

In fact, the ξ^i are coordinates on each tangent space $T_a M$ but, by using translations in this vector space, it is possible to replace the differentials $d\xi^i(v)$ by their value at the origin where they coincide with the dx^i (this is the content of Formula (III.55)).

The tangent linear map to $d(A_{|T_q M})$ associates to the vector $w = \sum_{i=1}^{n} w^i \partial / \partial x^i$ in $T_q M$ the vector $\sum_{i,j}^{n} (\partial^2 A / \partial \xi^j \partial \xi^i) \, w^j dx^i$, an element of T_q^*M. It is thus of rank n precisely when the matrix of second derivatives of A in the vertical directions is invertible, a condition for which we were seeking an intrinsic expression. Hence:

Proposition IX.53. *A necessary and sufficient condition for a Lagrangian A defined on a configuration space M to be regular is that, in a natural chart (x^i, ξ^i) of TM, the matrix $(\partial^2 A / \partial \xi^i \partial \xi^j)$ is invertible.*

Example IX.54. Let $A = \frac{1}{2} \sum_{i,j=1}^{n} g_{ij} \xi^i \xi^j$ be the energy associated to a bilinear form g defined on a configuration space M and whose expression in the chart $x = (x^1, \cdots, x^n)$ is $g = \sum_{i,j=1}^{n} g_{ij} \, dx^i \, dx^j$.

The map $d(A_{T_q M})$ evaluated at a point v has local expression given by $\sum_{i,j=1}^{n} g_{ij} v^j dx^i$. It is thus of maximum rank exactly *when the bilinear form g defines a duality between $T_q M$ and T_q^*M*, i.e. when g is non-degenerate.

This is obviously the case if g is a scalar product, but there are other interesting cases such as the case of *Lorentzian* metrics which are basic in general relativity.

IX.55. Among the Lagrangians which are *not regular* are those which are *positively homogeneous of degree* 1. In fact, we can express the homogeneity relation for the Lagrangian A by the relation $A(e^t \dot{q}) = e^t A(\dot{q})$, from which, by taking the derivative of this relation at $t = 0$, we obtain the *Euler identity* $\partial_D A = A$, where D denotes the *Liouville vector field* which, in any parametrisation (x^i, ξ^i) of TM, is written $D = \sum_{i=1}^{n} \xi^i \, \partial / \partial \xi^i$. Its flow is formed by dilations in each tangent space.

Differentiating the identity $\partial_D A = A$ restricted to each tangent space, we obtain that $\sum_{i=1}^{n} \xi^i \, \partial^2 A / \partial \xi^i \partial \xi^j = 0$, which means precisely that the Lagrangian cannot be regular, D lying in the kernel of the tangent linear map of $v \mapsto d(A_{|T_{\pi_M(v)}M})$.

Corollary IX.56. *The extremals $\tau \mapsto \gamma(\tau)$ of a regular Lagrangian A are regular curves. They are solutions of the Euler–Lagrange equations which can be written, in the natural extension (x^i, ξ^i) of any coordinate system (x^i), in the form*

$$(\text{IX.57}) \qquad \frac{d^2 \gamma^i}{d\tau^2} + \sum_{j=1}^{n} \alpha^{ij} \left(\sum_{k=1}^{n} \frac{\partial^2 A}{\partial x^k \partial \xi^j} \frac{d\gamma^k}{d\tau} - \frac{\partial A}{\partial x^j} \right) = 0,$$

(where (α_{ij}) denotes the inverse of the matrix $(\partial^2 A / \partial \xi^i \partial \xi^j)$), i.e., as second order differential equations which are solved for the second derivatives.

A solution is determined by its position and velocity at the initial moment.

Proof. To obtain the equations of the extremals in the form (IX.57), it is enough to apply the inverse of the matrix of second derivatives to Formula (IX.49).

As a second order differential equation defined on an open subset of $\mathbb{R}^n$, with differentiable coefficients and solved for the second derivatives, can be written as an ordinary differential equation of first order on an open subset of $\mathbb{R}^{2n}$, also with differentiable coefficients, by taking the first derivatives as auxiliary variables, its solutions are determined by their initial value and initial first derivative, whence the second part of the corollary. $\qquad \square$

Exercise IX.58. Show that *the extremals of the action associated to a Lagrangian A, assumed to be regular, are the projections onto M of the integral curves of an intrinsically defined vector field X_A on TM whose expression in a local chart (x^i, ξ^i) at a point $v = (q^i, v^i)$ is*

$$X_A(v) = \sum_{i=1}^{n} v^i \frac{\partial}{\partial x^i} + \sum_{i,j}^{n} \alpha_{ij} \left(\frac{\partial A}{\partial x^j} - \sum_{k=1}^{n} \frac{\partial^2 A}{\partial x^k \partial \xi^j} v^k \right) \frac{\partial}{\partial \xi^i}$$

where (α_{ij}) denotes the inverse of the matrix $(\partial^2 A / \partial \xi^i \partial \xi^j)$.

Definition IX.59. If A is a Lagrangian defined on the extension by velocities TM of a configuration space M, the *energy* E^A which is associated to it is the observable $E^A = \partial_D A - A$ where D denotes the Liouville vector field (which generates the scaling transformations in each tangent space).

Example IX.60. Let us show how to obtain from this formalism the equations of Newtonian mechanics for a particle of mass m in a 3-dimensional Euclidean vector space $(V, (\, | \,))$ in the presence of a potential φ.

The trajectories of the particle are extremals of a Lagrangian A which, by using the natural isomorphism from TV onto $V \times V$, in which a point v is written as (q, χ), is given by $A(q, \chi) = \frac{1}{2} m \, (\chi | \chi) - \varphi(q)$. We then have $E^A(q, \chi) = \frac{1}{2} m \, (\chi | \chi) + \varphi(q)$ because of the homogeneity of degree 2 and 0 respectively of the two terms featuring in the Lagrangian, and thus E^A coincides with the *total energy* in the sense of mechanics, which justifies the terminology we have used.

This Lagrangian is regular: the potential term, which depends only on the position of the particle, does not affect the regularity; while the quadratic term, which does not depend on the point of the space being considered, has derivative $\flat$, where $\flat$ denotes the *linear isomorphism* from V onto its dual V^* defined by the scalar product.

The Euler–Lagrange equations given by (IX.57) then take the form

$$m\,\gamma''(\tau) = -(\mathrm{Grad})\varphi(\gamma(\tau))$$

since $\mathrm{Grad} = \sharp \circ d$ where $\sharp$ denotes, as usual, the isomorphism from V^* onto V defined by the Euclidean scalar product, namely the inverse of $\flat$.

We have recovered, of course, the classical *equations of Newtonian mechanics*.

Proposition IX.61. *Along the curve of velocity-vectors $\dot\gamma$ of an extremal curve γ of the action functional $\mathbf{A}$ induced by a Lagrangian A, the energy is conserved, i.e., we have $(dE^A/d\tau)(\dot\gamma(\tau)) = 0$.*

Proof. Let us evaluate $\partial_{\dot\gamma} E^A$ at time τ_0. For this, we take a chart x at $\gamma(\tau)$. Since γ is an extremal of the action, we have that

(IX.62)
$$\left(\frac{d}{d\tau}\frac{\partial A}{\partial \xi^i} - \frac{\partial A}{\partial x^i}\right)(\gamma^i(\tau), \frac{d\gamma^i}{d\tau}) = 0 \ .$$

We thus have

$$\frac{dE^A}{d\tau}\bigg|_{\tau=\tau_0} = \frac{d}{d\tau}\left(\left(\sum_{i=1}^{n}\frac{\partial A}{\partial \xi^i}\frac{d\gamma^i}{d\tau} - A\right)\circ \dot\gamma\right)_{\bigg|\tau=\tau_0}$$

$$= \sum_{i=1}^{n}\left\{\frac{d}{d\tau}\left(\frac{\partial A}{\partial \xi^i}\right)\frac{d\gamma^i}{d\tau} + \frac{\partial A}{\partial \xi^i}\frac{d^2\gamma^i}{d\tau^2}\right\} - \frac{d}{d\tau}\left(A(\gamma^i(\tau), \frac{d\gamma^i}{d\tau})\right)$$

$$= \sum_{i=1}^{n}\left\{\frac{d}{d\tau}\left(\frac{\partial A}{\partial \xi^i}\right)\frac{d\gamma^i}{d\tau} + \frac{\partial A}{\partial \xi^i}\frac{d^2\gamma^i}{d\tau^2} - \frac{\partial A}{\partial x^i}\frac{d\gamma^i}{d\tau} - \frac{\partial A}{\partial \xi^i}\frac{d^2\gamma^i}{d\tau^2}\right\} \ ,$$

whence the required property after using (IX.62).   □

Examples IX.63. Let us review briefly two important examples:

i) In the case where the Lagrangian A is associated to a Riemannian metric g by $A(v) = \frac{1}{2}g(v, v)$, we see that $E^A = A$, for since A is homogeneous of degree 2 it satisfies the Euler identity $\partial_D A = 2A$ (cf. IX.55). This justifies the term *"energy"* we have used for the functional associated to this particular Lagrangian.

Proposition IX.61 thus appears as a simple generalisation of Proposition IX.74.

ii) The case of Lagrangians which are homogeneous of degree 1, which we have seen in IX.55 cannot be regular, is singled out by the fact that their associated energy *vanish identically*. This is notably the case for the Lagrangian given by the length associated to a Riemannian metric.

IX.64. Among the classical variations on the theme of the Euler–Lagrange equations is that consisting of the determination of the extremals of a variational problem which is obtained from another by imposing constraints on the positions (often called *holonomic constraints* in mechanics).

These situations give other examples of the use of the formalism of Lagrange multipliers.

Project IX.65. Let N be a submanifold of a configuration space M on which a Lagrangian A is defined. Compare the Euler–Lagrange equations associated to the Lagrangian $A_{|TN}$ with those of the Lagrangian A and interpret the extra term.

IX.66. One question which arises naturally is the following: "*to what extent do the Euler–Lagrange equations depend on the Lagrangian?*" or, more precisely, "*are there distinct Lagrangians which define the same Euler–Lagrange equations, and which therefore have the same extremals?*".

The complete answer to this question, often called the *inverse problem of the Calculus of Variations*, is rather subtle. It is still the object of current research, but we give a partial answer in the form of Exercise IX.67. This exercise introduces the classical notion of a *gauge transformation*, the name usually given by physicists to the operation of passing from one Lagrangian to another with the same extremals. This definition has been extensively amplified in the past decade and has taken on a precise mathematical meaning, but an exhaustive presentation of these ideas would take too long.

Exercise IX.67. Show that, if one adds to a Lagrangian A defined on a configuration space M a closed differential 1-form λ on M, viewed as an observable on TM, then the new Lagrangian $A + \lambda$ has the same extremals as A.

IX.68. We cannot leave this section on the Euler–Lagrange equations without mentioning the special form which the equations take when the curves considered lie in a 1-dimensional space or are described in such a way that one can reduce the problem to one involving real functions of a real variable. This will lead us to consider the classical problems of the Calculus of Variations.

IX.69. If we work in dimension 1, the configuration spaces to be considered are the real line $\mathbb{R}$ (or one of its subintervals) and the circle S^1. In these two cases, the extension by velocities is trivial and the Lagrangian A can be considered simply to be a function of two variables: the *position* and the *velocity*.

In problems which reduce to describing the extremal curves as functions of a real variable (for example, because they are assumed to be the graphs in $\mathbb{R}^2$ of functions of a real variable f), the Lagrangian A becomes a function of three variables: two to describe the point of the graph $(u, f(u))$ and one more for the velocity-vector of the curve which is determined by the derivative $f'(u)$ of the function f.

The most classical examples of such Lagrangians are:

i) the *length* of the graph of a function f, given by the formula

$$f \mapsto \int_{u_0}^{u_1} \sqrt{1 + (f'(\tau))^2}\, d\tau \; ;$$

ii) the *time taken to traverse a trajectory under gravity* which, assuming the trajectory is the graph of a function f in a vertical plane, leads to the functional

$$f \mapsto \int_{u_0}^{u_1} \sqrt{\frac{1 + (f'(\tau)^2)}{f(\tau) - f(u_0)}} \, d\tau$$

where the points $(u_0, f(u_0))$ and $(u_1, f(u_1))$ joined by the curve are fixed.[176]

In both cases, the Euler–Lagrange equations take the simplified form

(IX.70) $$\frac{d}{du} \frac{\partial A}{\partial f'} - \frac{\partial A}{\partial f} = 0 \,.$$

Exercise IX.71. Show that the Euler–Lagrange equations really do take the form (IX.42) with the notations explained in IX.69.

Remark IX.72. The functional "length of a graph" is particularly interesting when it is considered in conjunction with the area which the curve encloses. To seek the curve of minimal length which encloses a given area is called the *isoperimetric problem*.

This is an example of a *constraint* imposed on an infinite-dimensional configuration space which can be solved by resorting to the technique of Lagrange multipliers, suitably generalised.

D. The Geometry of Geodesics

In this section, we study the problem of *shortest paths*, i.e., in a configuration space equipped with a Riemannian metric (which allows us to speak of the length of a tangent vector, and thus of the "length" and "energy" functionals), we look for the curves of minimal length joining two points.

We show in particular that *shortest paths always exist on a compact configuration space.*

IX.73. As we have already remarked, the equation for the critical points of the *length* functional (or the *energy* functional) do not allow one to distinguish between those curves which minimise the length and those which are merely extremals. To avoid any confusion over this minimisation property, we call any extremal of the length functional a *geodesic*.

> *In this section, we fix a configuration space M and a Riemannian metric g on M, and denote by* **L** *and* **E** *respectively the* length *and* energy *functionals associated to g.*

---

[176] The extremal curves of this problem are called *brachistochrone* curves and their study played an important role in the birth of the Calculus of Variations.

Proposition IX.74. *If $\gamma : [\alpha, \beta] \to M$ is a geodesic, every curve obtained from it by reparametrisation is also a geodesic.*

The curves which are extremals of the energy are geodesics whose velocity-vector has constant length, and which are thus parametrised proportionally to arc-length.

Proof. We have established the invariance of the length under reparametrisation in IX.26. Hence, we know that, if γ is a geodesic and ψ a diffeomorphism of $[\alpha, \beta]$, then $L(\tilde{\gamma}) = L(\gamma)$ where $\tilde{\gamma} = \gamma \circ \psi$. This implies that $\tilde{\gamma}$ belongs to a critical level of the functional L, but this does not imply a priori that $\tilde{\gamma}$ is also a critical point of L. We must make use of the special manner in which $\tilde{\gamma}$ is obtained from γ, namely by the action of the group of diffeomorphisms of $[\alpha, \beta]$.

For this, if $\tilde{\Gamma} : [\alpha, \beta] \times] - \epsilon, \epsilon [\to M$ is any variation of $\tilde{\gamma}$, we can associate to it a variation Γ of γ by putting $\Gamma(\tau, s) = \tilde{\Gamma}(\psi^{-1}(\tau, s))$ so that $\tilde{\gamma}_s = \gamma_s \circ \psi$. Again by IX.26, we have $L(\gamma_s) = L(\tilde{\gamma}_s)$ for all $s \in] - \epsilon, \epsilon [$. Consequently, the derivative at 0 of $s \mapsto L(\tilde{\gamma}_s)$ is equal to that of $s \mapsto L(\gamma_s)$, and hence is equal to zero. This proves that $\tilde{\gamma}$ is a geodesic.

Since the Lagrangians defining the length and the energy are powers of each other, we consider the more general problem of comparing the Euler–Lagrange equations for a Lagrangian A and for one of its powers $B = A^k$. In a local chart (x^i, ξ^i), we have

$$\frac{d}{d\tau}\left(\frac{\partial B}{\partial \xi^i}\right) - \frac{\partial B}{\partial x^i} = kA^{k-1}\left(\frac{d}{d\tau}\left(\frac{\partial A}{\partial \xi^i}\right) - \frac{\partial A}{\partial x^i}\right) + k(k-1)A^{k-2}\frac{dA}{d\tau}\frac{\partial A}{\partial \xi^i} ;$$

this implies that the Euler–Lagrange equations for A and $B = A^k$ coincide provided that $dA/d\tau = 0$ along the extremal. This is precisely the assumption we have made. We thus obtain the stated result by taking A to be successively the norm of the velocity-vector and its square.

It only remains to show that the length of the velocity-vector is constant along an extremal of the energy. This follows from a direct calculation of the derivative of the function $g(\dot{\gamma}, \dot{\gamma})$ along the extremal curve, using Formula (IX.57). $\square$

Remark IX.75. The existence of an infinite-dimensional invariance group for the "length" functional $\mathbf{L}$, namely the *group of diffeomorphisms of the interval* $[\alpha, \beta]$ on which the curve is defined, which should a priori be considered an interesting geometric property, has unpleasant analytic consequences.

Indeed, the fact that the Lagrangian of $\mathbf{L}$ is homogeneous of degree 1 (which is the reason for its invariance under reparametrisation) implies its non-regularity, and thus the fact that we cannot write the equations for the geodesics in the form of a second order differential equation solved for the second derivatives. The *"energy"* functional, on the other hand, is regular since the matrix of second derivatives coincides with that of the Riemannian metric g.

Theorem IX.76. *In a configuration space M equipped with a Riemannian metric g, a curve γ is a geodesic parametrised proportionally to arc length if its expression in a local chart (x^i) satisfies the system of ordinary differential equations*

$$\frac{d^2\gamma^i}{d\tau^2} + \sum_{j,k}^{n} \Gamma_{jk}^{i}\frac{d\gamma^j}{d\tau}\frac{d\gamma^k}{d\tau} = 0, \quad 1 \leq i \leq n,$$

where

$$\Gamma_{jk}^{i} = \frac{1}{2}\sum_{m=1}^{n} g^{im}\left(\frac{\partial g_{mj}}{\partial x^k} + \frac{\partial g_{mk}}{\partial x^j} - \frac{\partial g_{jk}}{\partial x^m}\right),$$

the matrix (g^{ij}) being the inverse of the positive-definite symmetric matrix (g_{ij}).

Proof. The equations for the geodesics in the particular case we are considering are obtained from Formula (IX.57) by using the fact that we have, in this case, $2\,A(v) = \sum_{i,j}^{n} g_{ij}\chi^i\chi^j$ if $v = (q^i, \chi^i)$ in the coordinate system (x^i, ξ^i).

The inverse of the matrix of second derivatives of A becomes the inverse of the symmetric matrix (g_{ij}).   □

IX.77. The quantities $\Gamma_j{}^i{}_k$ which appear in IX.76 are traditionally called the *Christoffel*[177] *symbols* of the Riemannian metric g. They play an important role in Riemannian geometry and lead to the notion of a *connection* (also called a *covariant derivative*) which is a central concept in gauge theories.

Theorem IX.78. *For any Riemannian metric defined on a compact, connected configuration space, there always exists at least one minimising geodesic joining any two points.*

Proof. We refer to paragraph 10 of (Milnor 1973) for the complete proof.

The principal arguments are the following. First of all one uses the fact that finding the shortest paths consists in integrating the vector field defining the Euler–Lagrange equations which we know, by conservation of energy IX.61, is tangent to the submanifold of TM formed by the vectors whose length is a constant, 1 for example. Since M is assumed to be compact, this submanifold is compact.

By VI.75, the vector field we are considering is complete, which means that the geodesic can be prolonged to infinity.

It then remains to prove that the set of points which can be reached by geodesics originating at a given point is both open and closed in M, and hence is the whole of M since M is assumed to be connected.   □

IX.79. There is a very interesting approach which brings families of geodesics into play. On a configuration space M equipped with a Riemannian metric g, we define a metric by taking, for any two points q_1 and q_2 of M, $d_g(q_1, q_2)$ to be the infimum of the lengths of the curves joining q_1 and q_2. (Exercise: Check that this is a distance.)

A fundamental theorem, due originally to H. Hopf, asserts that this infimimu is attained (necessarily by a geodesic) when all the geodesics are infinitely extendable. As we have seen in IX.78, this is, in particular, the case if M is compact. The distance d_g then makes M into a complete metric space.

To study the geometry of such a space, it is thus natural to consider the way in which the length of the geodesics realising the distance between their endpoints varies. This is precisely what Proposition IX. 80 does.

First Variation Formula for the Length IX.80. *Let $s \mapsto \gamma_s$ be a family of geodesics defined on the interval $[\alpha, \beta]$ in a configuration space M equipped with a Riemannian metric g, the parameter s being assumed to be proportional to arc-length. If X denotes the transverse vector of this variation of the geodesic $\gamma = \gamma_0$, we have*

$$\frac{d}{ds}\mathbf{L}(\gamma_s)_{|s=0} = g(X(\beta), \dot\gamma(\beta)) - g(X(\alpha), \dot\gamma(\alpha)) ,$$

---

[177] The German mathematician Elvin Bruno Christoffel (1829–1900) made contributions to invariant theory and the geometry of surfaces.

in other words, the infinitesimal variation of the length depends only on the angle which the transverse vector of the variation makes with the velocity-vector $\dot{\gamma}$ of the geodesic at its endpoints.

Proof. This is seen to be a special case of the First Variation Formula IX.33 once we observe that the integral term does not contribute since we are considering a family of geodesics.   □

IX.81. Another approach consists in studying the properties of the configuration space M, equipped with a Riemannian metric g, by *starting at some point* and "exploring" the space by means of geodesics defined in terms of their initial conditions at this point. This is the object of the following definition.

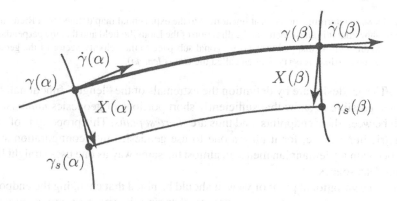

Fig. IX.82 Variation of a geodesic through geodesics

Definition-Proposition IX.83. *Let g be a Riemannian metric defined on a configuration space M. The map which to a tangent vector $\dot{q}$ at a point $q = \pi_M(\dot{q})$ in M associates the point with parameter 1 on the geodesic γ originating at q and having initial velocity-vector $\dot{q}$ is called the* exponential map *defined by the metric g and is denoted by $\exp_q$. It defines a local diffeomorphism from a neighbourhood of 0_q in $T_q M$ onto a neighbourhood of q in M.*

Proof. If we work in a local chart (x^i), a geodesic $\tau \mapsto \gamma(\tau)$ is a solution of a second order differential equation, solved for the second derivatives which, putting $\gamma' = v$, can be written locally as the system of equations

(IX.84)
$$\begin{cases} \dfrac{d\gamma^i}{d\tau} = v^i \\ \dfrac{dv^i}{d\tau} = \displaystyle\sum_{j,k}^n \Gamma^i_{\ jk} \, v^j v^k \end{cases} \quad (1 \leq i \leq n),$$

whence the fact that a geodesic is determined by its initial velocity-vector $\dot{\gamma}(0)$, i.e., by its initial position $\gamma(0)$ and its initial derivative $\gamma'(0)$ in a local chart.

Note, however, that the Local Existence Theorem IV.36 for solutions of a differential equation only guarantees the existence of solutions for small time intervals. Now, to be able to define $\exp_q$ at a tangent vector $\dot{q}$, the integral curve must be defined up to time 1. To show that $\exp_q$ is indeed defined in a neighbourhood of the zero vector in $T_q M$, we shall use the homogeneity of the system IX.84, which implies that the point at time τ on a geodesic having initial velocity-vector $\dot{q}$ is also the point at time 1 on the geodesic having initial velocity-vector $\tau \dot{q}$. This coincidence of integral curves follows directly from the Uniqueness Theorem IV.28 for integral curves, after checking that the curve being considered is indeed a solution.

The fact that the map $\exp_q$ defines a local diffeomorphism from a neighbourhood of 0_q in $T_q M$ onto a neighbourhood of q in M follows from the Local Inversion Theorem III.49, observing that $T_{0_q} \exp_q$ is the identity on $T_q M$, by the very definition of the initial condition used to solve the system of differential equations IX.84. $\square$

Remark IX.85. The homogeneity property used in the proof can also be used to show the following result: *"on a compact configuration space, the exponential map is defined on the whole tangent space, i.e., the geodesics are infinitely extendable."*

Project IX.86. Determine the tangent linear map to the exponential map defined by a Riemannian metric g. Show that it is an isometry in the direction of the Liouville field and that the perpendicular subspace to this field is mapped to the orthogonal subspace to the velocity-vector of the geodesic at the image point (this property is often called the *Gauss lemma*).

IX.87. The geodesics are by definition the extremals of the "length" functional. It is, however, possible to show that sufficiently short portions of geodesics *minimise* the length between their endpoints, and thus are *shortest paths*. This property is of great geometric importance, for it allows one to use geodesics in a configuration space equipped with a Riemannian metric in almost the same way as one uses straight lines in Euclidean spaces.

From the variational point of view, it should be noted that changing the endpoints of the curves being considered amounts to changing the space of curves on which the calculus of variations problem is posed, and thus to treat a different problem from the one we started with.

Project IX.88. Show that geodesics parametrised by arc-length minimise *locally* both the length and the energy (one should use the Gauss lemma IX.86).

Examples IX.89. In a configuration space equipped with a general Riemannian metric, it is not possible to solve the geodesic equation explicitly. In certain special geometric situations, however, this problem can be solved completely.

Among them, we mention the (trivial) case of *Euclidean spaces* whose geodesics are the straight lines, and that of *"round spheres"*, namely spheres with the metric they inherit from their embedding in Euclidean space, whose geodesics are the *great circles*. (Exercise: Check this by using a chart on the sphere and the geodesic equation.)

Among the examples which have been of great historical importance, we mention the *Poincaré half-plane* $H = \{z \mid z = x + iy,\ x,\ y \in \mathbb{R},\ y > 0\}$ equipped with the Riemannian metric $g = y^{-2}(dx^2 + dy^2)$. The geodesics of this space are all the *circles orthogonal to the real axis*, and the *straight lines parallel to the y axis*. (Exercise: Prove it). The geometry which it defines was the first known example of a space which violates *Euclid's parallel axiom*.

Project IX.90. Determine the geodesics on the sphere S^2 for the metric induced on it by an embedding as an ellipsoid in $\mathbb{R}^3$.

IX.91. It is traditional to call the flow of the vector field on TM introduced in Exercise IX.58 the *geodesic flow*; the geodesics are the projections onto M of its integral curves. This one-parameter group of diffeomorphisms of TM is a valuable tool for studying the geometry defined by the Riemannian metric.

Thus, one shows that this flow has chaotic behaviour in the extensions by velocities of all the compact quotients of the Poincaré half-plane by discrete groups of isometries (one then says that this flow is *ergodic*). Generalisations of this (simple) example serve as important reference points in the branch of mathematics which deals with these questions, which is called *ergodic theory*. The models which this theory treats are intimately related to fundamental questions in statistical mechanics.

IX.92. Among the classical problems in the study of geodesics is the search for *periodic geodesics*, i.e., geodesics $\tau \mapsto \gamma(\tau)$ such that, for at least two values τ_0 and τ_1 of the parameter, we have $\gamma(\tau_0) = \gamma(\tau_1)$ and $\dot{\gamma}(\tau_0) = \dot{\gamma}(\tau_1)$.

Very varied techniques have been put to work in dealing with this problem. They belong to group theory and to algebraic topology, as well as to geometry. However, the most powerful techniques derive from considerations related to the Hamiltonian approach which we present in Chapter X. From the analytic point of view, this problem is often approached by working in the space of differentiable maps from S^1 to the configuration space M.

IX.93. Only very special manifolds have all their geodesics closed, but they are nevertheless of great geometric interest (see (Besse 1978)). Besides the standard Riemannian metrics on spheres and real projective spaces (obtained as the quotient of the round metric on the spheres), it is remarkable that this is also the case of the most natural Riemannian metrics on complex projective spaces.

On this topic, we mention the following property: *"there exists an infinite parameter family of Riemannian metrics on the sphere S^2 all of whose geodesics are closed"* (the determination of such metrics can be viewed as a problem in differential equations which was studied a century ago by Darboux[178], and taken up by Zoll[179] and Funk[180] at the beginning of this XXth century).

On the other hand, it is a remarkable theorem in the 1960s due to Green[181] that only the standard Riemannian metric on the real projective plane $\mathbb{R}P^2$, has all its geodesics closed. This underlines once more the importance of the global form of a space for the study of its global properties.

---

[178] The French mathematician Garton Darboux (1842–1917) contributed to many areas of mathematics, in particular differential geometry and analysis. His books played an important role in making these theories better known. Borel, É. Cartan and Picard were his students.

[179] The German mathematician Otto Zoll was a student of Hilbert.

[180] The Austrian mathematician Paul Funk (1886–1969) contributed to the Calculus of Variations.

[181] The American mathematician Leon Green (1926–2009) made a number of contributions to differential geometry.

E. Motion of a Rigid Body

In this section we study the motion of a rigid body about a fixed point formulated as a variational problem whose configuration space is, as we have shown in Chapter V, the rotation group SO_3.

IX.94. The problem of determining the motion of a rigid body about a fixed point reduces to the study of the geodesics of certain special metrics on SO_3.

IX.95. In Chapter VII, we saw that the extension by velocities of SO_3 can be written as the product $SO_3 \times \mathcal{A}$ after identifying $T_I SO_3$ with the space $\mathcal{A}$ of skew-symmetric transformations for the standard scalar product which is the one we shall use from now on. This space can itself be identified with $\mathbb{R}^3$ by associating to a skew-symmetric matrix B the vector $a \in \mathbb{R}^3$ such that, for every vector v in $\mathbb{R}^3$, $B(v) = a \times v$, where $\times$ denotes the cross product (which assumes that $\mathbb{R}^3$ has been oriented).

By using these isomorphisms, we can view a trajectory $t \mapsto q(t)$ of a rigid body Σ as the curve $L_{q(t)^{-1}} \dot{q}(t)$ in $\mathcal{A}$.

To describe the motion of Σ by the variational approach, it is necessary to know the Lagrangian A of which the motion of Σ is an extremal. Consider first the case when Σ *is not subject to any external forces*. The Lagrangian of the solid in motion then reduces to its *kinetic energy* E_c which is a quadratic form in the velocity-vector of its trajectory in the configuration space. This form E_c is obtained by integrating the individual kinetic energies of the particles comprising the body.

IX.96. The velocity v of a point m of a rigid body with respect to a fixed reference frame in which O is at rest can be written $v = \omega \times \overrightarrow{Om}$, where ω denotes the instantaneous angular velocity of Σ. The kinetic energy density of a solid with respect to its mass density (which is fixed) can thus be written $\|\overrightarrow{Om} \times \omega\|^2$, from which we see that the value of the quadratic form E_c in a position corresponding to a point $q \in SO_3$ can be obtained from its value at another point q' simply by transporting it by the rotation which sends q to q'.

It follows that the form E_c describing the *inertia distribution* of Σ, can be viewed as a Riemannian metric on SO_3 invariant under left translations. We deduce from this that "*the motions of a rigid body are the geodesics of a left-invariant Riemannian metric on SO_3 which is determined by the inertia distribution of the body*".

IX.97. The symmetric linear transformation associated to E_c by the natural Riemannian metric on SO_3 is often called the *inertia operator* (or the *inertia tensor*). This metric is defined by left transporting to each tangent space the metric defined on $\mathcal{A}$ by the standard metric on $\mathbb{R}^3$ using the identification given in IX.95. The directions defined by the eigenvectors of this transformation are called the *axes of inertia* of Σ.

In an orthonormal basis (A_1, A_2, A_3) of $\mathcal{A}$ corresponding to the axes of inertia in which $\omega = \sum_{i=1}^{3} \omega^i A_i$, we have $E_c = \frac{1}{2}(I_1(\omega^1)^2 + I_2(\omega^2)^2 + I_3(\omega^3)^2)$ where the I_i are traditionally called the *principal moments of inertia*, although in our context this expression may lead to confusion (for the notion of a moment, see Chapter XI).

Exercise IX.98. Show that the instantaneous angular velocity ω of a rigid body with inertia operator I moving about a fixed point O is a solution of the non-linear differential equation

$$I(\frac{d\omega}{dt}) = I(\omega) \times \omega ,$$

which, in a basis in which I is diagonalised with eigenvalues the principal moments of inertia I_1, I_2 and I_3, can be written

$$\begin{cases} I_1 \dfrac{d\omega_1}{dt} = (I_2 - I_3)\, \omega_2\, \omega_3 \\[2mm] I_2 \dfrac{d\omega_2}{dt} = (I_3 - I_1)\, \omega_3\, \omega_1 \\[2mm] I_3 \dfrac{d\omega_3}{dt} = (I_1 - I_2)\, \omega_1\, \omega_2 \end{cases} ,$$

ω^1, ω^2 and ω^3 denoting the components of ω (this result is known as the *"Euler equations of rigid body motion"*).

Remark IX.99. One of the consequences of the result contained in Exercise IX.98 is that *"the only stationary motions of a rigid body moving about a fixed point, not subject to any external forces and having distinct principal moments, are the rotations around one of its axes of inertia"*.

Even though they give all the information necessary for solving the problem, the description of the motion of a rigid body which one obtains by the methods we have outlined might be considered very (too?) abstract. Among the more concrete descriptions, that of the *inertia ellipsoid*, introduced by Poinsot[182], is classical.

By an (easy) modification of the Euler equations, one can see that *this ellipsoid rolls without slipping on a fixed plane*, the curves traced out by the points of contact on the plane and on the ellipsoid being called the *polhode* and the *herpolhode*. Note, however, that the inertia ellipsoid is not a truly *"concrete"* representation of the inertia quadratic form since it has no physical reality.

IX.100. One of the most realistic cases is provided by the *Lagrange top*, which is a rigid body having an axial symmetry (and thus having two of its principal moments of inertia equal) moving under gravity. It is possible to make a complete study of its motion by using a parametrisation of SO_3. The interested reader may consult pages 150 to 161 of (Arnol'd 1976), where the particular cases of practical importance are discussed in detail.

Let us mention, however, that the motion of the top decomposes into three periodic motions: a *rotation* (around its axis of symmetry), a *precession* and a *nutation* (motions of its axis of rotation, the first equatorial, the second azimuthal).

---

[182] Louis Poinsot (1777–1859), French mathematician and specialist in mechanics, was opposed to the analytic school of the first half of the 19th century. He gave a geometric treatment of a number of fundamental questions in mechanics.

It is only because of the presence of a *sufficient number of conserved quantities* that the solution can be carried through completely. Displaying such quantities is precisely the object of Chapter XI. Thus, we shall not enter into a discussion of them here.

F. Surfaces of Stationary Area

In this section we present very briefly a variational problem in several variables; this means that the objects we are varying are no longer curves, but configuration spaces of dimension greater than 1 (we consider here the case of *surfaces*).

IX.101. For simplicity, we shall only consider surfaces which are *submanifolds of Euclidean space* $\mathbb{R}^3$.

By analogy with the case of curves whose endpoints are fixed, we assume that the space S of surfaces on which a functional will be defined is the space of embeddings of the disc $\mathbf{D}^2$ defined by $\mathbf{D}^2 = \{(x, y) \mid x, y \in \mathbb{R}, \ x^2 + y^2 < 1\}$ whose boundary circle is sent to a fixed curve γ in $\mathbb{R}^3$.

A point of the space S which is of interest to us is thus a differentiable map $\sigma : \mathbf{D}^2 \longrightarrow \mathbb{R}^3$, assumed to be of rank 2 everywhere (i.e., such that we have $(\partial\sigma/\partial x) \times (\partial\sigma/\partial y) \neq 0$), and satisfying also a *boundary condition* on the edge of the disc. As we have already indicated in Chapter VIII, a tangent vector at a point σ of S is a vector field X defined along σ.

IX.102. The functional which we are going to consider is the *area denoted* by $\mathbf{S}$.

Since the variational problem we envisage involves objects which are maps from 2-dimensional spaces, the appropriate notion of a Lagrangian is defined on the set of *2-planes* of $\mathbb{R}^3$. Here, this will be the *volume element* induced on the tangent plane at a point σ by the first fundamental form g, i.e., the determinant with respect to a g-orthonormal basis of $T_m\sigma$ (abusing notation, the map σ is consistently identified with its image in $\mathbb{R}^3$) of the tangent plane basis formed by $\partial\sigma/\partial x$ and $\partial\sigma/\partial y$, so that

$$\mathbf{S}(\sigma) = \int_{\mathbf{D}^2} |\partial\sigma/\partial x \times \partial\sigma/\partial y| \ dx \, dy \ .$$

One shows that the functional $\mathbf{S}$ is invariant under reparametrisations of the disc $\mathbf{D}^2$. (Exercise: Actually prove this.)

Just as in the case of curves, it is also possible to study the "*energy*" functional $\mathbf{E}$, which is defined as follows

$$\mathbf{E}(\sigma) = \frac{1}{2} \int_{\mathbf{D}^2} \|T_{x,y}\sigma\|^2 \ dx \, dy \ ,$$

a formula which presupposes that one has fixed a Riemannian metric on the disc $\mathbf{D}^2$ to evaluate the norm of the tangent linear map $T\sigma$. The physical model which justifies this terminology is that of the *energy stored at the interface between two media*, often called the "*surface tension energy*", of which the most picturesque example is provided by soap films.

IX.103. As in the case of curves, these two functionals are closely related, but now the situation is more complicated. In the study of these relations, the fact that the source space is 2-dimensional enters in a decisive way, but a discussion of this point would take us too far afield.

We mention only that these relations enter in a fundamental way in the study of string theory, being studied by theoretical physicists.

IX.104. It may be useful to see how to write the functional S in the classical partial derivative notation due to Monge[183]. Recall that, for a function z of two variables x and y, it is traditional to put

$$p = \frac{\partial z}{\partial x}, \quad q = \frac{\partial z}{\partial y}, \quad r = \frac{\partial^2 z}{\partial x^2}, \quad s = \frac{\partial^2 z}{\partial x \partial y}, \quad t = \frac{\partial^2 z}{\partial y^2}.$$

In our case, if we write σ as a graph $\sigma(x, y) = (x, y, z(x, y))$, we obtain the expression $\|(\partial\sigma/\partial x) \times (\partial\sigma/\partial y)\| = \sqrt{1 + p^2 + q^2}$ for the area, so that

$$S(\sigma) = \int_{D^2} \sqrt{1 + p^2 + q^2} \, dx \, dy.$$

IX.105. To understand the geometry of a surface σ in Euclidean space $\mathbb{R}^3$, it is useful to "thicken" it, i.e., to consider the family σ_r of surfaces obtained by taking the points which are a distance r from σ. By the First Variation Formula for the Length IX.80 in $\mathbb{R}^3$, it is easy to see that, for r sufficiently small, one can define a map $r \mapsto m_r$, where m_r denotes the point of σ_r which is closest to a fixed point m in $\sigma = \sigma_0$. Moreover, the curve $r \mapsto m_r$ is normal to σ at m. We define in this way a normal variation of σ.

On the *parallel* surfaces σ_r to σ, the Euclidean metric induces a metric which we have called their *first fundamental form* g_r. Since, for r sufficiently small, the map $m \mapsto m_r$ defines a diffeomorphism ψ_r from σ to σ_r, we obtain a curve of metrics $r \mapsto \psi_r^*(g_r)$ on σ, which allows us to define a new fundamental invariant attached to the surface σ.

Definition IX.106. Let σ be a surface in $\mathbb{R}^3$ with first fundamental form g and let (σ_r, g_r) be the parallel surfaces. The field of symmetric bilinear forms h given by the negative of the derivative at $r = 0$ of the curve of metrics $r \mapsto \frac{1}{2}\psi_r^*(g_r)$ is called the *second fundamental form* of σ.

---

[183] Gaspard Monge (1746–1818), French mathematician, has a body of work which embraces physics, chemistry and technology as well as mathematics. He was the founder of descriptive geometry... and his teaching methods were used from the foundation of the École Polytechnique. He also led a very active political life, both under the Revolution and the Empire.

Remark IX.107. The appearance of the numerical coefficient $-\frac{1}{2}$ is necessary to recover the classical formulas for the second fundamental form which come from other definitions.

The presentation we have adopted has a strong mechanical connection. If one thinks of the surface $\sigma = \sigma_0$ as the middle slice of a thin slab, the surfaces σ_r are deformed according to their embedding in the ambient space. Their infinitesimal rate of deformation is exactly the second fundamental form of the middle slice.

IX.108. On the surface σ, we thus have a scalar product g, its first fundamental form, and a symmetric bilinear form h, its second fundamental form. It is natural to consider the operator associated to h by using the duality which g defines. This operator is often called the *shape operator* or Weingarten[184] operator of the surface σ. It has two fundamental invariants, its *trace*, traditionally denoted by H and called the *mean curvature* of the surface, and its *determinant*, traditionally denoted by K and called its *Gaussian*[185] *curvature* of the surface σ.

The name *mean curvature* comes from the fact that H is the sum of the inverses of the radii of curvature at the point considered. The Gaussian curvature K is their product. The "*Theorema egregium*" of Gauss asserts that "*the Gaussian curvature depends only on the first fundamental form, regarded as a field*". In fact, it is possible to express K as a non-linear function, which seems complicated at first sight, of the first and second partial derivatives of the coefficients of g in a parametrisation of the surface σ and of the coefficients of σ itself (see pages 36 and 37 of (Dombrowski 1979) to admire the formula).

IX.109. For a surface given as the graph over the (x, y) coordinate plane, we give the expressions for the above quantities in the notation of Monge. For the first two fundamental forms g and h, we have

$$\begin{cases} g = (1 + p^2)\, dx^2 + 2pq\, dx\, dy + (1 + q^2)\, dy^2 \,, \\ h = (1 + p^2 + q^2)^{-\frac{1}{2}}(r\, dx^2 + 2s\, dx\, dy + t\, dy^2) \,, \end{cases}$$

from which one obtains the following formulas for the mean curvature H and the Gaussian curvature K:

(IX.110)
$$H = \frac{(1 + q^2)r - 2pqs + (1 + p^2)t}{(1 + p^2 + q^2)^{\frac{3}{2}}} \,,$$

(IX.111)
$$K = \frac{rt - s^2}{(1 + p^2 + q^2)^2} \,.$$

---

[184] Julius Weingarten (1836–1910), German mathematician, contributed to the development of differential geometry.

[185] Carl Friedrich Gauss (1777–1855), German astronomer, mathematician and physicist, is generally considered to be the greatest mathematician of the first half of the 19th century. Two of his works, the "*Disquisitiones arithmeticae*" and the "*Disquisitiones circa superficies curvas*", are still referred to, both because of the profundity of their results and their rigorous style. His work in geodesy played an important guiding role in his work on the foundations of the geometric concepts of a curve and a surface.

Exercise IX.112. Prove Formulas (IX.110) and (IX.111).

Theorem IX.113. *A surface σ is of stationary area if and only if its mean curvature vanishes at every point of the surface.*

Proof. See pages 382 and 383 of (Dubrovin, Novikov and Fomenko 1982). ☐

IX.114. By Formula (IX.111), we see that the analytic expression of stationarity for the *"area"* functional of a surface which is the graph in $\mathbb{R}^3$ of a function f of two variables x and y is given by the non-linear partial differential equation

$$\left(1 + (\frac{\partial f}{\partial y})^2\right) \frac{\partial^2 f}{\partial x^2} - 2 \frac{\partial f}{\partial x} \frac{\partial f}{\partial y} \frac{\partial^2 f}{\partial x \partial y} + \left(1 + (\frac{\partial f}{\partial x})^2\right) \frac{\partial^2 f}{\partial y^2} = 0 ,$$

which seems a priori rather formidable.

IX.115. The general problem of describing the surface(s) of stationary area spanning a given contour is known as the *"Plateau[186] problem"*. Even though, to obtain a solution, one has only to dip this contour, realised in the form of a stiff wire, into soapy water, this problem has even today not been completely solved. However, one knows a certain number of situations where it has been solved in a satisfactory way.

Among the simple configurations giving rise to surfaces of stationary area, there is of course the *"plane"* configuration where the boundary is a plane circle. The disc which the circle bounds in this plane is then of stationary area. More interesting from a geometric point of view is the configuration of two coaxial circles (which we consider even though it violates our convention that the surface should have a connected topology). If the two circles are rather close, one can find a connected surface of revolution of stationary area spanning the two circles, namely the *catenoid*, which can be obtained by rotating a hanging chain around a line parallel to the tangent at its lowest point.

There are, of course, many other examples of surfaces of stationary area which have been discussed, but for more details... and beautiful pictures, we refer to (Hildebrandt and Tromba 1987).

Remark IX.116. As in the case of curves, the comparison for surfaces of the variational problems defined by the area and the energy is of great interest.

Not being invariant under a change of parametrisation, the *energy functional singles out* a particular one, but it is not as easy as in the case of curves to formulate the property which characterises it in purely geometric terms. Moreover, this parametrisation depends explicitly on the chosen Riemannian metric on the surface, a choice of which is less innocuous than that of an element of length on a curve where it depends only on a function in one variable.

---

[186] after the Belgian physicist Joseph Antoine Ferdinand Plateau (1801–1883) who, by studying the surface tension of liquid films, became interested in surfaces of minimal area.

IX.117. The search for minimal surfaces is closely related to the theory of functions of a complex variable via the *Gauss map*. In particular, there exist formulas due to Weierstrass involving complex integration which allow one to construct minimal surfaces starting from two arbitrarily given holomorphic functions.

Unfortunately, this approach does not allow one to decide if the surfaces thus constructed are embedded or only immersed. It is to solve problems of this kind that D. Hoffman and Meeks have used computer graphics. Their results are eventually stated in purely mathematical terms, but the *exploration* by graphical means has played a crucial role leading to an infinity of new examples of minimal surfaces of finite total curvature which are embedded, surfaces of which the only examples up to 1985 were the plane and the catenoid.

G. Historical Notes

IX.118. The fundamental equations of the Calculus of Variations were established by Euler and Lagrange. They are described in full generality, and in a way which is still comprehensible today, in the "*Méchanique Analitique*" of Lagrange, of which the first edition appeared in 1788. In particular, the passage illustrated in Fig. IX.119 can be found there.

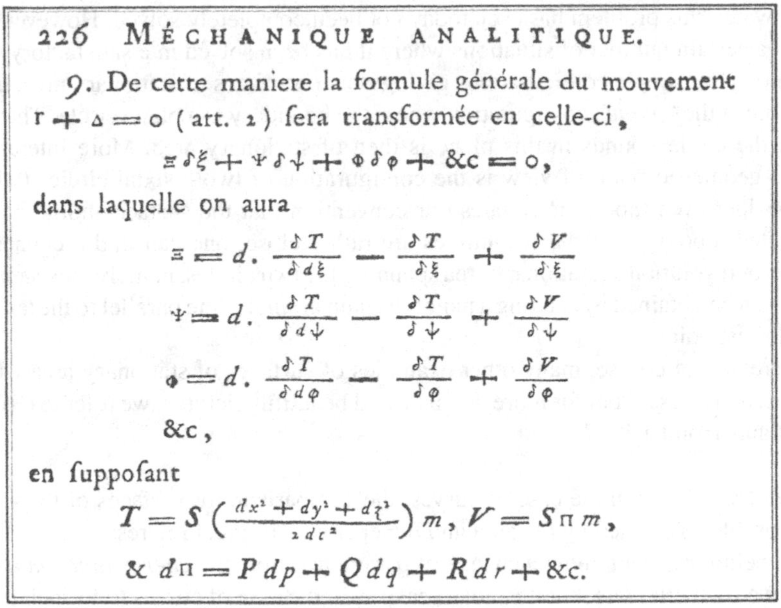

226   MÉCHANIQUE   ANALITIQUE.

9. De cette maniere la formule générale du mouvement
$\Gamma + \Delta = 0$ (art. 2) fera transformée en celle-ci ,

$$\Xi\,\delta\xi + \Psi\,\delta\psi + \Phi\,\delta\varphi + \&c = 0,$$

dans laquelle on aura

$$\Xi = d \cdot \frac{\delta T}{\delta\, d\xi} - \frac{\delta T}{\delta\xi} + \frac{\delta V}{\delta\xi}$$

$$\Psi = d \cdot \frac{\delta T}{\delta\, d\psi} - \frac{\delta T}{\delta\psi} + \frac{\delta V}{\delta\psi}$$

$$\Phi = d \cdot \frac{\delta T}{\delta\, d\varphi} - \frac{\delta T}{\delta\varphi} + \frac{\delta V}{\delta\varphi}$$

&c ,

en fuppofant

$$T = S \left(\frac{dx^2 + dy^2 + dz^2}{2\, dt^2} \right) m , \quad V = S\,\Pi\,m ,$$

$$\& \; d\Pi = P\,dp + Q\,dq + R\,dr + \&c.$$

Fig. IX.119 The Euler–Lagrange equations in the version of Lagrange in 1788

IX.120. The first modern exposition of these equations can be found in the *"Mathematical Foundations of Quantum Mechanics"* of Mackey[187]. But this manifestly intrinsic formulation of Lagrangian mechanics had been in mathematical circulation since the beginning of the 1960s (the names of Palais and J. Klein[188] should be mentioned in connection with these developments). The rather subtle structure of the extension by velocities of a configuration space remains an obstacle to the widespread use of this approach.

IX.121. We have seen that problems in the Calculus of Variations in one variable lead to second order differential equations. Problems in the Calculus of Variations in several variables, such as that of surfaces of extremal area, give rise to second order partial differential equations which are often non-linear and which have a special form (one often says that these systems can be put in *divergence* form). Their mathematical structure is much more elaborate, and is still not completely elucidated today.

---

[187] George Whitelaw Mackey (1916–2006) was an American mathematician who made major contributions to operator algebras, group representation and mathematical aspects of quantum mechanics.

[188] The French mathematician Joseph Klein (1914–1997) defended a thesis entitled *Espaces variationnels et mécanique* under the supervision of Lichnerowicz.

Chapter X
The Hamiltonian Viewpoint

In this chapter we present the *Hamiltonian approach* to variational problems which is, in a sense, *dual* to the Lagrangian approach which was the subject of Chapter IX.

To this end we introduce in Sect. A the *extension by momenta* of a configuration space, which is the analogue for cotangent vectors of the extension by velocities. This new configuration space has a special geometry which stems from the existence of a universal 2-form called the *Liouville symplectic form*.

By using the Liouville form, it is possible to associate, to any observable on the extension by momenta, a vector field, its *symplectic gradient*, whose definition we give in Sect. B. The differential equation associated to the symplectic gradient of the *Hamiltonian* (the name given to the energy of a system when it is viewed as an observable on the extension by momenta) is the general expression of *Hamilton's equations*, which are an extension of the equation we met in Chapter IV.

Section C relates the preceding developments to the Euler–Lagrange equations of a variational problem which we presented in Chapter IX. The correspondence is set up by using the *Legendre transform*. The formulations of the *Principle of Least Action* in the two approaches, Lagrangian and Hamiltonian, are compared.

In the Hamiltonian approach, it is possible to see directly the dynamical properties of the observables, i.e., their time evolution, by means of *Poisson brackets*. This operation, introduced in Chapter IV, extends to configuration spaces as we explain in Sect. D.

© The Author(s), under exclusive license to Springer Nature Switzerland AG 2022
J.-P. Bourguignon, *Variational Calculus*, Springer Monographs in Mathematics,
https://doi.org/10.1007/978-3-031-18307-2_10

A. The Extension by Momenta and its Symplectic Structure

We make the collection of all linear forms on the tangent spaces to a configuration space into a new configuration space, its *extension by momenta*.

We show that there are several special, and universal, multilinear forms on this space which give it an interesting geometry. One of them, the *Liouville 2-form*, generalises to the case of configuration spaces the antisymmetric bilinear form defined in Chapter IV on the product of a vector space and its dual.

X.1. The Hamiltonian viewpoint, which is the subject of this chapter, involves in a fundamental way observables depending on both the positions and the *momenta* (i.e., the cotangent vectors at the point considered).

This leads us to make the collection of all cotangent vectors into a configuration space, just as we constructed the extension by velocities starting with the notion of a tangent vector at a point.

X.2. Thus, we form $T^*M = \bigcup_{q \in M} T_q^*M$, the disjoint union of all the cotangent spaces of M. We have, of course, a natural projection map, which we *denote* by ϖ_M, which associates to any element $\lambda \in T_a^*M \subset T^*M$ its point of attachment a.

We make T^*M into a configuration space as we did in Proposition IX.4 for the extension by velocities TM. For this, if x is a chart at a point a of M with domain an open set U, we define its natural extension[189] T^*x at $\alpha \in T_a^*M$ by putting, for an element $\lambda \in T^*U (= \varpi_M^{-1}(U))$,

$$T^*x(\lambda) = (x^1(\varpi_M(\lambda)), \cdots, x^n(\varpi_M(\lambda)), \langle \lambda, \frac{\partial}{\partial x^1} \rangle, \cdots, \langle \lambda, \frac{\partial}{\partial x^n} \rangle),$$

in other words, the last n coordinates are the components of the form λ in the basis of coordinate differentials of the chart x.

As to the notation for the $2n$ coordinates on the extension by momenta arising from the natural extension of a chart x, we proceed as for the extension by velocities. The n coordinates of a point of the base are, of course, denoted by $(x^1, \cdots, x^n)$, and we *agree* to *denote* by $(X_1, \cdots, X_n)$ the other n coordinates, which correspond to the evaluation of a cotangent vector on the natural basis $(\partial/\partial x^i)_{1 \leq i \leq n}$ (and this is why we have placed the indices on these functions on T^*M at the bottom). In summary, the natural extension of the chart x will be denoted by (x, X).

The traditional notation for these n additional coordinates, dual to the $\dot{q}^i$, is p_i; this is also the notation usually used to denote *momenta* in mechanics, and we shall see the reason for this coincidence in Chapter XI.

X.3. By using these maps, we can define a topology on T^*M and, in fact, a configuration space structure as Proposition X.4 asserts.

---

[189] Note that the notation we have used is not altogether orthodox. It is *not true* that for every map $f : M \to N$ it is possible to define an extension of it that maps T^*M to T^*N. Covariant objects, on the contrary, are those which are adapted to taking inverse images. But here we use explicitly the fact that x is invertible, by taking the inverse image under its inverse.

Proposition X.4. *The natural extensions to* T^*M *of the charts on an n-dimensional configuration space M make* T^*M *into a 2n-dimensional configuration space, called its* extension by momenta.

Proof. This is in every way analogous to that of Proposition IX.4. Given two charts x and y of M, it suffices to determine the change of coordinates map $\tau_{T^*y}^{T^*x}$ between their natural extensions.
    We obtain

$$\tau_{T^*y}^{T^*x}(x, X) = (\tau_y^x(x), {}^t(T\tau_x^y)(X))$$

by comparing terms on both sides using the formula in X.2.   □

Exercise X.5. Show that a differential form λ defined on a configuration space M is differentiable if and only if $\lambda : M \mapsto T^*M$ is a differentiable map.

Exercise X.6. Show that the natural map $\varpi_M : T^*M \longrightarrow M$ is a differentiable submersion.[190]

X.7. In the presentation we have adopted, the extension by velocities came first because it seemed the most natural and the most concrete. Moreover, this is the way things happened historically.
    Nevertheless, it is possible to deduce everything we have done by considering the extension by momenta first and proceeding as follows. One can construct a cotangent vector as the term of first order in the expansion of an observable, so that the cotangent space at a point a can be identified with the quotient of the ideal consisting of the observables which vanish at a by its square (i.e., the space generated by the products of two observables which vanish at a).
    It might seem that this point of view does not fit into our context, but it has proved to be very powerful for discussing spaces more general than configuration spaces, for example spaces with singularities.

X.8. We start with an n-dimensional configuration space M and we consider its extension by momenta T^*M. We propose to show that, as a configuration space, T^*M has a very special geometry which is the key to the Hamiltonian approach to variational problems, the subject of this chapter.

Throughout this chapter, we work in local charts (x^i, X_i)
of the extension by momenta T^*M *of a configuration space M*
which are the extensions of local charts (x^i) *of M.*

Definition-Proposition X.9. *Let M be a configuration space. The differential 1-form* α^M *on* T^*M *which, at a point* λ, *is equal to* $\alpha^M(\lambda) = \varpi_M^*(\lambda)$, *is called the* Liouville 1-form.
    In any natural coordinate system (x^i, X_i) *on* T^*M, *we have*

(X.10) $$\alpha^M = \sum_i^n X_i \, dx^i \ .$$

Proof. Let $\lambda = \sum_i^n \lambda_i \, dx^i$ be the expression of a form λ in the natural basis of the chart, so that $X_i(\lambda) = \lambda_i$. In the chosen chart, the natural projection ϖ_M is given by $(\varpi_M)_x^{T^*x}(x^i, X_i) = (x^i)$. It follows from the equation defining α^M that $\sum_i^n \lambda_i \, dx^i$ is its value at $\lambda(= (x^i, X_i))$, as stated.   □

---

[190] hence the name *cotangent bundle* given to T^*M by mathematicians.

Remark X.11. The intrinsic and universal character of the Liouville 1-form α^M is usually established by using the special form (X.10) which it takes in any natural coordinate system.

The abstract definition we have given, although having the advantage of being manifestly intrinsic, does not at first sight give any indication of the great geometric importance of the object thus defined, or that it is actually the "magic" form $\sum_{i=1}^n p_i \, dq^i$ of mechanics, as Formula (X.10) indicates.

X.12. To be in a position to discuss the Hamiltonian approach, we still need one more ingredient which we have already touched upon in Chapter VIII.

X.13. We have already met 1-forms and alternating n-forms in a configuration space of dimension n. In an analogous way, one can consider, purely algebraically, *alternating* 2-forms, i.e., bilinear forms which change sign when their arguments are interchanged.

The notation used to signify this antisymmetry is the following: if λ and ν are two linear forms on a vector space E, then one constructs an *alternating* form, *denoted* by $\lambda \wedge \nu$, depending bilinearly on them, by putting, for any two vectors v and w, $(\lambda \wedge \nu)(v, w) = \lambda(v) \, \nu(w) - \lambda(w) \, \nu(v)$. Thus, if (ϵ^i) denotes a basis of the dual of E, an alternating 2-form ω is written $\omega = \sum_{i,j}^n \omega_{ij} \, \epsilon^i \wedge \epsilon^j$, the symbol $\wedge$ reminding us that the coefficients are antisymmetric.[191]

X.14. There exists an operation on exterior differential forms which generalises the differential of observables.

X.15. In fact, starting with a differential 1-form λ whose expression in a parametrisation (x^i) is $\lambda = \sum_i^n \lambda_i \, dx^i$, taking the derivatives of its coefficients and trying to define a bilinear form on a pair of vectors v and w such that $v = \sum_i^n v^i (\partial/\partial x^i)$ and $w = \sum_i^n w^i (\partial/\partial x^i)$, would lead us to $\sum_{i,j}^n v^i \, w^j (\partial \lambda_i/\partial x^j)$.

But, *this bilinear form is not intrinsically defined*, for if we start with the expression for λ in another chart (y^α), we find that the bilinear form becomes, using Greek indices to denote the components in the new natural frame, $\sum_{\alpha,\beta}^n v^\alpha \, w^\beta (\partial \lambda_\alpha/\partial y^\beta)$; consequently, recalling that we have $v^\alpha = \sum_i^n v^i (\partial y^\alpha/\partial x^i)$ on the one hand, and $w^\beta = \sum_j^n w^j (\partial y^\beta/\partial x^j)$ on the other hand, and finally $\lambda_\alpha = \sum_i^n \lambda_i (\partial x^i/\partial y^\alpha)$, we find the unwanted term $\sum v^\alpha \, w^\beta \lambda_i (\partial^2 x^i/\partial y^\alpha \partial y^\beta)$ which has no reason to be zero, proving the naivety of our tentative definition.

X.16. Even though it was unsuccessful, the preceding discussion contains exactly what is needed to arrive at the correct notion.

---

[191] In fact, $\wedge$ is the manifestation in our situation of an *exterior product*, so-called by contrast with the *inner product*, which we shall meet a little later (the name "inner product" has also been used in the past to mean a *scalar product*). To any pair of elements of a vector space, this operation associates an alternating 2-vector (a vector in the dual of the space of alternating bilinear forms). A complete discussion of this fundamental product in multilinear algebra, as well as that of the tensor product (from which it is derived), would lead us too far away from the main topic of our exposition.

X. 17. In fact, if the change of coordinates map is at least of class C^2 (we have always assumed that the changes of coordinates are infinitely differentiable), then by the Schwarz Lemma III.97, the matrix of second derivatives $(\partial^2 x^i / \partial y^\alpha \partial y^\beta)$ is symmetric in α and β. Consequently, the alternating bilinear form

$$(v, w) \mapsto (d\lambda)(v, w) = \sum_{i,j}^{n} v^i w^j \left(\frac{\partial \lambda_i}{\partial x^j} - \frac{\partial \lambda_j}{\partial x^i} \right)$$

is well-defined, since with this definition the unwanted term encountered earlier on comparing the expressions in two charts makes no contribution. The form $d\lambda$ is called the *exterior differential* of the differential 1-form λ. With the notations introduced above, the exterior differential 2-form $d\lambda$ can be written

$$(X.18) \qquad d\lambda = \sum_{1 \le i < j \le n} \left(\frac{\partial \lambda_j}{\partial x^i} - \frac{\partial \lambda_i}{\partial x^j} \right) dx^i \wedge dx^j .$$

(Recall that the notation $\wedge$ implies that there is an antisymmetry here.)

X.19. The way we have introduced the exterior differential d ensures that it behaves well with respect to diffeomorphisms, i.e., that, for any diffeomorphism ψ defined on a configuration space M and taking values in a configuration space N, and for any differential 1-form λ on N, we have

$$(X.20) \qquad\qquad \psi^*(d\lambda) = d(\psi^* \lambda) .$$

Formula (X.20) still holds for any differentiable map ψ. (Exercise: Prove it.)

Proposition X.21. *The exterior differential of the differential of an observable vanishes, i.e., $d \circ d = 0$.*

Proof. Since the statement we wish to establish is intrinsic, it suffices to prove it in a chart x. The verification then follows directly from the Schwarz Lemma III.97. In fact, if φ is an observable, we have $d\varphi = \sum_i^n \partial \varphi / \partial x^i \, dx^i$ so that

$$d(d\varphi) = \sum_{i,j}^{n} \left(\frac{\partial^2 \varphi}{\partial x^i \partial x^j} - \frac{\partial^2 \varphi}{\partial x^j \partial x^i} \right) dx^i \wedge dx^j$$

which is 0 by the symmetry of the second derivatives. $\square$

Proposition X.22. *Let M be a configuration space. The exterior differential 2-form $\omega^M = d\alpha^M$ is non-degenerate called the* Liouville symplectic 2-form. *In any chart (x^i, X_i),*

$$(X.23) \qquad\qquad \omega^M = \sum_{i} dX_i \wedge dx^i ,$$

in other words, if $V_1 = (v_1^i, (V_1)_i)$ and $V_2 = (v_2^i, (V_2)_i)$ are tangent vectors to T^*M,

$$\omega^M(V_1, V_2) = \sum_{i=1}^{n} \left((V_1)_i \, v_2^i - (V_2)_i \, v_1^i \right) .$$

Proof. Starting from the expression X.10 for the Liouville form α^M in a chart (x^i, X_i), we see from X.18 that the Liouville 2-form indeed has the stated form.

If $V = \sum_i^n (v^i \, \partial/\partial x^i - V_i \, \partial/\partial X_i)$ is tangent to T^*M, $\omega^M(V, .) = \sum_i^n (V_i dx^i + v^i dX_i)$ is zero only if $v^i = V_i = 0$ for $1 \le i \le n$, since the forms $dx^1, ..., dx^n, dX_1, ..., dX_n$ constitute a basis of $T_\lambda T^*M$ at each point λ of T^*M, hence ω^M is non-degenerate.   □

Exercise X.24. Show that the Liouville symplectic form ω^M is characterised by the following property: "*for any differential 1-form λ on a configuration space M, viewed as a section of the projection $\varpi_M : T^*M \longrightarrow M$, we have $\lambda^*(\omega^M) = d\lambda$.*"

X.25. It is possible to give a general definition of a *symplectic differential 2-form* on a configuration space. For this, it is first necessary to extend the exterior differential to 2-forms by following the same path we followed to extend it from 0-forms (observables) to differential 1-forms. A *symplectic form* on a configuration space is then simply "*a non-degenerate exterior differential 2-form whose exterior differential is zero*" (one says that it is "*closed*").

Giving a symplectic form on a configuration space defines a geometry which, according to a theorem of Darboux (see Exercise X.26), has no local invariants (except its dimension). The search for global invariants of such configuration spaces has been developed only recently.

Giving a Riemannian metric on a configuration space gives rise, on the contrary, to rich local invariants (already introduced by Riemann around the middle of the XIXth century and developed by Ricci and É. Cartan in the early XXth) such as the curvature tensor and its covariant derivatives.

Nevertheless, there is a geometry which ties together Riemannian and symplectic geometry. It is naturally defined on complex analytic subvarieties of complex projective spaces. It is called *kählerian geometry* after Kähler.[192] Very important properties of algebraic varieties can be proved using this geometry, which is very rich. The methods which it introduces into algebraic geometry, such as Hodge theory, are often qualified by the term *transcendental*, as opposed to the purely algebraic methods which use only the properties of polynomials.

Project X.26. Let N be a configuration space equipped with a symplectic form ω. Show that, in a neighbourhood of any point, there exists a chart $(x^1, \cdots, x^{2m})$ in which ω takes the form $\sum_1^m dx^i \wedge dx^{i+m}$. (This property, which asserts the local equivalence of all symplectic forms, is called *Darboux's theorem.*)

X.27. As is usual whenever a new notion has been introduced, we must consider the transformations which preserve it. To this end, we make the following definition.

Definition X.28. Let M be a configuration space. A diffeomorphism Ψ of T^*M such that $\Psi^*(\omega^M) = \omega^M$ is called a *canonical transformation.*

X.29. We begin by studying the linear canonical transformations. Thus, in a vector space E of even dimension equipped with a non-degenerate alternating 2-form ω, we consider the linear transformations l which satisfy the identity $l^*\omega = \omega$.

The subgroup of the general linear group thus defined is called the *symplectic group* and is usually *denoted* by $\mathrm{Sp}_\omega E$ (in the case of $\mathbb{R}^{2m}$ equipped with the standard alternating form $\omega_0 = \sum_i^m \epsilon^i \wedge \epsilon^{i+m}$, this group is denoted by Sp_m).

---

[192] The German mathematician and philosopher Erich Kähler (1906–2000) had many major contributions to several fields: algebra, non-linear partial differential equations and differential geometry. He is one the main developers of ideas of É. Cartan. In a visionary paper he introduced the concept of a *Kähler metric*, at the crossroad of Riemannian and symplectic geometries, which proved to be extremely important in differential geometry, complex analysis and finally in theoretical physics.

Proposition X.30. *The natural extension $T^*\psi$ to T^*M of a diffeomorphism ψ of a configuration space M is a canonical transformation.*

Proof. Let $x = (x^1, \cdots, x^n)$ be a chart in a neighbourhood of a point a in M. Since ψ is a diffeomorphism of M, $y = x \circ \psi$ is a chart in a neighbourhood of the point $\psi^{-1}(a)$. By construction, the expression of the map ψ in the charts x in the source and y in the target (which we denote as usual by ψ_y^x) is the identity, so that $\psi_y^x(x^1) = y^1, ..., \psi_y^x(x^n) = y^n$. In the natural extensions $T^*x = (x^i, X_i)$ and $T^*y = (y^\alpha, Y_\alpha)$ of the charts x and y to T^*M, we thus find that $(T^*\psi)_{T^*x}^{T^*y}$ is also the identity, so we also have $Y_1 = X_1, ..., Y_n = X_n$.

Since, in the natural extension of the chart y, ω^M can be written in the form $\sum_\alpha^n dy^\alpha \wedge dY_\alpha$ by Proposition X.22, the expression of the inverse image exterior differential 2-form $(T^*\psi)^* \omega^M$ in the chart x is $\sum_i^n dx^i \wedge dX_i$, i.e., we have proved that $(T^*\psi)^* \omega^M = \omega^M$. $\quad\square$

Remark X.31. i) Recall once again that a differentiable map $f : M \to N$ does not in general induce a map from T^*M to T^*N, but rather, by taking the inverse image, a map f^* from T^*N to T^*M. Here, we have assumed that f is a diffeomorphism, which allowed us to consider T^*f by identifying it with $(f^{-1})^*$.

ii) Furthermore, one should not be surprised by the tautological nature of the preceding proof. It depends only on the fact that the Liouville 2-form is *universally* defined on configuration spaces; in other words, its definition does not depend on which extension of the local chart of the configuration space M is used to express it.

X.32. As we have seen in Chapter VIII, the space of vector fields on a configuration space N can be considered to be the tangent space at the identity of the group $\mathcal{D}N$ of diffeomorphisms of N.

Now that we have the symplectic form ω^M at our disposal, it is natural to ask what is the tangent vector to a curve $t \mapsto \Psi_t$ originating at Id_{T^*M} which consists of canonical transformations. If $X = (d\Psi_t/dt)_{|t=0}$, on differentiating the identity $\Psi_t^*(\omega^M) = \omega^M$ and using the definition of the Lie derivative, we obtain $\mathcal{L}_X \omega^M = 0$.

A vector field satisfying this condition is called a *Hamiltonian vector field* (or an *infinitesimal canonical transformation*).

Exercise X.33. Show that the flow of a Hamiltonian vector field consists of canonical transformations.

B. The General Form of Hamilton's Equations

We exploit the geometry defined by the Liouville 2-form on the extension by momenta, in particular the duality. To any observable on the extension by momenta is associated a vector field, its *Hamiltonian* vector field. The differential equation which it defines is called the system of *Hamilton's equations* associated to the observable.

The conservation law for the Hamiltonian is established, and the special properties of the flows of Hamiltonian vector fields are examined.

X.34. First of we examine all the consequences of the algebraic property of non-degeneracy of the symplectic form ω^M.

X.35. Being non-degenerate, ω^M defines at each point λ in T^*M (a space which we shall temporarily denote by N) a *duality* between $T_\lambda N$ and $T_\lambda^* N$ by associating to each vector v in $T_\lambda N$ the 1-form $i_v \omega^M$ which, evaluated on an arbitrary element w in $T_n N$, is equal to $(i_v \omega^M)(w) = \omega^M(v, w)$, and which is called the *interior product* of ω^M by v.

We shall mainly use the inverse of this operation, which associates to a 1-form λ, i.e., an element of T^*N, the vector v such that $i_v \omega^M = \lambda$. If the 1-form λ is the differential $d\varphi$ of an observable φ, the image vector is called the *Hamiltonian vector field* associated to φ, or its *symplectic gradient*, and denoted by Ω_φ. In a natural chart (x^i, X_i), we have

$$\Omega_\varphi = \sum_i^n ((\partial \varphi / \partial X_i)\, \partial / \partial x^i - (\partial \varphi / \partial x^i)\, \partial / \partial X_i).$$

X.36. Let $\varphi : T^*M \longrightarrow \mathbb{R}$ be an observable. The differential equation on T^*M which defines the vector field $-\Omega_\varphi$ is called the system of *Hamilton's equations* associated to φ. The function is then called the *Hamiltonian function* (or simply the *Hamiltonian*) of the problem. An integral curve η of the vector field $-\Omega_\varphi$ thus satisfies the equation $d\eta / d\tau = -\Omega_\varphi(\eta(\tau))$.

For a Hamiltonian φ, *Hamilton's equations can be written*, in a natural chart (x^i, X_i), *in the classical form*

$$(X.37) \qquad \begin{cases} \dfrac{dx^i}{d\tau} = \dfrac{\partial \varphi}{\partial X_i} \\[2mm] \dfrac{dX_i}{d\tau} = -\dfrac{\partial \varphi}{\partial x^i} \end{cases}, \quad 1 \leq i \leq n.$$

X.38. We shall now show that these equations describe the evolution of a system which is the solution of a variational problem. The main point is clearly to identify the observable which will be the Hamiltonian. In a model of physical origin, it can be identified with the *energy* of the system being modelled (we shall come back to this point in IX.59). We thus see that Hamilton's equations are in some ways extremely simple,[193] the difficulty at this stage being to understand their variational nature.

To guide us in this task, we first establish two fundamental properties of Hamilton's equations.

---

[193] On this point, it is interesting to quote Lagrange who, in his *"Second mémoire sur la théorie de la variation des constantes arbitraires dans les problèmes de mécanique"*, which appeared in 1809, said of these equations *"qu'elles sont, comme l'on voit, sous la forme la plus simple qu'il soit possible"*. As we shall see in Sect. E, he only treated infinitesimal variations of the motion by this method.

Proposition X.39. *Every observable φ on the extension by momenta T^*M of a configuration space M is constant along the integral curves of its symplectic gradient Ω_φ, i.e., $\partial_{\Omega_\varphi}\varphi = 0$.*

Proof. This is in every way similar to that of IV.75.   □

Exercise X.40. Prove Proposition X.39 by performing a calculation in local coordinates.

Remark X.41. It is worth emphasising the differences which exist between the notions of the gradient of an observable with respect to a Riemannian metric (the classical notion of gradient) and with respect to a symplectic form such as ω^M.

When we evaluate the derivative of an observable φ in the direction of its (Riemannian) gradient $\mathrm{Grad}\,\varphi$, we have $\partial_{\mathrm{Grad}\,\varphi}\varphi = g(\mathrm{Grad}\,\varphi, \mathrm{Grad}\,\varphi) \geq 0$ and this quantity is zero only at the critical points of φ. This relation models a *dissipative* phenomenon.

If we do the same thing for the symplectic gradient Ω_φ, we obtain $\partial_{\Omega_\varphi}\varphi = 0$ as we have pointed out in Proposition X.39. Such a relation models a *conservative* phenomenon, as we shall discuss in detail in Chapter XI.

When applied to models of physical or mechanical origin, when the Riemannian metric g defines the kinetic energy of a system for example, these properties have fundamental mechanical or physical consequences.

Proposition X.42. *The flow (Φ_t) of the symplectic gradient of an observable φ, defined on the extension by momenta T^*M of a configuration space M, consists of canonical transformations.*

Proof. Even though this proposition is fundamental, we only give the idea of its proof which depends on the use of the Lie derivative via an identity, often called the *"Fundamental Identity of the Calculus of Variations"*, according to which, *for any vector field Y on T^*M, we have*

$$(\text{X.43}) \qquad\qquad \mathcal{L}_Y\omega^M = d(i_Y\omega^M)\,,$$

(for a proof, see (Abraham and Marsden 1978), for example).

It then suffices to note that the dual 1-form, with respect to ω^M, of the symplectic gradient Ω_φ is by definition the differential of φ, and hence, by Proposition X.21, Formula (X.43) asserts that Ω_φ is a Hamiltonian vector field. Since the flow of a Hamiltonian vector field is formed by canonical transformations (cf Exercise X.33), we have the result stated.   □

Remark X.44. In fact, starting with the fundamental identity (X.43), one can show much more, namely that *"a vector field Y on the extension by momenta T^*M of a configuration space M is a Hamiltonian vector field if and only if Y is the dual (with respect to the Liouville symplectic form) of a closed differential form"*.

X.45. Proposition X.42 is in fact valid in any configuration space equipped with a symplectic form (this is a direct consequence of Exercise X.46). This is the main justification for including the hypothesis of being closed in the definition of a symplectic form. In this chapter and in Chapter XI, we shall see numerous consequences of this property, which is the foundation of the success of the Hamiltonian approach.

In the proof of Proposition X.42, it is clear that the only property of Ω_φ which was used is that the dual 1-form is *closed*. The distinction between the closed 1-forms and the 1-forms which are differentials of observables is related to the global structure of the configuration space N on which the symplectic form is defined in the following way: *"every Hamiltonian vector field on N is a symplectic gradient if the first cohomology space of N with real coefficients vanishes."* This is an illustration of how global topological properties of a configuration space can be determined by using the exterior differential forms defined on it. The discovery of the possibility of such a relation goes back to É. Cartan, but the most general version was given by de Rham.

Project X.46. Establish the general form of the Fundamental Identity of the Calculus of Variations, according to which *"for every vector field X and every exterior differential k-form α on a configuration space N, $\mathcal{L}_X \alpha = i_X d\alpha + d i_X \alpha$"*.

Project X.47. Show that *"the volume element $(\omega^M)^n$ defined by the Liouville 2-form is invariant under the flow of every Hamiltonian vector field"*, a property known as *"Liouville's theorem"*.

X.48. The result presented in Project X.47 has very important consequences. If we consider, for example, a *bounded* domain D in the extension by momenta, the volume of the image of D under the flow of a Hamiltonian vector field, evaluated with respect to $(\omega^M)^n$, is constant. Starting from this result, one can establish, for example, the *"recurrence theorem"* of Poincaré according to which the time evolution of every point of the extension by momenta passes through any given neighbourhood of its initial position infinitely many times.

X.49. By definition, a canonical transformation Ψ preserves the Liouville 2-form so that, if (x^i, X_i) is a natural chart on T^*M, we have

$$\omega^M = \sum_i^n d(X_i \circ \Psi) \wedge d(x^i \circ \Psi) .$$

This formula implies that Hamilton's equations retain their simple form (X.37) if one considers the Hamiltonian as a function of the new variables $(x^i \circ \Psi, X_i \circ \Psi)$. Note that this need no longer be the extension to T^*M of a chart on M if Ψ is not the extension to T^*M of a diffeomorphism of M (this is the case, for example, for the flow induced by a Hamiltonian ψ).

X.50. One often refers to this property by saying that Hamilton's equations are *invariant under canonical transformations*. The word "invariance" is used here in a formal sense: it is simply the *appearance* of the equations which does not change.

C. Relation with the Lagrangian Approach

In this section, we set up the correspondence (whenever possible) between the Euler–Lagrange equations and Hamilton's equations. This will lead us to a variational interpretation of Hamilton's equations (which is lacking so far) and to understand the role of the *energy* associated to a Lagrangian.

X.51. We begin with the simplest model we treated with the Euler–Lagrange equations: that of a particle of mass m which moves in a potential V defined on the space $\mathbb{R}^3$. At time τ, its trajectory γ has velocity-vector $\dot\gamma(\tau) = (\gamma(\tau), \gamma'(\tau))$.

In this case, if we define the *total energy* E of the particle by setting

$$E(\dot\gamma) = \frac{1}{2} m\left(\gamma'(\tau)|\gamma'(\tau)\right) + V(\gamma(\tau)),$$

we see that the *energy E is conserved during the motion* since

$$dE/d\tau = m\left(\gamma''(\tau)|\gamma'(\tau)\right) + \left(\mathrm{Grad}V(\gamma(\tau))|\gamma'(\tau)\right) = 0$$

because of the equation of motion $m\,\gamma''(\tau) = -\mathrm{Grad}V(\gamma(\tau))$.

In this particular case, the energy is thus a candidate for the Hamiltonian by Proposition X.39, provided that we can *"transfer"* from the extension by velocities TM to the extension by momenta T^*M.

X.52. We must therefore use a *correspondence between TM and T^*M*, and *construct for a general Lagrangian a conserved quantity* like the total energy in the special case just considered.

Definition X.53. Let A be a Lagrangian defined on the extension by velocities TM of a configuration space M. The *Legendre transform* associated to A is the map $\Lambda_A : TM \longrightarrow T^*M$ defined by $\Lambda_A(v) = d(A_{|T_qM})(v)$ (where $T_v^*(T_qM)$ is identified with T_q^*M by translation by the vector $-v$).

In the extensions Tx and T^*x of a chart $x = (x^1, \cdots, x^n)$ on M to TM and T^*M respectively, we have $(\Lambda_A)^{Tx}_{T^*x}(x^i, \xi^i) = (x^i, \partial A/\partial \xi^i)$.

Proposition X.54. *The Lagrangian A is regular if and only if the Legendre transform associated to it is a local diffeomorphism.*

Proof. The proof is almost a restatement of IX.50 and IX.52.

In fact, the Legendre transform Λ_A coincides with the map introduced in IX.50. Moreover, requiring that it has maximum rank, say $2n$ when the source space is also of dimension $2n$, is equivalent by the Local Inversion Theorem III.70 to it being a local diffeomorphism. $\qquad\square$

Remark X.55. By its construction, the Legendre transform Λ_A has the property of commuting with the natural projections π_M and ϖ_M from TM and T^*M to M respectively, i.e., we have $\varpi_M \circ \Lambda_A = \pi_M$. It is made so as to transform the impulse into the Liouville 1-form.

Note, however, that even if the Lagrangian A is regular, there is no guarantee that Λ_A gives a bijection between TM and T^*M. Of course, this is the most pleasant situation, since we then have a *global diffeomorphism between TM and T^*M*: in this situation, the Lagrangian is sometimes called *hyperregular* (for an important and natural example, see X.63).

Examples X.56. i) In the case of a Riemannian metric g, the expression for the Legendre transform is particularly simple since, for $v \in T_qM$, $\Lambda_A(v) = \flat_{g(q)}(v)$ where $\flat_{g(q)}$ is the *linear* isomorphism from T_qM to T_q^*M defined by the scalar product $g(q)$. This transformation is evidently a global diffeomorphism which means that A is a hyperregular Lagrangian.[194]

ii) The Legendre transform associated to a Lagrangian which is homogeneous of degree 1 clearly cannot even be a local diffeomorphism, since the Liouville vector field lies in the kernel of the tangent linear map of Λ_A, as the calculation contained in IX.55 shows.

Theorem X.57. *When the Lagrangian A is hyperregular, the Legendre transform Λ_A induces a bijection between the integral curves of the symplectic gradient of the Hamiltonian $H = E^A \circ (\Lambda_A)^{-1}$ and the curves of velocity-vectors of the extremals of the action* **A** *functional associated to A.*

Proof. It is possible to give a completely intrinsic proof of this fundamental theorem (cf. (Abraham and Marsden 1978), pages 218-219), but we have chosen to give here a more explicit proof.[195]

Since the Lagrangian A is assumed to be hyperregular, the Legendre transform Λ_A is a diffeomorphism from TM to T^*M, whose inverse we denote by Λ_A^{-1}. We remark that these two transformations commute with the natural projections π_M and ϖ_M from TM and T^*M to M.

Having chosen a coordinate system x, it is useful to introduce the maps Λ_A^X and Λ_A^ξ defined respectively by $\Lambda_A^X(v) = (X_i(\Lambda_A(v)))$ and $\Lambda_A^\xi(\lambda) = (\xi^i(\Lambda_A^{-1}(\lambda)))$. For $\lambda \in T^*M$ we have

$$H(\lambda) = \sum_i^n X_i(\lambda)\,(\Lambda_A^\xi)^i(\lambda) - A(\Lambda_A^{-1}(\lambda)).$$

We show that the correspondence we wish to establish holds for the curves whose projection onto M passes through a point q. Take a chart $x = (x^1, \cdots, x^n)$ at q. Let γ be a solution curve of the Euler–Lagrange equations. We show that the curve $\tau \mapsto (\gamma^k(\tau), (\partial A/\partial \xi^i)(\gamma^k(\tau), d\gamma^k/d\tau))$ satisfies Hamilton's equations.

For this, we calculate along γ

$$-\frac{\partial H}{\partial x^i} = -\sum_{j=1}^n X_j\left(\frac{\partial \Lambda_A^\xi}{\partial x^i}\right)^j + \frac{\partial A}{\partial x^i} + \sum_j^n \left(\frac{\partial \Lambda_A^\xi}{\partial x^i}\right)^j \frac{\partial A}{\partial \xi^j} = \frac{\partial A}{\partial x^i}$$

which is equal to $d(\partial A/\partial \xi^i)/d\tau$ since γ is an extremal of **A**. Since, by definition of the Legendre transform, $X_i(\Lambda_A(x^i, \xi^i)) = \partial A/\partial \xi^i$, we have thus verified "half" of Hamilton's equations.

To establish the other half, we calculate along γ

$$\frac{\partial H}{\partial X_i} = (\Lambda_A^\xi)^i + \sum_j^n X_j\left(\frac{\partial \Lambda_A^\xi}{\partial X_i}\right)^j - \sum_j^n \frac{\partial A}{\partial \xi^j}\left(\frac{\partial \Lambda_A^\xi}{\partial X_i}\right)^j = \frac{d\gamma^i}{d\tau}$$

since $\xi^i(\dot\gamma) = d\gamma^i/d\tau$ and $X_j \circ \Lambda_A = \partial A/\partial \xi^j$.

To complete the proof, we note that, the Lagrangian being regular, for any point of TM, there exists a solution of the Euler–Lagrange equations having this point as initial condition since these curves are solutions of a second order differential equation which is solved for its second derivatives. The Legendre transform being a diffeomorphism, we have thus exhibited a solution curve of Hamilton's equation, passing through any point of T^*M, which is the image of the curve of velocity-vectors of a solution of Lagrange's equations: by the Uniqueness Theorem IV.28 for the integral curves originating at a given point, this establishes the desired correspondence.   □

---

[194] Note that the positive character of the scalar product was irrelevant in our discussion, the only important property being the fact that g is non-degenerate. This remark can be used in general relativity, where one is given a *Lorentzian* metric in every tangent space.

[195] which is not to say that it is more illuminating.

Remark X.58. When the Lagrangian is not hyperregular, several difficulties occur. Starting from a Lagrangian A, without the inverse of the Legendre transform at our disposal, we can no longer define a *Hamiltonian* as an observable on T^*M given as image of the energy E^A. At best, the Hamiltonian will only be defined on the image of TM under Λ_A, and even then possibly with some ambiguity. It was to deal with this situation that Dirac developed his *theory of constraints*. In this setting, it is again possible to establish some kind of correspondence between extremals of the action A associated to the Lagrangian A and integral curves of the symplectic gradient.

X.59. One can also start directly from an observable H on T^*M, which is taken to be the *Hamiltonian*, and define an analogue λ_H of the Legendre transform (cf. Exercise X.60). One can then define a new observable L^H which, when transferred to TM, will be the Lagrangian of the problem. This will all make sense if the Hamiltonian is hyperregular, i.e., if λ_H is a diffeomorphism from T^*M onto TM.

Exercise X.60. Let H be an observable on the extension by momenta T^*M of a configuration space M. Show that $\sum_{i=1}^{n}(\partial H/\partial X_i)\,\partial/\partial x^i$ is a tangent vector to M independent of the chart x used to express it, and hence that it is intrinsic.

X.61. *Variational principles* have played a fundamental role in the evolution of mathematical models in mechanics and physics, as we have already emphasised in the preface to this course.

The approach we have followed in Chapter IX led us to introduce the extremals of a variational Principle of Least Action which is due to Hamilton. The original formulation of this principle is due to Maupertuis[196].

There are other variational formulations which should be mentioned. One of these is *Fermat's Principle*, which is one of the foundations of geometrical optics. There are others, associated with Gauss and Jacobi, which are more closely related to geometrical considerations.

X.62. We now make several comments on a variational formulation which incorporates the Hamiltonian viewpoint by introducing the idea of a *duality between time and energy* which is important for many developments in physics. It comes close, in fact, to Maupertuis' original formulation.

X.63. Instead of considering arbitrary curves joining two points, we restrict ourselves to curves for which the energy of the velocity-vector is a constant e (i.e., the analogue of curves parametrised by arc-length of a Riemannian metric on the configuration space). Moreover, we assume that e is a regular value of E as a function from TM to $\mathbb{R}$.

---

[196] Pierre-Louis Moreau de Maupertuis (1698–1759), French mathematician and astronomer, formulated the Principle of Least Action in 1744. There was a priority dispute over the discovery of this principle, and he was accused of having plagiarised Leibniz. He introduced the theory of gravitation into France in 1732, and participated in its verification by leading an expedition to Lapland in 1736 designed to obtain experimental evidence of the polar flattening of the Earth.

We then have the following statement: *in the product-space of curves γ and parametrisations ψ respecting the energy constraint $E(\gamma \circ \psi) = c$, where $c \in \mathbb{R}$, a curve γ and the identity reparametrisation are critical for the functional*

$$(\gamma, \psi) \mapsto \int_{\psi(\alpha)}^{\psi(\beta)} (D.A)(\dot{\gamma}(\tau))\, d\tau$$

if and only if γ is a solution of the Euler–Lagrange equations. (Exercise: Prove it.)

Remark X.64. In the preceding formulation, we have introduced a constraint on the energy. This leads us to *consider new configuration spaces obtained as submanifolds of the extension by velocities or by momenta of a configuration space.* To be able to continue to do Variational Calculus with tools analogous to those we have used up to now, some of the notions we have introduced, such as that of a *symplectic structure*, must be generalised.

The Hamiltonian approach lends itself to this much more readily than the Lagrangian viewpoint (as we will unfortunately not be able to show!). In fact, the latter approach is effective only when the constraints depend only on the positions.

Note also that from a variational point of view, the introduction of a constraint on the energy makes it necessary to introduce an additional degree of freedom on "*the law of time*". This suggests a pairing between energy and time which we shall meet again in Sect. D.

X.65. It might be useful to formulate our study of the relation between the Euler–Lagrange equations and those of Hamilton in purely mathematical terms. The geometries which are defined on the tangent bundle and on the cotangent bundle are, in fact, very different in nature.

The special character of the internal geometry of TM appears above all when one studies closely the structure of TTM. This space is fibred in *two* different ways over TM : by the natural projection π_{TM} as a tangent bundle and by the tangent linear map $T\pi_M$ of the projection of TM onto M. This singles out certain special elements, the *jets of curves*, and allows one to define an *involution* of TM which is of great geometric interest. These considerations lead to picking out from among the vector fields on TM certain special fields which define second order differential equations (one example is given by the Euler–Lagrange equations). Given a Lagrangian, one can find such a field which is related in a complicated way to the Lagrangian (and this is where the non-triviality of the Euler–Lagrange equations lies). In the case where the Lagrangian is regular, there is an underlying symplectic structure *which depends on the Lagrangian*. (For more details, see (Abraham and Marsden 1978) or (Besse 1978).)

As far as the cotangent bundle is concerned, it is the presence of Liouville's universal differential 2-form which gives it its special structure. All the variational properties arise from symplectic duality once one is given an observable. This leads to the idea that any manifold equipped with a symplectic form is just as good as a cotangent bundle for treating questions related to mechanics. This point of view has been exploited extensively in the passage from classical mechanics to quantum mechanics.

X.66. There is an analogue of the Hamiltonian approach for problems involving objects in several variables, such as surfaces or more general fields. Unfortunately, it is much more complicated. It is necessary to work in spaces related to *spaces of jets of maps*. The symplectic structure is replaced by a more elaborate mathematical structure, sometimes called a *multisymplectic* structure. The complexity of these objects is without doubt one of the reasons why this approach has not had success comparable with that which we have exploited in the case of curves, i.e. objects depending on one real variable.

D. Poisson Brackets of Observables

The purpose of this section is to extend the notion of *Poisson bracket* to configuration spaces. This operation can be defined on the algebra of observables on the extension by momenta of a configuration space.

We study notably its behaviour under canonical transformations.

X.67. The *Poisson bracket*, suitably generalised to the setting of configuration spaces, turns out to be a very powerful tool in the dynamical study of variational problems.

Definition-Proposition X.68. *Let φ and ζ be two observables on the extension by momenta T^*M of a configuration space M. Their Poisson bracket, denoted by $\{\varphi, \zeta\}$, is the observable $\partial_{\Omega_\varphi} \zeta$, which can also be written $-\partial_{\Omega_\zeta} \varphi$, and which in a natural chart (x^i, X_i) has the expression*

$$(X.69) \qquad \{\varphi, \zeta\} = \sum_{i=1}^{n} \left(\frac{\partial \varphi}{\partial x^i} \frac{\partial \zeta}{\partial X_i} - \frac{\partial \varphi}{\partial X_i} \frac{\partial \zeta}{\partial x^i} \right).$$

The Poisson bracket is an alternating bilinear map from $\mathcal{F}M \times \mathcal{F}M$ to $\mathcal{F}M$.

Proof. This is (almost) a copy of the argument given in IV.65.

Formula (X.69) follows directly from the defining relations $i_{\Omega_\varphi} \omega^M = d\varphi$ and $\omega^M = \sum_{i=1}^{n} dX_i \wedge dx^i$. The bilinearity of the Poisson bracket also follows from this.

The alternating character of the Poisson bracket is an immediate consequence of the fact that

$$\partial_{\Omega_\varphi} \zeta = \langle d\zeta, \Omega_\varphi \rangle = \omega^M(\Omega_\varphi, \Omega_\zeta) = -\omega^M(\Omega_\zeta, \Omega_\varphi),$$

from which the skew-symmetry follows by interchanging φ and ζ.   □

Remark X.70. It is clear that the Poisson bracket as we have defined it is an extension of the definition given in IV.65.

The novelty is, of course, the fact that the symplectic form ω^M is universally defined on the extension by momenta of a configuration space.

Project X.71. Show that the Poisson bracket satisfies the *Jacobi identity*: for observables φ_1, φ_2 and φ_3,

$$\{\varphi_1, \{\varphi_2, \varphi_3\}\} + \{\varphi_2, \{\varphi_3, \varphi_1\}\} + \{\varphi_3, \{\varphi_1, \varphi_2\}\} = 0.$$

Remark X.72. In fact, in 1808 Lagrange introduced quantities which are closely related to Poisson brackets and which are known as *Lagrange brackets*. To any pair of vector fields they associate an observable. (For more details on this topic, we refer the interested reader to (Abraham and Marsden 1978), page 196.)

Proposition X.73. *If Ψ is a canonical transformation defined on the extension by momenta T^*M of a configuration space M, then, for any pair of observables (φ, ζ) on T^*M,*

$$(X.74) \qquad \{\varphi \circ \Psi, \zeta \circ \Psi\} = \{\varphi, \zeta\} \circ \Psi.$$

Proof. By definition, if Ψ is a canonical transformation, $\Psi^* \omega^M = \omega^M$. Moreover, we know by Eq. (X.20) that $\Psi^* d\varphi = d(\varphi \circ \Psi)$. It follows that $\Omega_{\varphi \circ \Psi} = (T\Psi)(\Omega_\varphi)$. Formula (X.74) is now clear. In fact,

$$\{\varphi, \zeta\} \circ \Psi = (\partial_{\Omega_\varphi} \zeta) \circ \Psi = \partial_{T\Psi(\Omega_\varphi)}(\zeta \circ \Psi) = \partial_{\Omega_{\varphi \circ \Psi}}(\zeta \circ \Psi) = \{\varphi \circ \Psi, \zeta \circ \Psi\}$$

which is precisely the relation to be proved. $\square$

Proposition X.75. *Let φ be an observable on the extension by momenta T^*M of a configuration space M. Along an integral curve $\tau \mapsto \eta(\tau)$ of the vector field Ω_H defined by a Hamiltonian H, φ satisfies the equation $d(\varphi \circ \eta)/d\tau = \{H, \varphi\} \circ \eta$.*

In particular, φ is a constant of the motion defined by the Hamiltonian H if and only if $\{H, \varphi\} = 0$.

Proof. This is almost tautologous. By the Chain Rule VI.26, we have $d\varphi/d\tau = \partial_{\Omega_H} \varphi$ which is by definition $\{H, \varphi\}$.

The second part follows immediately. $\square$

X.76. Proposition X.75 contains as a special case Proposition X.61 since the Poisson bracket is alternating, and hence $\{H, H\} = 0$.

X.77. Proposition X.75 generalises to quantum mechanics. By the correspondence principle, the algebra of observables must be replaced by the algebra of operators on a Hilbert space of states. In this new algebra, it is the *commutator of operators* which plays the role of the Poisson bracket, hence the fact that in quantum mechanics observables must commute with the Hamiltonian to be conserved.

Remark X.78. The reader will find an example of the use of the Poisson bracket in connection with the search for first integrals in XI.16.

E. Historical Notes

X.79. Even in the case of mechanical models, it is essential to study configuration spaces equipped with a symplectic form which are more general than extensions by momenta of configuration spaces (notably because, in the solution of a problem, one wants to take into account the first integrals of the motion, as we explain in Chapter XI). This is the starting point of a special branch of differential geometry, called *symplectic geometry*.

   This geometry was already developed in the XIXth century; indeed, then Darboux obtained one of the fundamental theorems of the subject, asserting that this geometry has no local invariants. This theory la dormant for some time, while the interests of mathematicians turned towards global questions, probably because of the "size" of its invariance group, which is infinite-dimensional. Nevertheless, with the development of quantum mechanics, it has received new stimulation from physics.

X.80. The new lease of life which symplectic geometry is experiencing today is largely due to the desire for a mathematical formulation of the *"correspondence principle"* in quantum mechanics. But it seems that the work of Sophus Lie dating back to the end of the XIXth century dealing with analogous questions has been largely ignored.

Another reason for its rejuvenation is the appearance of this geometry as a tool for solving questions of analysis related to the study of the propogation of singularities of certain partial differential equations. This marriage has given birth to a new point of view on these questions of analysis: today one speaks of *microlocal analysis*.

X.81. Even today, only a few results are available on the global classification of symplectic configuration spaces; even so, one should mention the work of Calabi[197] and the considerable work of Gromov[198].

---

[197] The Italian-American mathematician Eugenio Calabi, born in 2023, contributed to many fields in mathematics, from differential and algebraic geometry to analysis (in particular to the theory of partial differential equations). Besides important work on minimal surface theory, he shaped the fields of affine and Kähler geometries by proving several uniqueness results and constructing important examples. He was also the first to prove a global result in symplectic geometry.

[198] Mikhail Leonidovich Gromov, born in 1943, is a Russian-French mathematician whose works revolutionised geometry at large by showing how working with less regular spaces can provide key information about more regular spaces. He in particular made new linkages between group theory and geometry by defining metric structures on groups. His view on estimates also impacted the theory of partial differential equations and image analysis as well as many other fields, also indirectly some aspects of theoretical physics. He has been a major contributor to the development of global symplectic geometry. Later in his career, he developed a strong interest in theoretical biology. He received the Wolf Prize in 1993, the Kyoto Prize in 2002 and the Abel Prize in 2008.

Chapter XI
Symmetries and Conservation Laws

In this final chapter, we study the relation between the presence of symmetries and the existence of conserved quantities in a variational problem. This theme plays a very important role both conceptually and from a practical point of view, because of the simplifications it allows in calculations.

In Sect. A, we give a mathematical presentation of the notion of *symmetry* by introducing the *action* of a Lie group on a configuration space together with various examples of the classical situations where the notion arises. Particular attention is paid to the case of the action induced on the extensions by velocities and by momenta by the action of a group on a configuration space.

We show in Sect. B how quantities which are conserved during the time evolution of a system generated by a vector field appear, and this leads us to the notion of a *first integral*. In the presence of such integrals, it is often possible to simplify the study of the evolution of a system; this introduces, as we shall see in several examples, additional reasons for considering configuration spaces other than open subsets of vector spaces.

Section C is devoted to giving a general definition of the *moment* associated to the action of a group on a configuration space. Of course, this generalises the notion of the moment of a force about a point, and it can be formulated very simply in the Hamiltonian approach. We deduce the theorem of E. Noether which asserts that the moment is a conserved quantity for a motion whose Hamiltonian is invariant under the group action.

When we wish to make successive reductions by using first integrals, the Poisson brackets of these observables must be zero (one also says that they are *in involution*). This is the topic of Sect. D in which we also present several properties of systems having sufficiently many first integrals in involution, the *integrable systems*, which makes it possible to solve them completely.

© The Author(s), under exclusive license to Springer Nature Switzerland AG 2022
J.-P. Bourguignon, *Variational Calculus*, Springer Monographs in Mathematics,
https://doi.org/10.1007/978-3-031-18307-2_11

A. Group Actions and Symmetries

We formalise the notion of symmetry in the form of the *action* of a group on a configuration space.

Various examples of geometric interest are examined as well as the extensions of an action to the extensions by velocities and by momenta of an action on a configuration space.

XI.1. Many attempts have been made to formulate clearly the concept of a "*system with symmetries*". Today we have a precise mathematical formulation which interprets such a system as one which is left invariant by a group of transformations. This formulation allows one to obtain all the fundamental results in a simple and systematic way, and also to carry out calculations completely.

In our setting, where the objects introduced are assumed to be differentiable, we shall be interested in groups which are also configuration spaces, i.e., *Lie groups*,[199] of which we have already met certain examples before.

Definition XI.2. Let M be a configuration space and G a Lie group. A differentiable map $\mu : G \times M \longrightarrow M$ is called an *action* of G on M if, for all g, $g' \in G$ and all $q \in M$, $\mu(g, \mu(g', q)) = \mu(g.g', q)$ (one also says that G *acts* on M).

In general, we *write* $\mu(g, q) = g.q$.

Exercise XI.3. Show that an action of a Lie group G on a configuration space M can be defined equivalently as a homomorphism from G to the group $\mathcal{D}M$ of diffeomorphisms of M.

Examples XI.4. Let us show that this definition encompasses very diverse classical situations.

i) In a vector space E, we consider the group $Gl\,E$ of invertible linear transformations, which is an open subset of the vector space $L(E)$, and is thus a configuration space in a natural way. This is a Lie group since the composition of endomorphisms, which is the multiplication in this group, is a bilinear operation. Having chosen a basis of E, we can represent any linear transformation by a matrix in this basis, and the inverse, as is well-known, is a rational function of the entries of the matrix which is differentiable on the open subset of matrices with non-zero determinant. The map which to an element l of $Gl\,E$ and an element v of E associates the element $l(v)$ is then an action in the sense of XI.2. It is clearly differentiable and satisfies the composition relation tautologically.

ii) For a slight generalisation of the preceding example, one can consider E as an *affine space* (by forgetting the origin) and the group $Af\,E$ of invertible affine

---

[199] We recall that a configuration space G is called a *Lie group* if there is a group law μ defined on it such that the product, viewed as a map from $G \times G$ to G, is differentiable, and such that taking the inverse is also a differentiable map from G to G.

transformations, obtained by adjoining the translations and their composites with linear transformations.

iii) For a more substantial example, consider the action of the rotation group SO_n on the sphere S^{n-1} considered as a submanifold of $\mathbb{R}^n$. There are two obvious ways to check that this is an action: to see how to express a rotation in the stereographic charts (which at first sight looks rather unpleasant), or to use the fact that the action of SO_n on $\mathbb{R}^n$ is itself differentiable, and hence is so in any chart, and to note that the distance r to the origin is left invariant by the group of rotations and that S^{n-1} is the submanifold of $\mathbb{R}^n$ defined by the constraint $r = 1$.

iv) To conclude with a more general example, we consider two actions which are naturally defined on any Lie group G: the *right regular action* μ_r and the *left regular action* μ_l of the group G on itself, which are defined as follows. For $g \in G$, it is traditional to denote by R_g (resp. L_g) the diffeomorphism $\mu_r(g)$ (resp. $\mu_l(g)$) which to $g' \in G$ associates the element $\mu_r(g)(g') = g'g$ (resp. $\mu_l(g) = g'g^{-1}$). There is another action of G on itself which turns out to be very important, namely the *adjoint action*, which to an element $g \in G$ associates the diffeomorphism $g' \mapsto gg'g^{-1}$. In contrast to the preceding two actions, which leave no point of the group invariant, it leaves the *centre* of the group invariant. (Recall that this is the subgroup of G consisting of the elements of G which commute with all elements of G.)

Exercise XI.5. Show that the operation which to an invertible linear transformation L of $\mathbb{R}^{n+1}$ and a line d in $\mathbb{R}^{n+1}$ associates the image $L(d)$ of this line is an action of $\mathbb{R}Gl_{n+1}$ on $\mathbb{R}P^n$.

Project XI.6. Show that the action of the complex general linear group $\mathbb{C}Gl_{m+1}$ on the complex projective space $\mathbb{C}P^m$, defined in a way analogous to that described in Exercise XI.5 for real projective spaces, is holomorphic.

XI.7. When a Lie group G operates on a configuration space M by an action μ, certain special vector fields on M, which we call the *velocity-vector fields of the action* μ, are singled out in the following way.

XI.8. If X is a tangent vector to G at the identity which is the velocity-vector of a curve γ on G, then, at any point q of M, the velocity-vector at 0 of the curve $t \mapsto \gamma(t).q$ on M defines a tangent vector $X_\mu(q)$ at q. The fact that $q \mapsto X_\mu(q)$ is differentiable follows from the Chain Rule III.46.

In Example i), which is linear, the vector field L_μ associated to a linear transformation L from E to itself is simply the field $v \mapsto (v, L(v))$ if we identify TE with $E \times E$ as usual. To take a specific example, if we consider $\mathbb{R}^n$ with its standard metric, we can consider the vector fields associated to the skew-symmetric matrices, which can be identified with the tangent vectors at the identity of the group SO_n (cf. IX.95). Thus, for $n = 3$, the velocity-vector field $(a_3)_\mu$ of infinitesimal rotations about the x^3-axis can be written

$$(a_3)_\mu(x^1, x^2, x^3) = x^2 \frac{\partial}{\partial x^1} - x^1 \frac{\partial}{\partial x^2}.$$

In Example iii), specialised to dimension 2, the velocity-vector fields X_μ of the rotations around the polar axis with angular velocity ω can be written in geographical coordinates (θ, ψ) $X(\theta, \psi) = \omega\, \partial/\partial\theta$. (Exercise: Prove it.)

Remark XI.9. The preceding development, which will be used a lot in the remainder of this chapter, only allows one to make use of the identity component of the group which acts. If the group acting is discrete, the identity is an isolated point of the group and consequently the tangent space at the identity is 0. Thus, the action of such a group element does not give rise to a velocity-vector field on the configuration space M.

The presence of discrete, even finite, symmetry groups is nevertheless interesting, since it may allow one to work on a smaller configuration space obtained by considering the orbits of the action of the discrete group, provided certain conditions are satisfied (the group should have no fixed points if it is finite, for example). This is, for example, what we did in VII.54 in our variational approach to the study of the eigenvectors of a symmetric linear transformation, when we restricted the quadratic form which it defines to the sphere S^{n-1}. Because of the invariance under the antipodal map which, being an involution, defines an action of the group $\mathbb{Z}_2$ with 2 elements, we were able to regard it as a variational problem on $\mathbb{RP}^{n-1}$ where we could be sure that this function had at least n critical points.

Proposition XI.10. *Let μ be an action of a Lie group G on a configuration space M. The map $\mu^T : G \times TM \longrightarrow TM$ which to (g, v) associates $T_{\pi_M(v)}(\mu(g))(v)$ defines an action of the same group G on the extension by velocities TM.*

In an analogous way, the map $\mu^{T^*} : G \times T^*M \longrightarrow T^*M$ which to (g, λ) associates $T^*_{\varpi_M(\lambda)}(\mu(g))(\lambda)$ defines an action of G on the extension by momenta T^*M.

Proof. The differentiability of the maps μ^T and μ^{T^*} is an immediate consequence of the Chain Rule III.46. Thus, it only remains to check that these maps are homomorphisms from G to the groups of diffeomorphisms of TM and T^*M respectively. From the calculation of the tangent linear map of the inverse of a differentiable map, it follows that the inverse of $\mu^T(g)$ is $\mu^T(g^{-1})$; the same is true for μ^{T^*}. The fact that $\mu^T(g.g') = \mu^T(g) \circ \mu^T(g')$ again follows from the Chain Rule III.46.   □

Exercise XI.11. Let μ be an action of a Lie group on a configuration space M. Determine the velocity-vector fields of the extensions μ^T of the action to TM and μ^{T^*} of the action to T^*M in terms of the velocity-vector fields X_μ on M.

Remark XI.12. It follows from Proposition X.30 that $\mu^{T^*}(g)$ is a canonical transformation for any $g \in G$. This remark will be fundamental in Sect. C. One sometimes says that μ^{T^*} is a *"symplectic action"* on the symplectic space (T^*M, ω^M).

XI.13. In the last few years, theoretical physicists (and, following them, other physicists and mathematicians) have been talking about *supersymmetry*. This notion appeared in order to be able to express certain transformations which exchange fermions and bosons. It is not possible to present the basic mathematical ideas which allow one to give a precise meaning to this notion without going into *spinor geometry*.

Even today, one cannot say whether these ideas, which have already shown their power in mathematics as well as in physics, allow one finally to create a *supergeometry* whose foundations can replace those of ordinary differential geometry. Among its characteristic features is the presence of *anticommuting* variables, in addition to the ordinary variables which serve as the charts of a configuration space. Among the points where a proper mathematical theory is lacking for the moment is the passage from an infinitesimal point of view (*superalgebras*) to a finite point of view (*supergroups*).

B. First Integrals and Conservation Laws

In this section, we review the properties of the quantities which are constant throughout a motion generated by a vector field and the use which can be made of them, notably in the case of vector fields arising in the Calculus of Variations.

XI.14. Even though we shall carry out the study of first integrals of vector fields defined on configuration spaces in complete generality in this section, we have in mind the application to the case of vector fields which appear in Variational Calculus, such as those which give rise to Hamilton's equations which we studied in Chapter X.

XI.15. Among the vector fields which we are particularly interested in are the symplectic gradient vector fields defined on extensions by momenta.

In Proposition IX.61, we have seen that these vector fields always exhibit at least one first integral: the *Hamiltonian* which defines them. In this special case, one speaks of the *Law of conservation of energy*, hence the fact that systems which can be modelled in this way are called *conservative*.

For these systems, there are very powerful methods for constructing first integrals which make use of the geometry of the extension by momenta. Proposition XI.16 gives one example.

Proposition XI.16. *Let Ω_H be the symplectic gradient of an observable H defined on the extension by momenta T^*M of a configuration space M. If φ and ζ are two first integrals of Ω_H defined possibly on only an open set of T^*M, their Poisson bracket $\{\varphi, \zeta\}$ is also a first integral of Ω_H where it is defined.*

Proof. It is completely analogous to that of IV.77.   □

XI.17. First integrals of a vector field can be used to *reduce the number of degrees of freedom* of the configuration space on which the vector field is originally defined, hence its dimension.

In fact, we know that the integral curves of the vector field all lie in level hypersurfaces of the first integrals. If we find sufficiently many first integrals, we can "*trap*" the integral curves in the intersection of the level hypersurfaces of the first integrals.

This can simplify the study of the phase portrait of the vector field provided that this intersection is non-singular. This condition is satisfied, in particular, when we consider p first integrals φ_i, $1 \le i \le p$, and the value taken by the map $\phi : T^*M \longrightarrow \mathbb{R}^p$ on the integral curve we wish to study is a regular value of ϕ. (See Sect. D for an application of this principle.)

XI.18. Let us give a specific example.

If we use Hamilton's equations to find the geodesics of a Riemannian metric g on a 2-dimensional configuration space M, we must find the integral curves of the vector field $-\Omega_H$ (where $H = \frac{1}{2}g \circ \sharp_g$) on T^*M, which is a 4-dimensional space. Since we know that H is conserved along the integral curves, and we are really interested in geodesics as geometric curves (i.e., forgetting the parametrisation) in T^*M, we can work on the submanifold $S^*M = H^{-1}(1)$. (Exercise: Show that this really is a submanifold.)

At every point λ of S^*M, the vector field $-\Omega_H$ is tangent to S^*M. We are thus reduced to studying a vector field on a 3-dimensional space. If we can find another quantity conserved along the integral curves of this vector field, we can reduce to the study of a vector field on a *surface*, which is much simpler.

Remark XI.19. The theorem of E. Noether[200] presented in Sect. C will give us a systematic method for deducing conserved quantities from the existence of symmetries of a system.

C. The Notion of a Moment and the Theorem of E. Noether

We continue the discussion of first integrals begun in the preceding section, but this time relating it to group actions. We will (finally!) elucidate the notion of a *moment* and the reason why we have called T^*M the *extension by momenta* of M.

XI.20. We begin with the elementary example of a vector space E (considered first of all as a configuration space, but also as an affine space).

The group of translations, which is isomorphic to E, acts by $(\vec{v}, q) \mapsto q + \vec{v}$. It is almost tautologous to say that the velocity-vector field associated to the element $\vec{V}$ of E (viewed as the tangent space at the identity of the group of translations), which we denote by V_E, is the constant vector field equal to $\vec{V}$ in E (viewed this time as a configuration space). Thus, we have $V_E(q) = \vec{V}$ for all $q \in E$.

The extension of this action μ_0 to $T^*E \cong E \times E^*$ is thus simply given by $\mu_0^{T^*}(\vec{v}, (q, Q)) = (q + \vec{v}, Q)$, so the vector field V_{T^*E} on $T^*E \cong E \times E^*$ associated to $\vec{V}$ is $V_{T^*E}(q, Q) = (\vec{V}, 0)$. If we now consider the differential 1-form λ_V on T^*E associated, via the symplectic form ω^E, to the vector field V_{T^*E} (i.e., we have $\lambda_V = i_{V_{T^*E}} \omega^E$), we find $\lambda_V(q, Q) = d(V^*)$ where V^* denotes the function on E^*

---

[200] The German mathematician Amalie Emmy Noether (1882–1935) made numerous major contributions to invariant theory and representation theory. She was the elder daughter of the algebraist Max Noether and became a role model for women in mathematics.

which at the point Q takes the value $V^*(Q) = \langle Q, V \rangle$. More concretely, if we have chosen a linear coordinate system (x^i) on E (which makes it isomorphic to $\mathbb{R}^n$), and if we take $V = e_i (= \partial/\partial x^i)$, we find $\lambda_{e_i} = dX_i$.

If we now consider a motion in E governed by Hamilton's equations associated to a Hamiltonian E which is invariant under translation by the vector tV for all $t \in \mathbb{R}$ (i.e., the flow of the vector field V_{μ_0}), we find that $0 = \partial_{V_{T^*E}} H = \{V^*, H\}$, and hence that V^* is a first integral of the motion.

This conservation law is traditionally called the *conservation of linear momentum*. The function V^* is then called the *conjugate momentum* of V. In the case where one takes $V = \partial/\partial x^i$, we find that the function X_i on $T^*\mathbb{R}^n$ is precisely its conjugate momentum, which justifies calling T^*M the *extension by momenta* of M.

Definition XI.21. Let μ be an action of a Lie group G on the extension by momenta T^*M of a configuration space M by canonical transformations. An observable p on T^*M with values in $(T_eG)^*$ (where e is the identity element of the group) is called a *moment* of the action μ if, for any vector $X \in T_eG$ and for any point $\lambda \in T^*M$, we have

(XI.22) $$d\langle p, X \rangle(\lambda) = (i_{X_\mu}\omega)(\varpi_M(\lambda)) \, .$$

Remark XI.23. Of course, we should first check that Definition XI.21 agrees with the one used in XI.20 in the case of translations.

For this, it is only necessary to check that the function V^* actually depends linearly[201] on $V \in E$.

XI.24. Definition XI.21 deserves comments, since it may look very roundabout.

XI.25. The observable p which appears there is vector-valued, so that its differential dp can be considered, on each tangent space to T^*M, to be a linear map with values in $(T_eG)^*$. We have made it scalar-valued by evaluating p on a vector X in T_eG, so that the left-hand side of XI.22 is an ordinary differential form on T^*M.

Eq. (XI.22) imposes a condition on the action μ since it says that it is possible to find a Hamiltonian function for each of the velocity-vector fields of the action which *depends linearly on the chosen vector in T_eG*. This is why the actions for which a moment exists are sometimes called *Hamiltonian actions*. Whether the actions of a group have this property depends on the global structure of the group.

Example XI.26. Let H be an observable on the extension by momenta T^*M of a configuration space M. The flow $(\mathcal{H}_t)$ of its symplectic gradient Ω_H (assumed complete) defines an action of $\mathbb{R}$ on T^*M whose velocity-vector field is precisely Ω_H. The $\mathcal{H}_t$ are canonical transformations by X.42. Since, by definition of the symplectic gradient, the vector 1 in $\mathbb{R}$ (which is customarily denoted by d/dt) satisfies

$$dH = i_{\frac{d}{dt}}\omega \, ,$$

---

[201] If we wanted to be pedantic, we would say that, in this case, the moment is in fact the isomorphism from E to its bidual E^{**}.

the observable H is a *moment* for this action.

Proposition XI.27. *Let μ be an action of a Lie group G on a configuration space M. The observable $p : T^*M \longrightarrow (T_eG)^*$ defined for $X \in T_eG$ by $\langle p(\lambda), X \rangle = \lambda(X_\mu)$ is a moment for the extended action μ^{T^*}.*

Proof. Note first of all that, by Remark XI.12, the extension μ^{T^*} of the action μ is by canonical transformations. It suffices to show that p satisfies (XI.22). For this we evaluate $d\langle p, X \rangle$. Note that $\lambda(X_\mu)$ can be written as $\alpha^M(X_{\mu^{T^*}})$ by the definition of the Liouville 1-form. Consequently,

$$d\langle p, X \rangle = i_{X_{\mu^{T^*}}} d\alpha^M = i_{X_{\mu^{T^*}}} \omega \,,$$

the first equality coming from the invariance of α^M by the flow of $X_{\mu^{T^*}}$.   □

Examples XI.28. i) We return to the extension μ^{T^*} of the action of the rotation group SO_n on $\mathbb{R}^n$. As we have seen in XI.8, the velocity-vector field of this action is given, at the point v, by Av, where A is an antisymmetric matrix (identified with an element of $T_I SO_n \cong \mathcal{A}$). On $T^*\mathbb{R}^n \cong \mathbb{R}^n \times (\mathbb{R}^n)^*$, the observable p, taking its values in $\mathcal{A}$ and defined by $\langle p(v, \lambda), A \rangle = \langle \lambda, A(v) \rangle$, is a moment for μ^{T^*}.

If we specialise this example to the case $n = 3$, we can use the identification of $\mathcal{A}$ with $\mathbb{R}^3$ introduced in IX.95: if ω is the element of $\mathbb{R}^3$ associated to an antisymmetric matrix A, we have $A(v) = \omega \times v$ (where $\times$ denotes the vector product). We then find that p, considered as having values in $\mathbb{R}^3$, is given by $p(v, \lambda) = v \times \lambda$, in other words that it can be written in linear coordinates as

$$p(x^1, x^2, x^3, X_1, X_2, X_3) = (x^2 X_3 - x^3 X_2, x^3 X_1 - x^1 X_3, x^1 X_2 - x^2 X_1) \,.$$

One recognises this as the expression for the *angular momentum*.

ii) Let us return to the case of the sphere $\mathbf{S}^2$ developed in XI.8. A moment for the action of the circle $\mathbf{T}^1$ on $\mathbf{S}^2$ by rotations around the polar axis is given, in the extension of the geographical coordinates $(\theta, \psi, \Theta, \Psi)$, by the formula contained in Proposition XI.27, which in this case takes the form $p(\theta, \psi, \Theta, \Psi) = \Theta$. (For an application of this calculation, see XI.33.)

XI.29. It is particularly interesting to look for moments in the case where the configuration space is itself a group. The existence of moments which satisfy particular invariance properties is related to the algebraic structure of the group (in fact, only that of its tangent space at the identity, which inherits the structure of a Lie algebra from the adjoint representation).

Souriau has applied these considerations to the Galilean group, which is the invariance group of the classical mechanics of particles. One finds that a number appears which is attached to the system being studied, and which can be identified with its *mass*. Its mathematical nature is that of a cohomology class (in a rather algebraic sense). If we turn to the case of special relativity theory, whose invariance group is the Poincaré group, this invariant disappears since the space in which it should live reduces to 0. This phenomenon is a consequence of the global structure of the invariance group.

XI.30. We are now ready to state the conservation law related to the action of a group which illustrates the relationship between symmetries of a system and conservations laws associated to its motion. In its general form, this law is due to E. Noether.

E. Noether's Theorem XI.31. *Let H be a Hamiltonian defined on the extension by momenta T^*M of a configuration space M. Suppose that there is an action μ of a Lie group G on T^*M by canonical transformations which leaves H invariant. Then any moment $p : T^*M \longrightarrow (T_eG)^*$ for this action is a first integral of the motion which is defined by the Hamiltonian H.*

Proof. We must show that $\partial_{\Omega_H} p$ is zero. To make this quantity a scalar, we evaluate it on an element X in T_eG. We then have

$$\langle \partial_{\Omega_H} p, X \rangle = \langle d(\langle p, X \rangle), \Omega_H \rangle = (i_{X_\mu} \omega)(\Omega_H) = \omega(X_\mu, \Omega_H) = -\langle dH, X_\mu \rangle$$

which is zero because of the hypothesis that H is invariant under the action. (Note that, in the preceding formula, there are duality pairings of two different kinds; that between observables on T^*M and tangent vectors, and that in the vector space T_eG.) $\qquad\square$

Corollary XI.32. *If H is a Hamiltonian defined on the extension by momenta of a configuration space M on which a Lie group G operates, then the canonical moment p defined in XI.27 is a first integral of the motion defined by the Hamiltonian H.*

Proof. This is simply a combination of Proposition XI.27 and E. Noether's Theorem XI.31. $\qquad\square$

Examples XI.33. i) One direct application of E. Noether's Theorem XI.31 is the following classical result: *"For a system in $\mathbb{R}^3$ whose motion is governed by a Hamiltonian which is invariant under the rotations around an axis, the angular momentum about this axis is a constant of the motion."*

ii) Let us give an explicit example of an application of Corollary XI.32 in the case of the sphere S^2. We consider the Hamiltonian H on S^2 induced by a Riemannian metric g which we assume is invariant under the rotations around the polar axis (for an *"abstract"* discussion of this example assuming only the existence of an action of T^1 on S^2, the interested reader should consult (Besse 1978), pages 95-96).

In the extension of the geographical chart (θ, ψ), this means that the Hamiltonian H induced by g can be written $H(\theta, \psi, \Theta, \Psi) = \Psi^2 + g^{\theta\theta}\Theta^2$, where $g^{\theta\theta}$ (which is the inverse of the coefficient $g_{\theta\theta}$ of g in the natural basis $(d\theta^2, d\theta d\psi, d\psi^2)$ of the symmetric bilinear forms) is a differentiable function defined on $[-\pi/2, \pi/2]$ which depends only on ψ and which satisfies certain boundary conditions at the endpoints of the interval because of the regularity of g over the whole of S^2. (Exercise: Find them.) The moment of the extension of the T^1-action to T^*S^2 has been given in XI.28.

E. Noether's Theorem XI.31 thus tells us that along an integral curve of Hamilton's equations for the Hamiltonian H, i.e., along a geodesic of the metric g, the coordinate Θ is constant, which means, by using the inverse Legendre transform, that the quantity $g_{\theta\theta}\, d\theta/dt$ is constant along any geodesic. This is *Clairaut's*[202] *first integral*, which allows one in many cases to simplify considerably the determination of the geodesics.

---

[202] Alexis Clairaut (1713–1765), French mathematician and astronomer, took part in the founding of the Calculus of Variations. At the age of 23, he accompanied Maupertuis on the expedition to Lapland which set out to measure the flattening of the earth.

Remark XI.34. There are generalisations of the conservation laws to quantities more complex than observables (even vector-valued ones like moments). Unfortunately, we cannot go into these developments even though they are a beautiful (and timely) illustration of the interaction between mathematics and other sciences.

The setting which turns out to be most suitable for these generalisations is that of the exterior differential calculus, which is mostly due to É. Cartan but of which certain aspects are due to Poincaré. One then speaks of *integral invariants*. We would have to develop, among other things, a theory of integration of exterior differential k-forms over k-dimensional submanifolds.

D. Observables in Involution and Integrable Systems

In this section, we complete our study of first integrals (whatever their origin) by examining the possible relations between them. This is made possible by using the Poisson bracket whose definition we have extended to configuration spaces.

Definition XI.35. Two observables φ and ζ on the extension by momenta T^*M of a configuration space M are said to be *in involution* if their Poisson bracket $\{\varphi, \zeta\}$ is identically zero.

Examples XI.36. Here are a few examples:

i) Among the examples of observables in involution are, of course, the *position variables* (by a "position variable", we mean an observable φ on T^*M which can be written as $\psi \circ \varpi_M$ where ψ is an observable on M). More generally, if x is a chart of M and (x^i, X_i) the corresponding coordinates on the extension by momenta, we have, for $1 \leq i, j \leq n$,

$$(XI.37) \qquad \{x^i, x^j\} = 0 , \quad \{x^i, X_j\} = \delta_{ij} , \quad \{X_i, X_j\} = 0 .$$

ii) Other examples arise naturally from the study of Hamiltonian systems because of the interpretation of the Poisson bracket with the Hamiltonian as the operator which describes the time evolution of the system (cf. X.75). If an observable φ is a first integral of the motion (the moment of a Lie group action, for example), then H and φ are in involution.

Note, however, that the components of a vector-valued first integral, such as, for example, the components of a moment, can fail to be in involution. This is, in particular, the case when the group G is non-Abelian, for the moment necessarily captures that part of the algebraic structure which the tangent space at the identity inherits from the adjoint representation.

One example of this situation is provided by the components of the angular momentum (p_1, p_2, p_3) (the moment of the action of the group SO_3 on $T^*\mathbb{R}^3$) which satisfy the relations

$$\{p_1, p_2\} = p_3, \quad \{p_2, p_3\} = p_1, \quad \{p_3, p_1\} = p_2.$$

Note, however, that this gives a beautiful example of an application of Proposition XI.16: *if two components of the angular momentum are first integrals of the motion of a Hamiltonian system in $T^*\mathbb{R}^3$, then the third is also a first integral.*

XI.38. Let us look again, in the light of Definition XI.35, at the statement of Darboux's theorem (cf. X.26) asserting that every symplectic structure is locally equivalent to the natural structure on a cotangent space. To prove this, it suffices to find a system of $2n$ variables which satisfy Eqs. (XI.37). When the problem is formulated in this way, it seems natural to try to define the variables inductively. This is the starting point of Darboux's proof, which continues by interpreting the Poisson brackets as first order partial differential equations for the observables.

XI.39. The importance of the notion of *observables in involution* is due essentially to the following fact: if one has k observables in involution on a $2n$-dimensional space (corresponding to an n-dimensional configuration space), the study of a Hamiltonian system defined by one of these observables can be reduced to that of a Hamiltonian system on a space of dimension $2n - 2(k - 1)$. This process, called *symplectic reduction*, has been used extensively for giving explicit solutions of Hamiltonian systems.

XI.40. If $\varphi_1, \ldots, \varphi_k$ are the k observables in involution, and are thus first integrals of the motion defined by any one of them, say φ_1, it is possible to study the motion on the subset C defined by the constraint $\phi \equiv (\varphi_1, \varphi_2, \cdots, \varphi_k) - (\alpha_1, \cdots, \alpha_k)$. If the α_i are regular values of the φ_i, this constraint is non-degenerate (and we shall assume this from now on), and C is a submanifold. One can then make $\mathbb{R}^k$ act on C by changing its ith coordinate according to the flow of the Hamiltonian defined by φ_i. The space of orbits of the subgroup $\{0\} \times \mathbb{R}^{k-1}$ of $\mathbb{R}^k$ on C is the required configuration space of dimension $2n - 2(k-1)$. On this new space, there is a naturally defined symplectic form induced by the Liouville symplectic form. (For more details, see (Abraham and Marsden 1978), page 298 et seq.)

Remark XI.41. This construction generalises when the conserved quantities are not necessarily in involution, but still give rise to a moment of a group action. Jacobi was the first to give such a construction in the case of the group SO_3, for applications to problems in celestial mechanics. This approach is now known as the *elimination of nodes*.

Definition XI.42 A Hamiltonian system on the extension by momenta of an n-dimensional configuration space M is said to be *completely integrable* if it has n first integrals in involution.

Example XI.43. We have already met an example of a completely integrable system in our study in XI.33 of the geodesics of a Riemannian metric invariant under the rotation group T^1. Such surfaces are often called *surfaces of revolution*. Jacobi contributed extensively to the study of such systems.

E. Historical Notes

XI.44. The study of *symmetries* has always excited great interest. It has been approached from a combination of philosophical and aesthetic, as well as scientific, points of view. To give just one example, we mention the full list of possible periodic tilings of the plane which are to be found in the frescoes of the Alhambra in Grenada.

One had to wait a long time for a precise mathematical formulation of this question, which uses the theory of groups.

XI.45. Celestial mechanics has been one of the great adventures of the human intellect, and its evolution is marked by discoveries which have often been landmarks in the history of thought. From our point of view, this is illustrated by the fact that the Law of conservation of angular momentum for the motion of a planet in a central force field was established by Kepler from observations of the planet Mars. He did not make use, and with good reason, of the tools we have put to work in this chapter.

The generalisation of such conservation laws to configuration spaces on which a group acts came only much later. It seems that this subject was initiated by Lie in his work on continuous groups of transformations, but this part of his work was ignored for a long time.

The importance of the dual of the tangent space to a Lie group for formulating the physical action for a system which admits this group as a symmetry group was brought to light in almost simultaneous work by Kirillov, Kostant, and Souriau. The former was also motivated by the representation theory of Lie groups.

XI.46. The problem of solving completely integrable systems is closely related to questions of an algebraic nature such as the theory of elliptic functions because of the appearance after reduction of tori as configuration spaces (i.e., spaces whose observables can be regarded as doubly periodic observables in the plane). This is the case, for example, in the problem of determining the geodesics on ellipsoids, which was solved by Jacobi.

This theme linking algebraic geometry and dynamics is flourishing again today thanks to studies of equations which have *soliton* solutions, such as the Korteweg[203]-de Vries[204] equation, for example, which models the motion of the water surface

---

[203] The Dutch mathematician Diederik Johannes Korteweg (1848–1941) was the first PhD holder of the University of Amsterdam for his dissertation entitled *On the Propagation of Waves in Elastic Tubes*.

[204] A student of Korteweg, Gustav de Vries (1866–1934) was a Dutch mathematician who initially studied physical chemistry.

in a shallow canal. One of the new features of this work is the appearance of infinite-dimensional configuration spaces (generalisations of tori, in fact). Another reason for mentioning these things in the conclusion of this paragraph is that they involve conservation properties of a rather special type, such as that of conserving the eigenvalues of certain linear operators, for example, which have been brought to light by Lax[205].

---

[205] The Hungarian-American mathematician Peter David Lax, born in 1926, has numerous contributions to pure and applied mathematics in the fields of fluid dynamics, scientific computing and conservation laws. At a very young age he worked for the Manhattan Project. The impact of his creative thinking is considerable. He received the Wolf Prize in 1987 and the Abel Prize in 2005.

Appendix: Basic Elements of Topology

1.1. A set M has the structure of a *topological space* if it is equipped with a family of subsets $(U_i)_{i \in I}$, containing $\emptyset$ and M, whose elements are called the *open sets* of M, and which is stable under taking *finite intersections* and *arbitrary unions*.

To define a topology on a set, one can give a distinguished family of subsets, the *base of the topology*, which one then completes by taking their intersections and unions according to the axioms which families of open sets satisfy.

Any subset X of a topological space M has a natural topology obtained by taking the family of open sets to be the *traces* $U_i \cap X$ of the open sets U_i of M. One then says that X is a *topological subspace* of M and that its topology is *induced* from that of M.

1.2. In a topological space, a set whose complement is open is called *closed*; the family of closed sets contains $\emptyset$ and M, and is stable under taking *finite unions* and *arbitrary intersections*, as one easily sees by taking complements in the properties satisfied by the family of open sets.

The smallest closed set containing a given subset A is called its *closure* and *denoted* by $\overline{A}$. It is easy to see that $\overline{A}$ is the intersection of all the closed sets containing A. 1.3. A subset V containing a point q in M is called a *neighbourhood*

of q if it contains an open set containing q. A point q' is called an *accumulation point* of a sequence $(q_n)_{n \in \mathbb{N}}$ of points of M if every neighbourhood of q' contains at least one point of the sequence.

A sequence $(q_n)_{n \in \mathbb{N}}$ in a topological space *converges* to a point q if every neighbourhood of q contains q_n for all except a finite number of values of n. 1.4. A

map f from a topological space M to a topological space N is said to be *continuous* if the inverse image under f of any open set of N is an open set of M. It is easy to check that the image under f of any convergent sequence in M is a convergent sequence in N.

A bijection from a topological space M to a topological space N which is continuous and has a continuous inverse is called a *homeomorphism*. The two spaces M and N are then said to be *homeomorphic*. Two spaces which are homeomorphic are indistinguishable from a topological point of view.

© The Author(s), under exclusive license to Springer Nature Switzerland AG 2022
J.-P. Bourguignon, *Variational Calculus*, Springer Monographs in Mathematics,
https://doi.org/10.1007/978-3-031-18307-2

1.5. The product of two topological spaces M_1 and M_2 has a natural topology, called the *product topology*: a base for this topology is given by the products of the open sets of each factor. This construction generalizes to an arbitrary number of factors.

1.6. A topological space M is said to be *connected* if it is not the union of two disjoint, non-empty open sets, which is the same thing as saying that M and $\emptyset$ are the only subsets of M which are both open and closed.

One can look for the connected subsets of a given topological space M, i.e. those which are connected in their subspace topology. It is then natural to introduce an equivalence relation on M by saying that two points are "equivalent" if there is a connected subset of M which contains both of them. The equivalence classes of this relation are called the *connected components* of M, and are thus the *maximal connected subsets* of M.

There is another notion of connectedness which is perhaps more intuitive in view of the role which the real numbers play in the foundations of analysis. This concerns the notion of *path connectedness*. A subset of a topological space is said to be *path connected* if any two of its points can be joined by the image of a continuous map from an interval of the real line (a "*path*"). In this way, one can also define an equivalence relation on a topological space whose equivalence classes are its *path connected components*.

References

Basic material on the topics of the notes

Abraham, R., Marsden, J. (1978) *Foundations of Mechanics*, 2nd edition, Benjamin, New York

Arnol'd, V.I. (1973) *Ordinary Differential Equations*, MIT Press, Cambridge, Massachusetts

Arnol'd, V.I. (1978) *Mathematical Methods of Classical Mechanics*, Graduate Texts in Mathematics 60, Springer, New York-Heidelberg-Berlin

Avez, A. (1983) *Calcul Différentiel*, Masson, Paris

Choquet, G. (1984) *Cours de Topologie*, 2nd edition, Masson, Paris

Deheuvels, R. (1981) *Formes Quadratiques et Groupes Classiques*, Presses Universitaires de France, Paris

Demazure, M. (1992) *Geometry, Catastrophes and Bifurcations*, Springer, New York-Berlin-Heidelberg-Tokyo

Dieudonné, J. (1968) *Foundations of Modern Analysis*, Academic Press, New York

Leborgne, D. (1982) *Calcul Différentiel et Géométrie*, Presses Universitaires de France, Paris

Schwartz, L. (1970) *Topologie générale et analyse fonctionnelle*, Hermann, Paris

Sternberg, S. (1965) *Lectures on Differential Geometry*, Prentice-Hall, Englewood Cliffs

Warner, F. (1983) *Foundations of Differentiable Manifolds and Lie Groups*, Graduate Texts in Mathematics 94, Springer, New York Berlin Heidelberg Tokyo

Books supporting some complements to the notes

Arnol'd, V.I. (1983) *Geometrical methods in the theory of ordinary differential equations*, Grundlehren der mathematischen Wissenschaften 250, Springer, New York-Heidelberg-Berlin

Dubrovin, B., Novikov, S.P., Fomenko, A. (1992) *Modern Geometry: Methods and Applications. Part I: The Geometry of Surfaces, Transformation Groups, and Fields*, 2nd edition, Graduate Texts in Mathematics 93, Springer, New York-Berlin-Heidelberg-Tokyo

© The Author(s), under exclusive license to Springer Nature Switzerland AG 2022   263
J.-P. Bourguignon, *Variational Calculus*, Springer Monographs in Mathematics,
https://doi.org/10.1007/978-3-031-18307-2

Dubrovin, B., Novikov, S. P., Fomenko, A. (1985) *Modern Geometry: Methods and Applications. Part II: The Geometry and Topology of Manifolds*, Graduate Texts in Mathematics 104, Springer, New York-Berlin-Heidelberg-Tokyo

More advanced books

Arnol'd V.I., Kowlov, V.V., Neishtadt A.I. (1988) *Dynamical Systems III: Mathematical Aspects of Classical and Celestial Mechanics*, Encyclopaedia of Mathematical Sciences Vol. 3, Springer, Berlin-Heidelberg-New York-London-Paris-Tokyo

Besse, A.I. (1978) *Manifolds all of whose Geodesics are Closed*, Ergebnisse der Mathematik und ihrer Grenzgebiete 93, Springer, New York-Heidelberg-Berlin

Gallavotti, G. (1983) *The Elements of Mechanics*, Springer, New York-Heidelberg-Berlin

Milnor, J. (1965) *Topology from the Differentiable Viewpoint*, University Press of Virginia, Charlottesville

Milnor, J. (1973) *Morse Theory*, Annals of Mathematics Studies 51, Princeton University Press, Princeton

Books having a historical interest

Caratheodory, C. (1935) *Variationsrechnung und partielle Differentialgleichungen erster Ordnung*, B.G. Teubner, Berlin; 2nd edition in English: *Calculus of Variations*, Chelsea, New York (1982)

Dombrowski, P. (1979) *150 Years After Gauss' "Disquisitiones generales circa superficies curvas"*, Astérisque 62, Soc. Math. France, Paris

Gelfand, I.M., Fomin, S.V. (1963) *Calculus of Variations*, Prentice-Hall, Englewood Cliffs

Goldstein, H. (1971) *Classical Mechanics*, Addison-Wesley, Reading, Massachusetts

Hildebrandt, S., Tromba, A. (1987) *Mathematics and Optimal Form*, Scientific American Library, W.H. Freeman, New York

Lanczos, C. (1970) *The Variational Principles of Mechanics*, 4th edition, University of Toronto Press, Toronto; reprinted by Dover, New York (1986)

Young, L.C. (1980) *Lectures on the Calculus of Variations and Optimal Control*, 2nd edition, Chelsea, New York

More recent books

Gallot, S., Hulin, D., Lafontaine, J. (2004) *Riemannian Geometry*, 3rd edition, Universitext, Springer

Jost, J. (2017) *Riemannian Geometry and Geometric Analysis*, 7th edition, Springer

Jost, J., Li-Jost, X. (1998) *Calculus of Variations*, Cambridge University Press

Zorich, V.A. (2016) *Mathematical Analysis II*, 2nd edition, Universitext Springer

Notation Index

© The Author(s), under exclusive license to Springer Nature Switzerland AG 2022
J.-P. Bourguignon, *Variational Calculus*, Springer Monographs in Mathematics,
https://doi.org/10.1007/978-3-031-18307-2

Subject Index

1-form
 differential, 158
 Liouville, 231

acceleration-vector, 209
accumulation point, 261
action
 functional, 203
 of a Lie group, 248
 regular, of a Lie group on itself,
  249
 transitive, 104
adjoint map, 67
admissible local coordinate systems,
 105
affine
 space, 4
 vector field, 93
algebra of observables, 109
alternating form, 191, 233
angular momentum, 254
area, 223
atlas, 106
axiom, Euclid's parallel, 218
axis
 of inertia, 220
 of rotation, 126

band, Möbius, 124
base of a topology, 261

basis
 dual, 28
 Hilbert, 46
 natural, associated to a chart,
  142
 unitary, 34
bidual, 31
bounded set, 6
Boy's surface, 164
bracket
 duality, 28
 Lie, 152
 Poisson, 97, 243
bundle, cotangent, 231

C^1-diffeomorphism, 70
catenoid, 225
Cauchy sequence, 22
Cayley–Klein parametrisation, 130
chart, 106
 adapted to a constraint, 161
Christoffel symbol, 216
Clairaut's first integral, 255
class
 first Chern, 202
 map of class C^2, 74
closed
 differential form, 160
 set, 261
closure of a set, 261

© The Author(s), under exclusive license to Springer Nature Switzerland AG 2022
J.-P. Bourguignon, *Variational Calculus*, Springer Monographs in Mathematics,
https://doi.org/10.1007/978-3-031-18307-2

Printed in the United States
by Baker & Taylor Publisher Services